AF751248

ENERGY EFFICIENCY – THE INNOVATIVE WAYS FOR SMART ENERGY, THE FUTURE TOWARDS MODERN UTILITIES

Edited by **Moustafa Eissa**

Energy Efficiency – The Innovative Ways for Smart Energy, the Future Towards Modern Utilities
http://dx.doi.org/10.5772/2590
Edited by Moustafa Eissa

Contributors

M.M. Eissa, S.M. Wasfy, M.M. Sallam, Joana Carla Soares Gonçalves, Denise Duarte, Leonardo Marques Monteiro, Mônica Pereira Marcondes, Norberto Corrêa da Silva Moura, Dionysis Xenakis, Nikos Passas, Ayman Radwan, Jonathan Rodriguez, Christos Verikoukis, Soib Taib, Anwar Al-Mofleh, Tomas Gil-Lopez, Miguel A. Galvez-Huerta, Juan Castejon-Navas, Paul O'Donohoe, Bjørn R. Sørensen, Dragan Šešlija, Ivana Ignjatović, Slobodan Dudić, H.M. Ramos, Luís F. C. Duarte, Elnatan C. Ferreira, José A. Siqueira Dias, Chenchen Yang, Feng Yang, Liang Liang, Xiping Xu, Seong-woo Woo, Jungwan Park, Jongyun Yoon, HongGyu Jeon, Luo Xianxi, Yuan Mingzhe, Wang Hong, Li Yuezhong, Rafaa Mraihi, Teuvo Aro, Said Ben Alla, Abdellah Ezzati, Ahmed Mohsen, Rodrigo Pantoni, Cleber Fonseca, Dennis Brandão, Giuseppe Procaccianti, Luca Ardito, Antonio Vetro', Maurizio Morisio, Glauber Brante, Marcos Tomio Kakitani, Richard Demo Souza

Published by InTech

Janeza Trdine 9, 51000 Rijeka, Croatia

Copyright © 2012 InTech

All chapters are distributed under the Creative Commons Attribution 3.0 license, which allows users to download, copy and build upon published articles even for commercial purposes, as long as the author and publisher are properly credited, which ensures maximum dissemination and a wider impact of our publications. However, users who aim to disseminate and distribute copies of this book as a whole must not seek monetary compensation for such service (excluded InTech representatives and agreed collaborations). After this work has been published by InTech, authors have the right to republish it, in whole or part, in any publication of which they are the author, and to make other personal use of the work. Any republication, referencing or personal use of the work must explicitly identify the original source.

Notice

Statements and opinions expressed in the chapters are these of the individual contributors and not necessarily those of the editors or publisher. No responsibility is accepted for the accuracy of information contained in the published chapters. The publisher assumes no responsibility for any damage or injury to persons or property arising out of the use of any materials, instructions, methods or ideas contained in the book.

Publishing Process Manager Dragana Manestar
Technical Editor InTech Prepress, Novi Sad
Cover InTech Design team

Additional hard copies can be obtained from orders@intechopen.com

Energy Efficiency – The Innovative Ways for Smart Energy, the Future Towards Modern Utilities, Edited by Moustafa Eissa
p. cm.
ISBN 978-953-51-0800-9

We are IntechOpen, the world's leading publisher of Open Access books

Built by scientists, for scientists

4,100+
Open access books available

116,000+
International authors and editors

120M+
Downloads

151
Countries delivered to

Our authors are among the
Top 1%
most cited scientists

12.2%
Contributors from top 500 universities

WEB OF SCIENCE™

Selection of our books indexed in the Book Citation Index
in Web of Science™ Core Collection (BKCI)

Interested in publishing with us?
Contact book.department@intechopen.com

Numbers displayed above are based on latest data collected.
For more information visit www.intechopen.com

For EU product safety concerns:
IN TECH d.o.o., Prolaz Marije Krucifikse Kozulić 3, 51000 Rijeka, Croatia,
info@intechopen.com or visit our website at intechopen.com.

Meet the editor

Dr. M. M. Eissa is a Professor at Helwan University. He authored more than 75 papers (29 IEEE and IET journal). In 1999, he was invited as a Visiting Research Fellow at Calgary University-Canada. From 2008 till 2010 he was a chair Prof. for "DSM and Energy Efficiency", project funded from Saudi Electricity Company. His research interests are digital relaying, application of WAM to power systems, PMU applications, smart grid and wireless application on power system. Dr. Eissa received "State country prize in the advanced technology science-Egypt 2002". He also owns two prizes from Helwan University; "Distinguished Researcher Award-2005" and "Incentive Researcher Award-2011". He is currently the PI for Smart Grid Frequency Monitoring Network, a project of Helwan NTRA.

Contents

Preface

The objective of this book is to present different programs and practical applications for energy efficiency in sufficient depth. The approach is given to transfer the long academic and practical experiences from researchers in the field of energy engineering to readers. The book is enabling readers to reach a sound understanding of a broad range of different topics related to energy efficiency. The book is highly recommended for engineers, researchers and technical staff involved in energy efficiency programs.

Energy efficiency is a relatively quick and effective way to minimize depletion of resources. It is the way for the future development of alternative resources. Effective energy efficiency programs can reduce a country's reliance on non-domestic energy sources, which can in turn improve national security and stabilize energy prices. Smart Energy is the philosophy of using the most cost effective long term approach to meeting the energy needs maintaining the lowest environmental impact and maximum efficiency.

The electric power delivery system is almost entirely a system with only modest use of sensors, minimal electronic communication and almost no electronic control. In the last 30 years almost all other industries in the world have modernized themselves with the use of sensors, communications, electrical and mechanical equipment and computational ability. For industries, many enormous improvements are produced in productivity, efficiency, quality of products and services, and environmental performance.

Smart grid is the way to achieve smart energy with optimized and high performance use of electrical and mechanical equipment, sensors, communications, computational capabilities, demand / load management and control in different forms, which enhances the overall functionality of the electric power delivery system. Traditional system becomes smart by sensing, communicating, applying intelligence, exercising control and through feedback, continually adjusting. This permits several functions in the power system and allows optimization of the use of generation and storage, transmission, distribution, distributed resources and consumer end uses. This is the way to ensure reliability and optimize or minimize the use of energy, mitigate environmental impact, manage assets, and reduce cost.

Improving energy efficiency will require concerted and effective policies and programs at the international and local levels in addition to extensive improvements in technology. This book provides some studies and specific sets of policies and programs that are implemented in order to maximize the potential for energy efficiency improvement. It contains unique studies that provide a multi-disciplinary forum for the discussion of critical issues in energy policy, science and technology, and their impact on society and the environment.

Moreover the book provides innovative ways of energy research by addressing key topics in this wide-ranging field; from different expert programs in the field related to electrical and mechanical equipment, load management and quality, to energy efficiency in sensors and software, measurement and auditing.

The book contains four main sections; Energy Efficiency with Load Management, Energy Efficiency-Equipment, Energy Efficiency-Measurement and Analysis, and Energy Efficiency Software and Sensors. Every section contains several chapters related to the topic of the section. More than 30 Scientists with academic and industrial expertise in the field of the energy efficiency have contributed to this book which aim was to provide sufficient innovative knowledge and present different energy efficiency policies from multi-disciplinary point of view.

Section 1: Energy Efficiency – Load Management

This section describes modified Intelligent Monitoring and Controlling System to high voltage customers. It also assembles the complete work of environmental design developed for the new research centre of Petroleum Companies. Energy Efficient Mobility Management for the Macrocell–Femtocell LTE Network is also presented. Finally this section defines the concept and the need for energy efficiency as a solution for energy management.

Section 2: Energy Efficiency – Equipment

This section describes high efficiency mix energy system design with low carbon footprint for wide-open Workshops. Ventilation fans are energy-demanding equipment that stands for a significant share of a building's total energy consumption. Improving energy efficiency of ventilation fans is thus important. In addition, different approach of controlling the static pressure difference of a fan is suggested. One of the important industry utilities that has to be encompassed by this energy policy are compressed air systems. The section is also concerned with the identification of the current state of energy efficiency in the production and usage of compressed air and possibilities for improvements that would yield the corresponding energy saving. Moreover, it presents an optimization model that determines the best hourly operation for one day, according to the electricity tariff, for a pumped storage system with water consumption and inlet discharge with wind turbines. Finally energy dissipation due to gas-liquid mixing as a function of different physical, geometric and dynamic variables of the system is enunciated in this section.

Section 3: Energy Efficiency – Measurement and Analysis

This section presents a comprehensive compilation of several state-of-the-art methods that can be used for the detailed electrical energy measurement in houses, with emphasis on the techniques which can provide a complete knowledge of the energy consumption of all appliances in a home. The section also introduces energy efficiency comparison study between the countries using a dynamic analysis. The reliability design and its direct effect on the energy efficiency are also discussed in this section. The steam in iron and steel plants is an important secondary energy. Accurate measurement of steam flow rate is of great significance for the rational use of steam and improving energy efficiency. However, due to the complex nature of steam and the low precision of instruments, the reliability of the measured data is low. That makes negative impact to the production scheduling. Here we have three data processing approaches proposed for the real-time flow rate measurements. In addition, energy consumption of transport sector depends on several factors, such as economic, fiscal, regulatory and technological factors. The investigation of the main driving factors of transport energy consumption changes requires analysis of the relationship between transport activity and economic growth. Finally this section also presents the problems of energy efficiency in transport sector, methods of determination of the contributing factors and the policy options to make the sector more sustainable.

Section 4: Energy Efficiency – Software and Sensors

An inefficient use of the available energy leads to poor performance and short life cycle of the sensors network. This section provides Hierarchical Adaptive Balanced Energy Efficient Routing Protocol to decrease probability of failure nodes and to prolong the time interval before the death of the first node and increasing the lifetime in heterogeneous Wireless sensor networks, which is crucial for many applications. Sensing and actuating nodes placed outdoors in urban environments so as to improve people's living conditions as well as to monitor compliance with increasingly strict environmental laws. Furthermore, in this section an application for urban networks using the IEEE 802.15.4 standard is presented, which is used for monitoring and control electric variables in a public lighting scenario. It also deals with the matter of finding relationships between software usage and power consumption. Two experiments have been designed, consisting in running benchmarks on two common desktop machines, simulating some typical scenarios and then measuring the energy consumption in order to make some statistical analysis on results. Finally the section also outlines wireless sensors network scenarios and analysis of the energy consumption of the devices.

Prof. M. M. Eissa
Faculty of Engineering at Helwan
Helwan University,
Egypt

Section 1

Energy Efficiency – Load Management

Load Management System Using Intelligent Monitoring and Control System for Commercial and Industrial Sectors

M.M. Eissa, S.M. Wasfy and M.M. Sallam

Additional information is available at the end of the chapter

http://dx.doi.org/10.5772/51850

1. Introduction

There are vast opportunities to improve energy use efficiency by eliminating waste through process optimization. Applying today's computing and control equipment and techniques is one of the most cost-effective and significant opportunities for larger energy users to reduce their energy costs and improve profits. An Energy Management Information System (EMIS) is an important element of a comprehensive energy management program. It provides relevant information to key individuals and departments that enable them to improve energy performance. Today it is normal for companies, particularly in process sectors, to collect huge amounts of real-time data from automated control systems, including Programmable Logic Controllers (PLCs), Supervisory Control and Data Acquisition (SCADA), etc. The captured data is shared and analysed in an orderly and precise way that identifies problem areas and provides solutions, this mass of data is merely information overload. Advances in information technology (IT), defined here as the use of computers to collect, analyse, control and distribute data, have developed rapidly. It is now common for managers and operators to have access to powerful computers and software. Today there are a number of techniques to analyse the factors that affect efficiency, and models are automatically generated based on "what if" scenarios in order to improve decisions to be taken.

The paper shows a very advanced technology for handling automatically more than 200 digital and analogue (i/p and o/p) parameters via intelligent monitoring and controlling system.

However, load management is the process of scheduling the loads to reduce the electric energy consumption and or the maximum demand. It is basically optimizing the

INTECH
open science | open minds

© 2012 The Author(s). Licensee InTech. This chapter is distributed under the terms of the Creative Commons Attribution License http://creativecommons.org/licenses/by/3.0), which permits unrestricted use, distribution, and reproduction in any medium, provided the original work is properly cited.

processes/loads to improve the system load factor. Load-management procedures involve changes to equipment and/or consumption patterns on the customer side. There are many methods of load management which can be followed by an industry or a utility, such as load shedding and restoring, load shifting, installing energy-efficient processes and equipment, energy storage devices, co-generation, non-conventional sources of energy, and reactive power control [1]-[3]. Meeting the peak demand is one of the major problems now facing the electric utilities. With the existing generating capacity being unmanageable, authorities are forced to implement load shedding in various sectors during most of the seasons. Load shifting will be a better option for most industries. Load shifting basically means scheduling the load in such a way that loads are diverted from peak period to off-peak periods, thereby shaving the peak and filling the valley of the load curve, so improving the load factor[4]-[6].

To encourage load shifting in industries, and thereby to reduce peak demand automatically, a new technology such as introduced here will be extended.

Also, power quality is of major concern to all types of industries, especially those operating with critical machinery and equipments. Poor quality of power leads to major problems like break-downs, production interruptions, excess energy consumption etc. Modern industries require automation of their operation enabling them to produce quality products and also for mass production. The conventional systems are being replaced by modern Power Electronic systems, bringing a variety of advantages to the users. Classic examples are DC & AC Drives, UPS, soft starters, etc. Power Quality Alarming and Analysis provides a comprehensive view into a facility's electrical distribution system. Power Quality can be monitored at the electrical mains or at any critical feeder branch in the distribution system such as described here. Devices in this category typically provide all of the parameters found in basic devices, plus advanced analysis capabilities [7]-[8]. These advanced analysis capabilities include using waveform capture to collect and view waveform shape and magnitude, providing harmonic analysis graphs, collection and storage of events and data, and recording single or multiple cycle waveforms based on triggers such as overvoltage or transients. With the ever-increasing use of sophisticated controls and equipment in industrial, commercial, and governmental facilities, the continuity, reliability, and quality of electrical service has become extremely crucial to many power users. Electrical systems are subject to a wide variety of power quality problems which can interrupt production processes, affect sensitive equipment, and cause downtime, scrap, and capacity losses. Momentary voltage fluctuations can disastrously impact production [7]-[8].

The proposed modified intelligent monitoring and controlling system will introduce monitoring, alarming, controlling, and power quality mitigation based on data collected and analyzed from the system. The original system can afford the following features:

- Complete information about the plant (circuit breakers status, source of feeding, and level of the consumed power).
- Information about the operating values of the voltage, operating values of the transformers, operating values of the medium voltage, load feeders, operating values of

the generators. These values will assist in getting any action to return the plant to its normal operation by minimum costs.

- Protective information such as the insulation of cables, temperatures of the generators. These parameters are used as a back up for the main protection.
- Information about the quality of the system (harmonics, current, voltages, power factors, flickers, etc.). These values will be very essential in case of future correction.
- Recorded information such case voltage spikes, reducing the voltage on the medium or current interruption.

2. Original system description

The hardware configuration of the original intelligent monitoring and controlling system is divided into two levels. The first level includes two workstations -1 and -2 with two different software programs are used for data handling and monitoring purpose. The second level includes the PLC for data collected that constituted from 10 digital meters and some smart sensors to cover many points in the system. Some digital meters are fed directly to the workstation-2 using different software for data handling. All other parameters such as breaker status, temperature, controllers, and cable insulators are fed through the PLC. Fig. 1 shows the overall structure of original intelligent monitoring and controlling system achieved at the Eastern Company in Egypt. The intelligent monitoring and controlling system uses the most recent technology of Profibus in data transferring. Workstaton-1 used the Wincc flexible software program for data handling received from the MV, Transformers and Generators. Workstation-2 used the Sicaro Q manger software program for data handling from the loads. Both workstations are linked through Ethernet network. One programmable logic controller S7-300 associated with 10 power meters for monitoring the MV, Transformers and Generators, Insulation relays, Temperature transducers for generators, and Circuit Breakers auxiliary points for all loads have been applied to workstation 1 through Profibus network-1. Workstation-2 associated with 12 power quality meters for monitoring all loads (Compressors, Pumps, Motors, Processes, etc.) via Profibus network-2. All system parameters are communicated using the Profibus technology. The output system is limited by given alarming and recommendation to the operator without doing any automatic actions for the system. The system components used in the system are produced from Siemens and can be described as:

PROFIBUS is the powerful, open and rugged bus system for process and field communication in cell networks with few stations and for data communication. Automation devices such as PLCs, PCs, HMI devices, sensors or actuators can communicate via this bus system. PROFIBUS is part of totally integrated automation, the uniform, integrated product and system range from Siemens for efficient automation of the entire production process for all sectors of industry. PROFIBUS can be used, for example, for the following applications: Factory automation, Process automation and Building automation. Different PROFIBUS versions are available for the various fields of application:

- Process or field communication (PROFIBUS DP) (for fast, cyclic data exchange with field devices). PROFIBUS PA (for intrinsic safety applications in process automation)

- Data communication (PROFIBUS FMS) (for data communication between programmable controllers and field devices).

Power Quality devices are installed at various measuring points in order to record a series of measurements of the required values for an analysis of the network quality. The devise can be installed on the load. In addition to all relevant measured variables, the meter can also record system disturbances, always when an upward or downward limit value violation has occurred. The recorded values can be called up and evaluated using a PC. Power Quality is available in 3 device versions with the following communication interfaces: RS232, RS485, and PROFIBUS-DP. Furthermore the device version Power Quality with PROFIBUS-DP interface opens up another area of application. Together with programmable control systems (PLCs), it can be used as a "sensor" for electrical measured variables. In the system achieved, the PROFIBUS-DP technology is used.

The Power Quality PAR parameterization software is executable under the Windows 2000/XP operating systems. The software allows you to define the device address, so that each device is uniquely identified and to configure the power quality for the communication protocol to be used (PROFIBUS DP) in order to prepare it for the measurement task. Fig. 2 shows the display of the currently transmitted measured values.

Power Meter is a power meter for panel mounting, with big graphic display and background illumination. The major application area is power monitoring and recording at MV and LV level. The major information types are measured values, alarms and status information. Power monitoring systems with Power Meter, a permanently installed system, enables continuous logging of energy-related data and provides information on operational characteristics of electrical systems. Power Meter helps identify sources of energy consumption and time of peak consumption. This knowledge allows you to allocate and reduce energy costs. Measured values include r.m.s values of voltages (phase-to-phase and/or phase-to-ground), currents, active, reactive and apparent power and energy, frequency, power factor, phase angle per phase, symmetry factor, harmonics of currents and voltages, total harmonic distortion. Ten maters are installed on the system and arranged on the incoming feeders, transformers and also on the generators. PROFIBUS-DP and Power Meter are connected in a master-slave operation mode. The communication parameters are loaded to the master station using the GSD file. The Power Meter supports data transmission rates from 9.6 kbit/s to 12 Mbit/s. The Measured values can be: Voltage, Current, Active power, Reactive power, Apparent power, Power factor, Active power factor, Phase angle, Frequency, Active energy demand, Active energy supply, Active energy total, Active energy total, Reactive energy, inductive, Reactive energy, capacitive, Reactive energy total, Apparent energy, Unbalance voltage, Unbalance current, THD voltage, THD current, Harmonic voltage, and Harmonic current.

SIMATIC S7-300 PLC: S7-300 programmable controller is made up of the following components:

- Power supply (PS)
- CPU
- Signal modules (SM)

- Function modules (FM)
- Communication processor (CP).

Several S7-300s can communicate together and with other SIMATIC S7 PLCs via PROFIBUS bus cables. Fig. 3 shows the components of the PLC. The Runtime application of the WinCC basic software offers all essential functions of a powerful SCADA-System. Using WinCC User Administrator, One can assign and control users access rights for configuration and runtime.

3. Original system operation

The application functions of the data collection and monitoring are all performed via two workstations, PLC and two different software programmes. The program can include data exchange communication protocol between the communication system and PLC, through digital power meters, breakers' status (On/Off), power quality monitoring, threshold for alarming. Figs. 4 and 5 show part of the system operation for monitoring feeder-2 and transformer-2 parameters for the system installed at Eastern Company in Egypt. Fig. 6 shows one of the event messages produced from the system. Fig. 7 shows the block diagram that demonstrates the various function components of IMCS. All of these components are programmed as functions of the system. Some of the system functionality can be described as;

3.1. Sensors, power meters and breakers status

Sensors and power meters communicate measurements and status information from the plant to the monitoring modules of the IMCS.

3.2. Monitoring

IMCS offers a wide range of options for monitoring the plant. Information can be monitored locally and centrally. Access to the IMCS is protected and the users must login to gain access to functionality. Information received from the plant can be monitored in different forms from data or trends mode. The monitoring includes; Alarms, Trends, Recommendations Status, Configuration utilities, Event messages, etc.

3.3. Control

The control of the system is limited while the system is based on monitoring purposes and given recommendation messages for the operator.

3.4. Data logging

The IMCS has the feature of data logging for some selected parameters such as (switching procedure during week; temperature, breaker status, transients, etc.) for further analysis.

3.5. Alarms, recommendation and event reporting

The main feature of the IMCS is producing the alarms, recommendation and event reporting functionality. The alarms, recommendation and event reporting are based on customers basis.

3.6. Predictive maintenance

Predictive maintenance is an essential part of the IMCS. The system gives all information about the power quality of the plant and cable insulation. This can be achieved through reports and alarms message produced from the system.

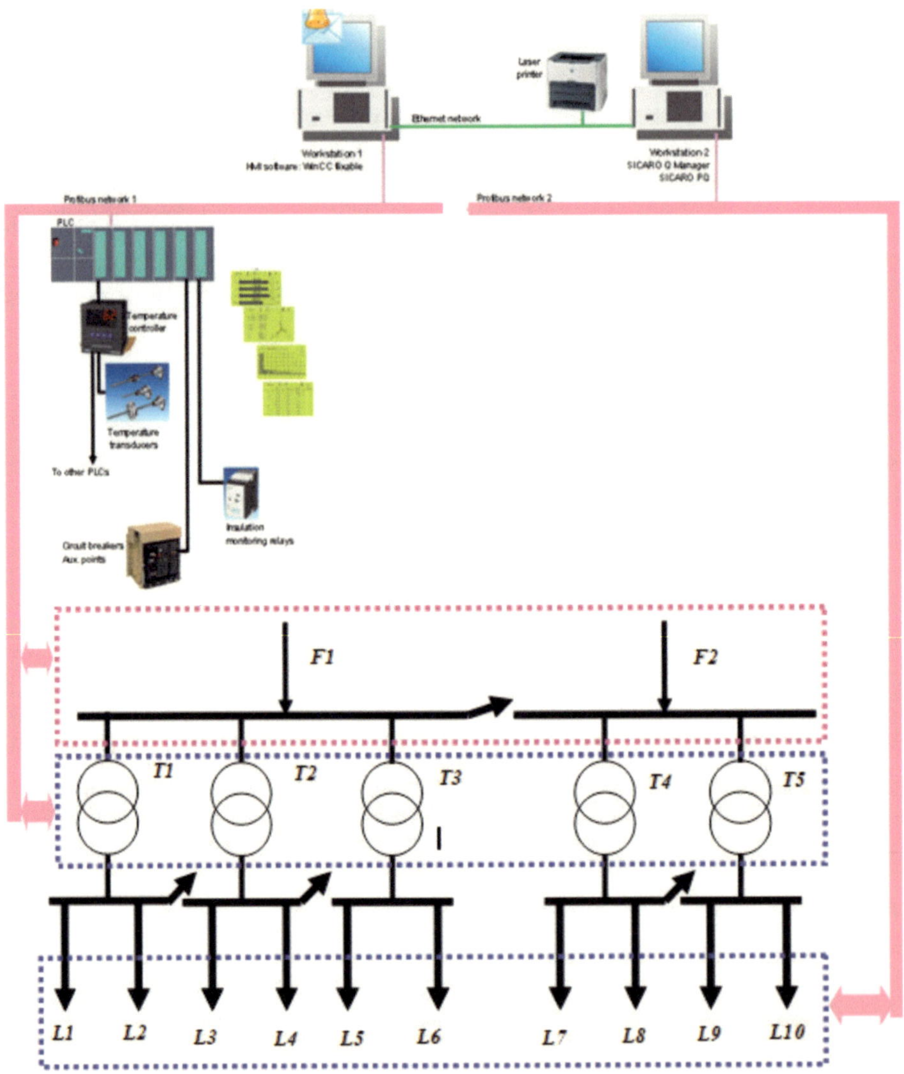

Figure 1. The overall structure of the original intelligent monitoring and controlling system.

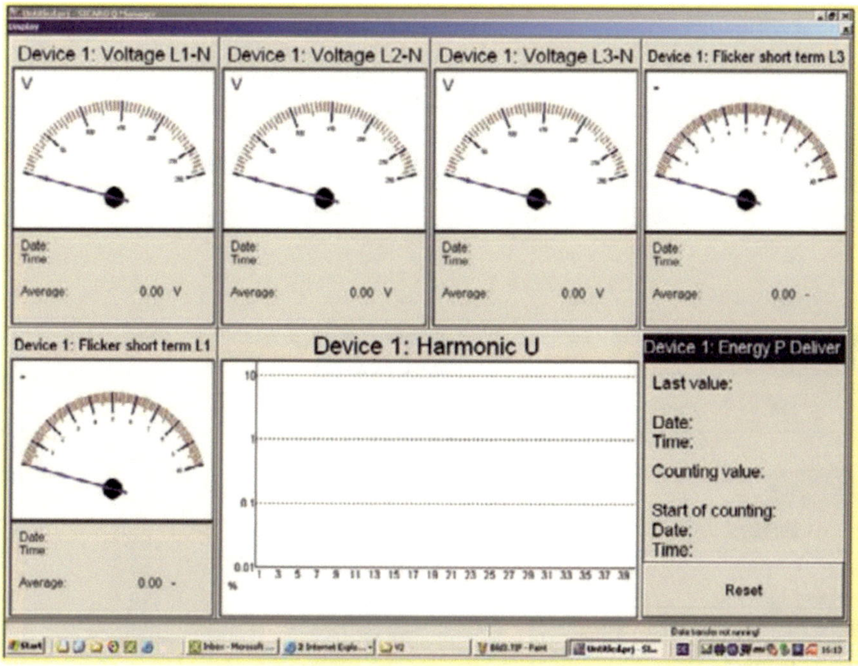

Figure 2. The display of the currently transmitted measured values of Quality Meter used in the system.

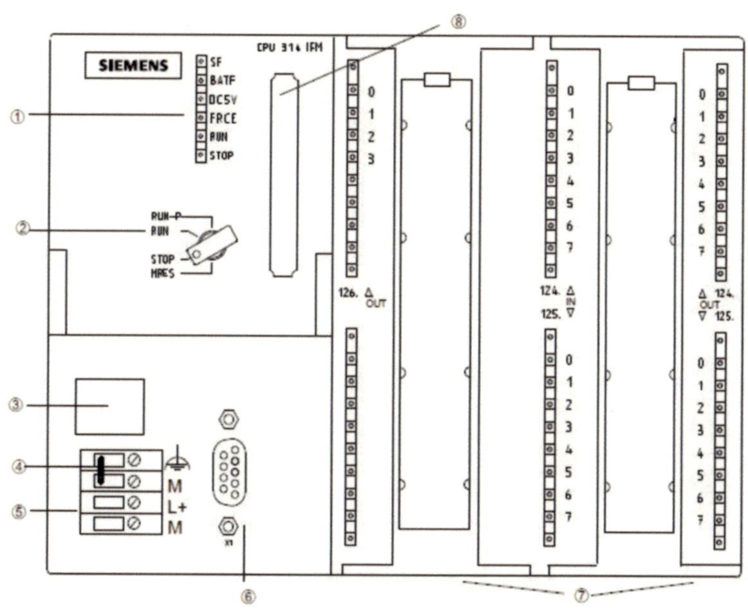

① Status and fault LEDs
② Mode selector
③ Compartment for backup battery or accumulator
④ Jumper (removable)
⑤ Terminals for power supply and functional ground
⑥ Multipoint Interface (MPI)
⑦ Integrated inputs/outputs
⑧ Slot for memory card (-5AE**10**- only)

Figure 3. The main components of the PLC.

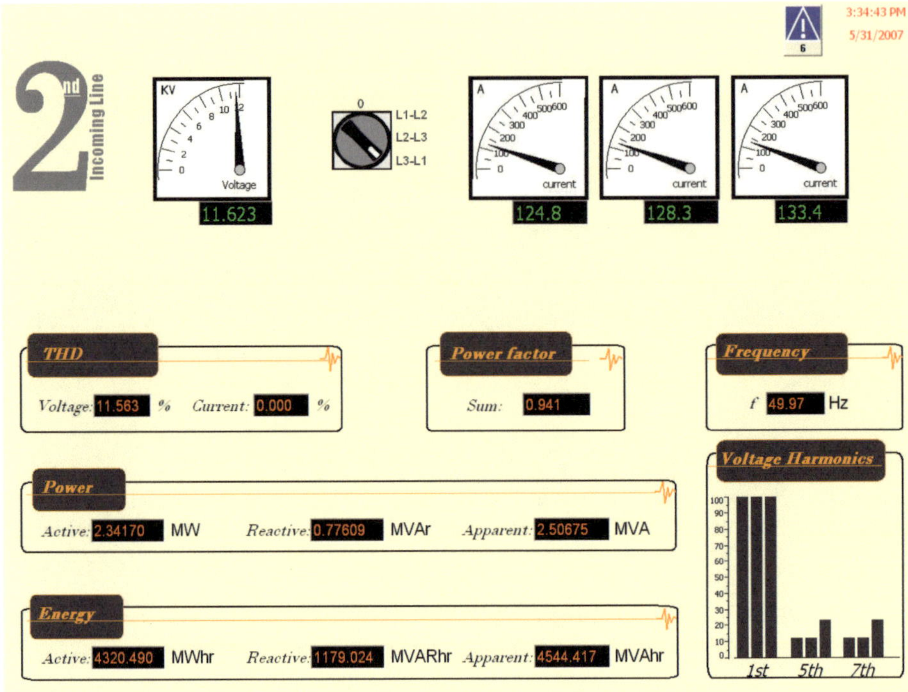

Figure 4. Part of the monitored data on fedder-2 at Eastern Company.

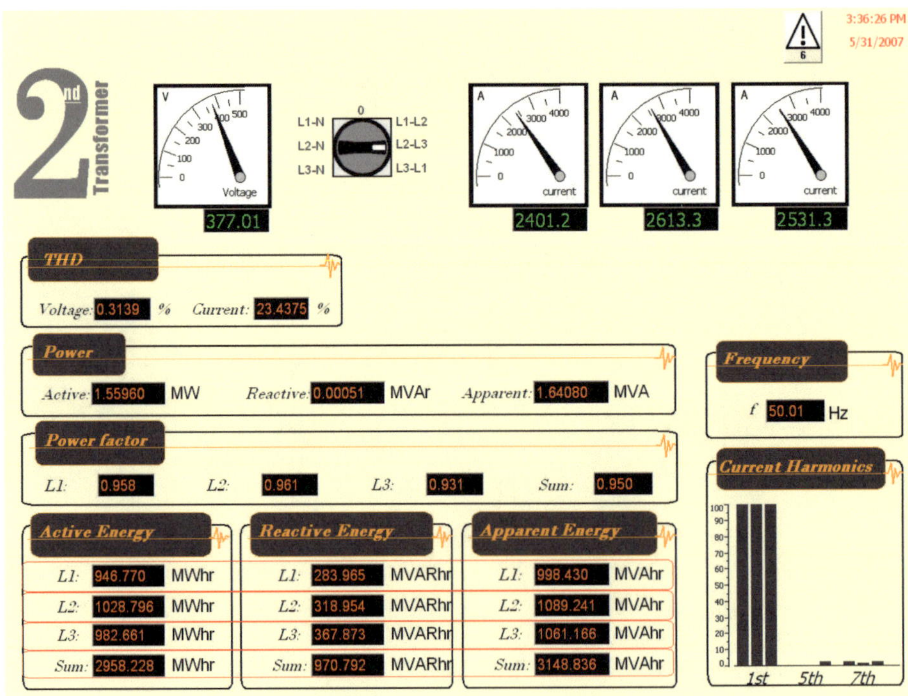

Figure 5. Part of the monitored data on Transfwer-2 at Eastern Company.

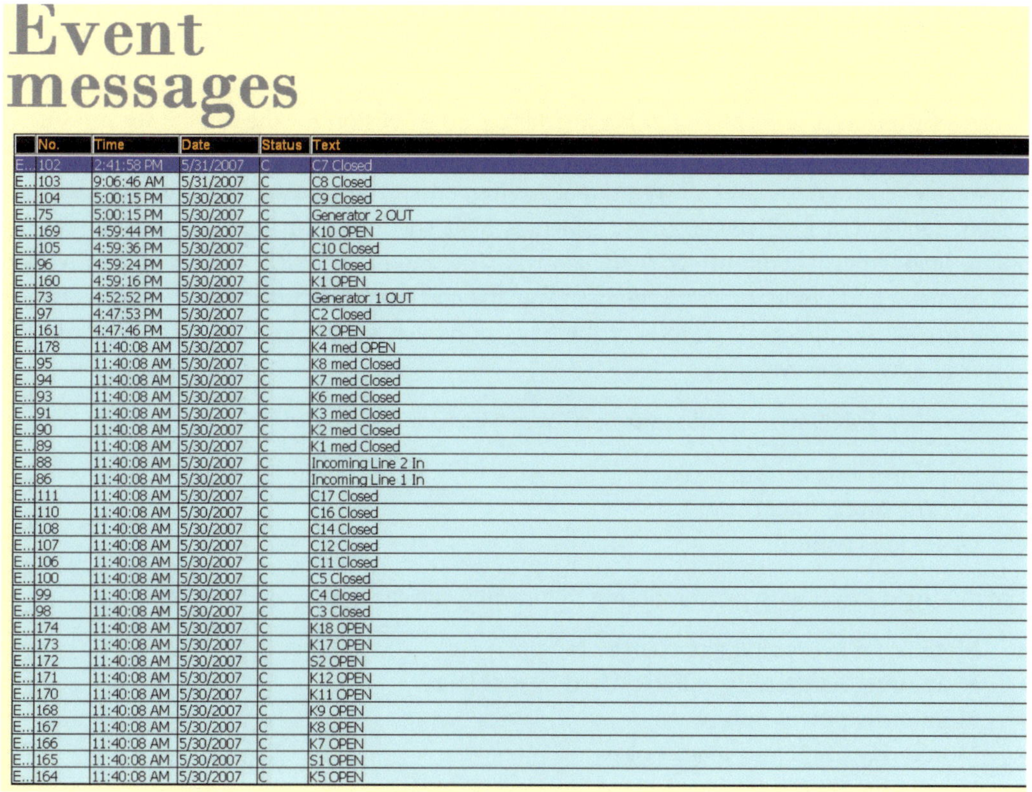

Event messages

	No.	Time	Date	Status	Text
E...	102	2:41:58 PM	5/31/2007	C	C7 Closed
E...	103	9:06:46 AM	5/31/2007	C	C8 Closed
E...	104	5:00:15 PM	5/30/2007	C	C9 Closed
E...	75	5:00:15 PM	5/30/2007	C	Generator 2 OUT
E...	169	4:59:44 PM	5/30/2007	C	K10 OPEN
E...	105	4:59:36 PM	5/30/2007	C	C10 Closed
E...	96	4:59:24 PM	5/30/2007	C	C1 Closed
E...	160	4:59:16 PM	5/30/2007	C	K1 OPEN
E...	73	4:52:52 PM	5/30/2007	C	Generator 1 OUT
E...	97	4:47:53 PM	5/30/2007	C	C2 Closed
E...	161	4:47:46 PM	5/30/2007	C	K2 OPEN
E...	178	11:40:08 AM	5/30/2007	C	K4 med OPEN
E...	95	11:40:08 AM	5/30/2007	C	K8 med Closed
E...	94	11:40:08 AM	5/30/2007	C	K7 med Closed
E...	93	11:40:08 AM	5/30/2007	C	K6 med Closed
E...	91	11:40:08 AM	5/30/2007	C	K3 med Closed
E...	90	11:40:08 AM	5/30/2007	C	K2 med Closed
E...	89	11:40:08 AM	5/30/2007	C	K1 med Closed
E...	88	11:40:08 AM	5/30/2007	C	Incoming Line 2 In
E...	86	11:40:08 AM	5/30/2007	C	Incoming Line 1 In
E...	111	11:40:08 AM	5/30/2007	C	C17 Closed
E...	110	11:40:08 AM	5/30/2007	C	C16 Closed
E...	108	11:40:08 AM	5/30/2007	C	C14 Closed
E...	107	11:40:08 AM	5/30/2007	C	C12 Closed
E...	106	11:40:08 AM	5/30/2007	C	C11 Closed
E...	100	11:40:08 AM	5/30/2007	C	C5 Closed
E...	99	11:40:08 AM	5/30/2007	C	C4 Closed
E...	98	11:40:08 AM	5/30/2007	C	C3 Closed
E...	174	11:40:08 AM	5/30/2007	C	K18 OPEN
E...	173	11:40:08 AM	5/30/2007	C	K17 OPEN
E...	172	11:40:08 AM	5/30/2007	C	S2 OPEN
E...	171	11:40:08 AM	5/30/2007	C	K12 OPEN
E...	170	11:40:08 AM	5/30/2007	C	K11 OPEN
E...	168	11:40:08 AM	5/30/2007	C	K9 OPEN
E...	167	11:40:08 AM	5/30/2007	C	K8 OPEN
E...	166	11:40:08 AM	5/30/2007	C	K7 OPEN
E...	165	11:40:08 AM	5/30/2007	C	S1 OPEN
E...	164	11:40:08 AM	5/30/2007	C	K5 OPEN

Figure 6. One of the output event messages produced from the system at Eastern Company.

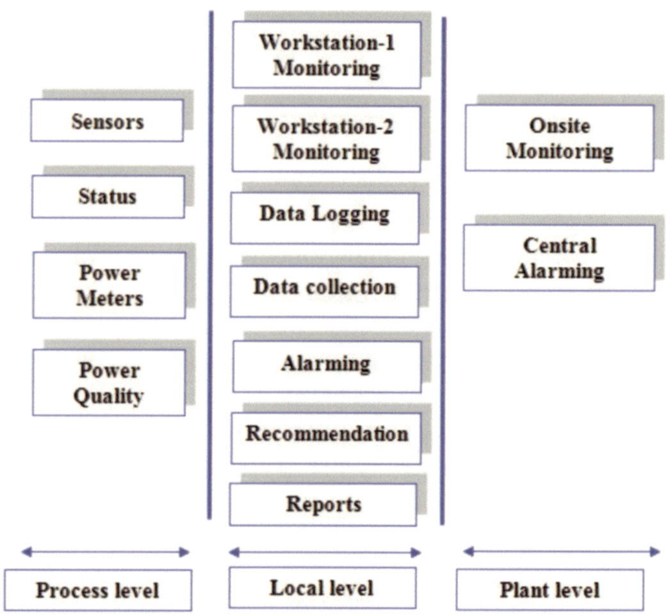

Figure 7. Block diagram of the IMCS components installed at Eastern Company.

4. System Modified for load management and quality mitigation schemes

The original system is used mainly for monitoring purposes with some recommendation messages produced from it. The paper introduces an extension for the system for producing many digital and analogue output signals from the PLC to control loads based on load management programs and power quality mitigation procedure. The proposed modified system can accept load management schemes load shedding during peak period, cycling on/off load control, and direct load control. The modified IMCS associated with load management and power quality schemes gives the customer the possibility of load reduction or control during the peak periods of the day, moreover, gives more information about the power quality of the system.

The modified IMCS with Load management can include:

- Load shedding during peak
- Cycling on/off load control,
- Direct load control.

The modified IMCS with power quality monitoring can include:

- Monitoring overvoltage or transients
- Monitoring harmonic graphs for feeders and loads
- Monitoring power failure
- Monitoring High frequency noise
- Monitoring Spikes
- Monitoring ground faults and deterioration insulation in cables

The IMCS can also monitor all the breakers status and temperatures of the stand-by generators. The following subsections explain some function components embedded in the modified IMCS

4.1. New generation of power quality meter

The new generation of the power monitoring device provides accurate knowledge of the systems characteristics with maximum, minimum and average values for voltage, current, power values, frequency, power factor, symmetry and THD. The SENTRON PAC4200 detects the values for active, reactive and apparent energy – both for high and low tariff. It measures ratings and power values via the four quadrants, i.e. power import and export are measured separately. The SENTRON PAC4200 also facilitates the detection of a measuring period's average values for active and reactive power. These values can be further processed into load curves in a power management system. Typically, 15-minute intervals are used for this purpose. PAC4200 also detects uneven harmonics from the 3rd to the 31st for voltage and current, the distortion current strength (Id), the phase angle and the asymmetry for voltage and current with reference to the amplitude and phase. For further processing of the measured data, the devices can be very easily integrated in superior automation and power management systems in the proposed system the meters are interfaced with SIMATIC PCS 7 powerrate and SIMATIC WinCC powerrate software packages. The Wincc powerrate software packages can handle very complicated schemes for load management.

4.2. Control

The purpose of the control module in the modified IMCS is to provide the control of the necessary parameters for each point in the plant. It provides the functionality required to control the load in case of peak period or in case of exceeding the threshold boundary. It offers control functionality, e.g. load shedding, on/off load control, direct load control, power quality control. It also provides the functionality required to control devices such as pumps, motors, compressors, on/off breakers, interlocking, power quality mitigation. Operation can be configured to be automatic. The modified system offers the facility for adjusting control parameters (e.g., set points, output quantity, tolerances, time delay) in order to achieve the desired condition for each program. Fig. 8 shows the new added function blocks for the modified intelligent monitoring and controlling system. The figure shows three main components; workstation-2 with control scheme, Control output module, new generation of the power quality meters compatible with Wincc powerrate software program. As given in Fig. 9, the proposed modified system uses one additional PLC interfaced with the new power quality meters located on the loads. The data is shared through Profibus network-2. Workstation-2 uses Wincc powerrate software program for programming and controlling purposes. Workstation-2 manipulates different load management programs through collected data received from the PLC. Many controlled output signals are produce from the PLC o/p modules. The proposed modified system avoids many of original system limitations by replacing the power quality meters interfaced directly with workstation-2. The proposed modified IMCS can manipulate the following programs;

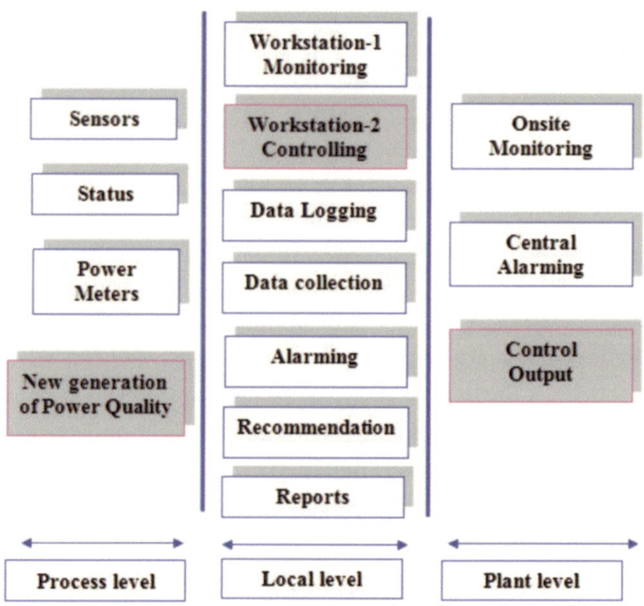

Figure 8. Block diagram of the modified IMCS function components.

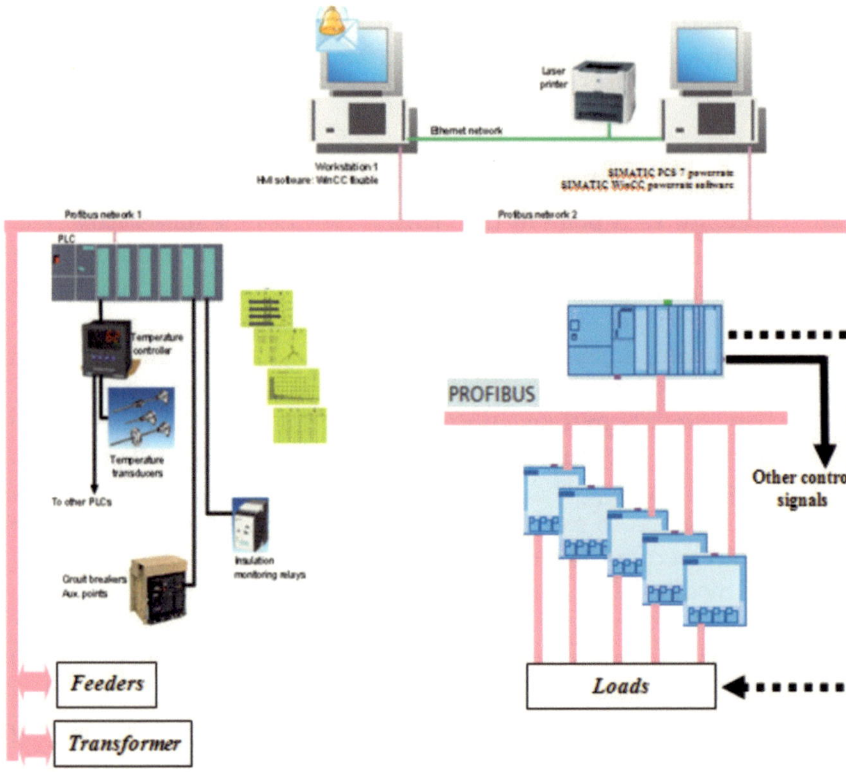

Figure 9. The overall structure of the proposed modified intelligent monitoring and controlling system.

4.2.1. Load management

A company's electric bill, in most countries, consists of two major components: demand charge and energy consumption charge. Demand charges are reset monthly and are based on the highest rate at which electricity is consumed during periods that are peak utility service hours. Demand charges are measured in kilowatts and, depending on the utility service provider, the highest consumption rate is measured in 15- or 30-minute intervals during peak hours or contracted value. Demand charges form a significant portion of a company's monthly electric bill. Peak load management strategies that lower a facility's demand during times when the peak demand is measured can result in significant facility cost savings, especially for commercial, industrial and governmental sectors. Fig. 10 shows the procedure of the load management that can be achieved automatically in the proposed modified IMCS. The scheme of load management can be built and programmed according to previous scenario using energy auditing. In such a case the customer can avoid any penalty from the utility and can save money as well. The procedure of the control can be run through workstation-2 and the readings of the loads received from PLC with power quality meters. The program can be run under Wincc powerrate software program. Firstly, the system will check the required load capacity to be shaved during peak value and then select the minimum diesel power generated that coves the required loads. The system will produce output signal through the PLC output module to the suitable generator/generators. In case of the load reduction the system disconnects the generator/-s sequentially according to the required load and get back the loads to the normal operation on their feeders.

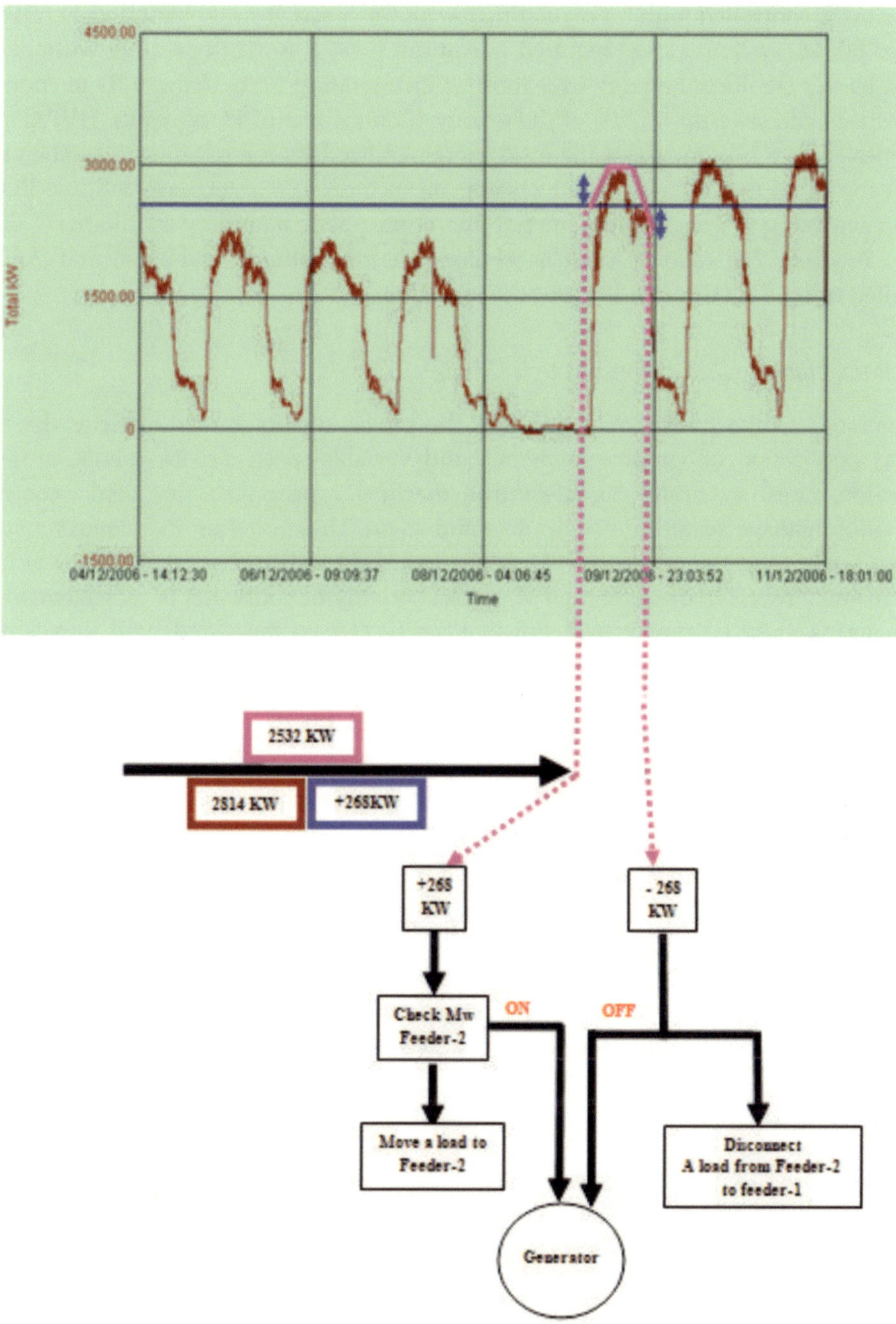

Figure 10. The load management control for avoiding the peak period.

4.2.2. Load reduction and energy saving using VFD for HVAC

Flow generating equipment like fans, pumps and compressors are often used without speed control. In stead, flow is traditionally controlled by throttling or using a valve or damper. When flow is controlled without regulating the motor speed, it runs continuously full speed. Because HVAC systems rarely required maximum flow, a system operating without speed control wastes significant energy over most of its operating time. Using VFD to control the motor speed can save up to 70% of the energy (Central line of Honeywell). HVAC system consumes a large percentage of the total energy utilized by the organization. The original IMCS can not do that. The extended system can include VFD interfaced with the PLC and can run according to the given parameters and atmospheric boundary conditions in the side control module. The devices such as chillers, motors (pumps and fans) and AHU are controlled using VFD through the proposed modified IMCS.

4.2.3. Power quality mitigation through IMCS

The demands to the power reliability and the power quality become stricter due to the popular application of variable-frequency and variable-speed drives, robots, automated production lines, accurate digital-control machines, programmable logic controllers, information manage systems in computers and so on. These devices and computer systems are very sensitive to the power-supply ripple and various disturbances. Any power quality problems may result in the reduction of product quality or confusion of management order which means great economic loss. Power Quality can be measured with power quality meters and analyzed with software in the control module. Power Quality events usually are infrequent, making them hard to detect and store without specialized equipment. Over a certain period, there may only be a few Power Quality events. The system is able to collect and analyze events on the basis for identifying the Power Quality problem. Analyzed data is the starting point for improvement. Data identifies potential sources of problems based on the timing of the events. This information can show the cause of the problem. Once the source is identified, improvement begins based on corrective actions. Corrective actions may include: changing motor starting procedures, replacing faulty switches or relays, filtering harmonic producing loads, or changing switching schedules for power factor correcting capacitors, changing control of the power filter. Once the corrections are in place, further monitoring will verify that the corrective action worked. The applied system such as IMCS can line with this specifications for power quality mitigation.

5. Conclusion

The application of good plant quality and demand control concepts requires an understanding of utility rates, auditing, and metering in addition to a basic knowledge of the process and load being controlled or shed. This process is professionally achieved in Eastern Company with 10MW capacity and 11kV/380V plant before installing the original intelligent monitoring and controlling system. The paper introduced an overview explanation for the original IMCS achieved at the Eastern Company. The paper also proposed modified intelligent monitoring and controlling system that has the feature of high speed data manipulation through the technology of Profibus and new generation of digital meters. The proposed modified system has the function of demand monitoring/load shedding scheme that operated in automatic mode. Also, Power Quality can be measured with power quality meters and analyzed with software in the control module. Data identifies potential sources of problems based on the timing of the events.

The feature of the energy consumption and power quality mitigation are significantly enhancing the power system operation. The proposed system has the features of

- Accurate load management during peak periods designed based on the plant requirements,
- Real time monitoring of the plant performance,
- Intelligent alarming capabilities for early power quality problems
- Predictive maintenance planning for cables and standby generators
- Energy management
- Power system efficiency
- Save money by avoiding any penalty from the utility
- Keep the voltage tolerance with the allowed limits
- Continuous monitoring for the power quality
- Load control and management with proper methodology
- Keep the consumption with the contracted limit
- Recoding information that can assist for any future development for the electrical network
- Catching any transient vents happened in the system
- Mitigation the plant through continuous monitoring of the power quality
- Fault diagnosis and alarming
- Identify and fix the causes of power disturbances to avoid recurrences
- Improved system efficiency

The authors are going to implement the modified components of the IMCS.

Author details

M.M. Eissa, S.M. Wasfy and M.M. Sallam
Helwan University at Helwan-Department of Elect. Eng., Cairo, Egypt

6. References

[1] M. R. Mcrae, R. M. Seheer and B. A. Smith, "Integrating Load Management Programs into Utility Operations and Planning with a Load Reduction Forecasting System," IEEE Trans., Vol PAS-104, No. 6, pp. 1321- 1325, June 1985.

[2] C. W. Gellings, "Interruptible Load Management into Utility Planning", IEEE Trans. Vol.PAS-104, No.8, pp.2079-2085, August 1985.

[3] C. W. Gellings, A. C. Johnson and P. Yatcko, "Load Management Assessment Methodology at PSE&G", IEEE Trans.,Vol. PAS-101, No.9, pp. 3349-3355, September, 1982.

[4] C. Alvarez, R.P. Malhame, A. Gabaldon, "A class of models for load management application and evaluation revisited", IEEE Transaction on Power Systems, Vol. 7, No. 4, pp. 1435, Novemebr-1992.

[5] L Ma Isaksen and N.W. Simons, "Bibliography and load management", IEEE Transactions on Power Apparatus and Systems 1981: PAS-100(5):1981.

[6] J.N. Sheen et al, "TOU pricing of electricity for load management in Taiwan power company", IEEE Trans on Power Systems 1994.

[7] Meier, Alexandra von (2006). Electric Power Systems: A Conceptual Introduction. John Wiley & Sons, Inc. ISBN 978-0-471-17859.

[8] Kusko, Alex; Marc Thompson (2007). Power Quality in Electrical Systems. McGraw Hill. ISBN 978-0071470759.

[9] M.M. Eissa. Demand Side Management Program Evaluation Based on Industrial and Commercial Field Data. Energy Policy 39 (October, 2011) 5961–5969.

Chapter 2

Environmental Design in Contemporary Brazilian Architecture: The Research Centre of the National Petroleum Company, CENPES, in Rio de Janeiro

Joana Carla Soares Gonçalves, Denise Duarte, Leonardo Marques Monteiro, Mônica Pereira Marcondes and Norberto Corrêa da Silva Moura

Additional information is available at the end of the chapter

http://dx.doi.org/10.5772/48420

1. Introduction

This paper assembles the complete work of environmental design developed for the new research centre of the Brazilian Petroleum Company, Petrobras, in the tropical city Rio de Janeiro, in Brazil (latitude 22.53S). The main objective is to make clear the relationship between architectural solutions, environmental strategies and quality of space, by presenting the criteria and methods applied to the architectural concept and technical assessment of four complementary areas of environmental design: outdoors comfort, daylight and natural ventilation in buildings and, ultimately, the energy performance of air conditioned spaces.

Undertaken by members of the Laboratory of Environment and Energy Studies (LABAUT) from the Department of Technology of the Faculty of Architecture and Urbanism of University of Sao Paulo (FAUUSP), the environmental design of the new research centre of Petrobras was a comprehensive project of research pro-design related to the environmental performance of contemporary buildings in one of the Brazilian's main cities, Rio de Janeiro.

The design project was the object of a national architectural competition held in 2004. The programme of activities is an extension of the existing research centre, including laboratory rooms, offices, a convention centre, restaurants, greenhouse spaces and other special facilities (e.g. energy generation and model testing of petroleum platforms).The total built area of the extension of the Petrobras Research Centre in Rio de Janeiro encompasses 66.700,78 m², built on 193.290,65 m² site at the Guanabara Bay in Rio de Janeiro (see figure 1),

INTECH
open science | open minds

© 2012 The Author(s). Licensee InTech. This chapter is distributed under the terms of the Creative Commons Attribution License http://creativecommons.org/licenses/by/3.0), which permits unrestricted use, distribution, and reproduction in any medium, provided the original work is properly cited.

resulting in a plot ratio of 34,5% [1]. Completely built in 2010, this research centre is the first building complex of its size and complexity in Brazil to integrate environmental principles at the very early stages of the design. However, it should be noticed that the environmental agenda was not a particularity of the second phase of the Petrobras Research Centre. Comfort issues including thermal comfort were already regarded in the conceptual ideas of the architect Sergio Bernardes's for the first phase of the Petrobras research centre from the 1970s [2].

Figure 1. Site location of the Petrobras Research Centre in the Guanabara Bay of Rio de Janeiro.

Alongside a series of functional requirements, the design brief aimed for environmentally responsive solutions related to the comfort of the occupants and buildings' energy efficiency, in which the use of daylight, natural ventilation and vegetation were mandatory. Moreover, the brief's environmental agenda included issues associated with water consumption and the environmental impact of building materials. A list of 10 items summarises the environmental brief, involving building's design and building services: 1. buildings' orientation according to solar radiation, 2. buildings' form according to principles of bioclimatic design, 3. appropriate materials to local environmental conditions, 4. window wall ratio (WWR) according to local environmental conditions and good use of daylight, 5. protection against solar radiation, 6. natural ventilation in buildings, 7. good use of daylight, 8. low environmental impact materials, 9. rain water harvesting and re-use of grey water and, 10. vegetation for local environmental benefits, such as ecological niches and biodiversity [1]. Those strategies had a rather generic approach, with no pre-established quantitative criteria or benchmarks, opening up the possibility for the creation of a bespoken environmental reference in the context of Brazilian contemporary architecture.

The winning architectural scheme was the proposal from Zanettini Arquitetura S.A., (co-authored by José Wagner Garcia), which was informed by creative contributions from the various complementary areas, including structural, mechanical and electrical engineering and landscape and environmental design, resulting in a truly conceptual holistic design proposal. The local warm-humid conditions of Rio de Janeiro had a major influence on the architecture of the winning design project, which was inspired in one hand by the local bioclimatic modernist architecture (specially from the period between 1930's and 1960's) and on the other hand, by contemporary environmental principles and methods as well as the possibilities of current construction technologies.

In terms of internal thermal environmental conditions the new buildings of the research centre encompassed totally naturally ventilated, mixed-mode and full-time air conditioned buildings as a function of buildings' use and the consequent environmental requirements. The naturally ventilated ones are the Operational Support Building and Utilities Centre all designed based on the architectural typology of the factory building. The main air conditioned and mixed-mode buildings are: the central Building (approximately 36.000 m^2), the Laboratories (approximately 33.000 m^2) and the Convention Centre (approximately 6.500 m^2), being those three functions located at the core of the architectural composition.

Apart from influencing the architectural design, the local climatic conditions also played an important role in re-establishing some of the basic environmental performance criteria, such as the definition of comfort parameters, energy consumption targets and daylighting levels, based on the warm-humid climate of Rio de Janeiro. As the architectural design progressed, environmental assessment evolved from the interpretation on principles and simplified analytical work to advanced simulation procedures, carried out over the first 9 months of the total design period which lasted 22 months (from November 2004 to September 2006), covering the integral part of the architectural design concept). Construction began in September 2006 and was completed in 2010.

Looking at the first stages of the design of the winning project for the Expansion of the Petrobras Research Centre in Rio de Janeiro, design of the architectural proposal designed by Zanettini Arquitetura S.A., co-authored by Arch. José Wagner Garcia, and supported by a diversified consultancy team, this work presents the environmental concepts and some of its qualitative and quantitative performance aspects, highlighting the role of the Environment and Energy Studies Group of the *Faculdade de Arquitetura e Urbanismo, Universidade de São Paulo* (Faculty of Architecture and Urbanism of the University of São Paulo).

The environmental studies of the expansion of Petrobras Research Centre in Rio de Janeiro had four main objectives: assess the thermal comfort in the open spaces created by the horizontal disposition of buildings on site; maximize the benefits of daylight, assess the thermal performance of free running buildings where natural ventilation was required as a function the programme; and finally assess the performance of architectural solutions for air-conditioned buildings, where active cooling was a design premise.

The new buildings of the expansion were classified in two groups: one which the main the spaces had to be naturally ventilated, being these the *Operational Support Building* and *Utilities Centre*, and the other where the artificial control of the thermal conditions by means of active cooling systems was a design premise, being these the *Central Building* (an office building), the *Laboratories* and the *Convention Centre*. In addition, three other buildings of the research centre had active cooling as a functional requirement: the two restaurants and the *Visualization Centre* (*Núcleo de Visualização e Colaboração*, NVC), (see figure 2) [1].

Figure 2. CENPES site planning, including the 1st phase of the Research Center towards the south and the expansion with new buildings on the north part of the site facing the bay, as presented in the winning proposal.

The initial requirement for active cooling in all office spaces of the Petrobras research centre for all year round could be associated with the air-conditioning cultural of working spaces (artificial cooling is an unquestioned factor in commercial buildings in most Brazilian cities, being definitely a common practice in Rio de Janeiro), rather than a climatic driven need. Challenging the supremacy of air conditioning in the context of office spaces in Rio de Janeiro, the efficiency of natural ventilation and the introduction of the mixed-mode strategy were critically evaluated for the various typologies and conditions of working environments within the new buildings of the research centre, being ultimately recommended in some particular cases. Initially a simplified analysis of the local climate suggested the possibility of natural ventilation in a typical office space for approximately 30% of the working hours over the year, which justified a more detailed analysis of the mixed-mode strategy for the final design proposal [3].

2. Environmental concept

The preliminary climatic diagnosis highlighted the importance of shading and light colors as well as the possibility of natural ventilation as the main passive strategies to reduce heat gains in buildings and improve thermal comfort both indoors and outdoors. Analysis were based in a reference climatic year, with hourly data, encompassing readings from 2000 to 2004 of the meteorological station situated at the International Airport of Rio de Janeiro, situated within 2Km from the site of the Petrobras Research Centre. Air temperatures were high, more than 29ºC for 10% of the year, and below 20ºC, for 10% of the year as well, combined with high relative humidity rates, more than 70% in 66% of the year [4].

In this context, the search for adequate building environmental strategies started at the concept stage, addressing thermal comfort in buildings and open spaces, daylighting, acoustics and the specific issue of cooling demand. A horizontal architectural composition of multiple buildings derived from the core objective of creating meeting areas in semi-outdoor spaces, With buildings connected by transitional spaces, site planning and architectural form were defined to respond to need of protection from solar orientation, versus the exposure to natural ventilation and views towards the bay. Double roofs and various shading devices, high-level openings and open circulation routes are some of the defining architectural features which are found in all key buildings of the expansion of the Petrobras Research Centre in Rio de Janeiro.

At the masterplanning scale, the main environmental strategy was to position the different functions of the programme in separate low-rise buildings, keeping people at the ground level, or close to it and in contact with the external environment. As buildings were interspersed by transitional spaces on a predominantly horizontal occupation of the site, a series of open and semi-opened areas of different environmental qualities, including sunny and shaded areas (or partially shaded), exposed to various wind directions, as well as different landscape projects were created between, around and within buildings [5].

The value of such transitional spaces to the overall design concept was primarily related to the possibility of comfortable outdoor spaces protected from the all year round inhibiting solar radiation of Rio de Janeiro, available for leisure, social interaction and working activities, in other words, introducing the outdoors experience in the daily routine of the occupants and visitors of the Petrobras Research Centre in the Guanabara bay. Furthermore, environmentally the transitional spaces also give the benefit of reducing the impact of solar gains in the thermal performance of buildings' internal spaces (being some of them artificially cooled). In summary, the main transition spaces of the complex are associated with the three main buildings: the terraces from the Central Building, the gardens between the wings of Laboratories and the central open atrium of the Convention Centre, which is the main access to the expansion of the Research Centre.

The building cluster formed by the main office building (the Central Building), the laboratories and the convention centre was conceived to be the core of the masterplan of the extension of the research centre (see figure 3) laboratories were allocated in parallel wings facing the north-south orientation, on the two sides of the main office building (which then looks at east and west towards the bay).

The emphasis given to the efficiency of space and functionality, specially with respects to the laboratories and their connections to the rest of the research centre, was a fundamental to the site planning of buildings both in the first phase as in the expansion of the Petrobras Research Centre, as it was the importance given to the transitional spaces in the overall environmental quality of the masterplan. Whilst in the first phase, the laboratories follow and radial displacement on the site, in the expansion project, parallel rows of laboratories oriented north-south are attached to the long linear main central building, as shown previously in figure 2.

The north-south orientation to the laboratories was chosen given the relatively minor exposition to the direct solar radiation, therefore the most favorable conditions to achieve good daylight (specially from the south) and minimize solar gains, considering that daylight was a fundamental requirement to the laboratories, where cellular office cells were designed to be naturally ventilated.

The north-south orientation of the laboratories' wings was also important to allow the penetration of the predominant south-east wind into the open and semi-open areas of the complex through the patios – the semi-open spaces between two parallel laboratory wings, whilst creating appropriate conditions for the installation of photovoltaic cells on the exposed areas of the roof of laboratories.

Figure 3. Physical model of the new masterplan for the expansion of the Research Center. Source: Zanettini Arquitetura S.A.

In the case of the Central Building, the orthogonal position in relation to the wings of laboratories resulted in the east-west orientation, which brought the challenges of providing solar control on the facades whilst allow for views and daylight. On the other hand, apart from giving the opportunity of views towards the bay, the east orientation facilitated exposure to the prevailing wind from south-east (comparable to the sea breeze) at the terrace level, where semi-open spaces totally protected from the sun were design to encourage outdoors working activities, leisure and social interaction. On the opposite orientation, the west façade looks at the first buildings of the Research Centre, built on the 70s.

Architecturally, the shading of windows and semi-open spaces coupled with the use of light colors on the external facades (primarily white) were primary strategies in response to the local climatic conditions. As a result, a series of opened circulation areas inside and between buildings, shaded from the direct sun but exposed to wind, alongside internal environments of diffuse daylight and protected views of the surroundings qualify the architecture of the buildings placed at the core of the new masterplan: The Central Building, Laboratories and Conventional Centre.

However, it is important to notice that the environmental qualities and the related architectural solution of the double roofs in the main office building and the laboratories were modified with the design development. In the case of the Central Building, the original permeable structure gave place to a more robust and closed roof (see figures 4 and 5), whilst in the laboratories, the space dedicated to capture daylight was taken by systems in response to the need for highly specialized technical installations (see figures 6 and 7). Despite the major changes in the design, in both cases the second roofs kept the original role of extra solar protection. The consequent differences in performance will be explored in the forthcoming topics.

Figure 4. Conceptual sketch of the Central Building with the permeable screen roof filtering sun, light and air flow creating environmental diversity.

In addition, the two utility buildings, Operational Support and Utilities Centre, follow the factory-like building typology, in which daylight and natural ventilation by stack effect are intrinsically related to the roof design and its orientation in relation to the sun and the winds (see figure 8).

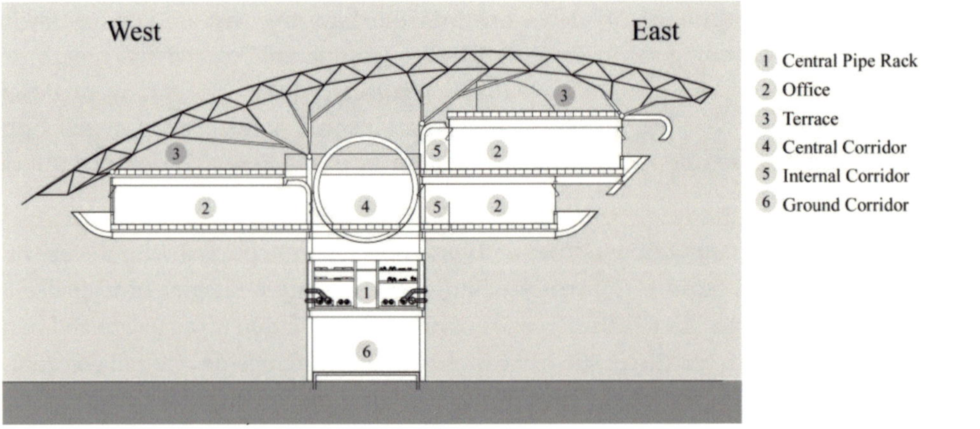

Figure 5. The final design of the Central Building with the insulated metal sandwich roof offering a higher protection against the sun and the rain.

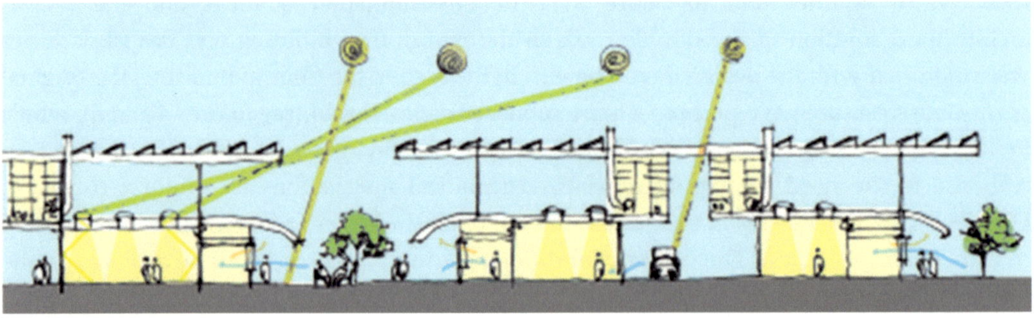

Figure 6. Conceptual sketch of the Laboratories showing the original concept of the double roof shading skylights.

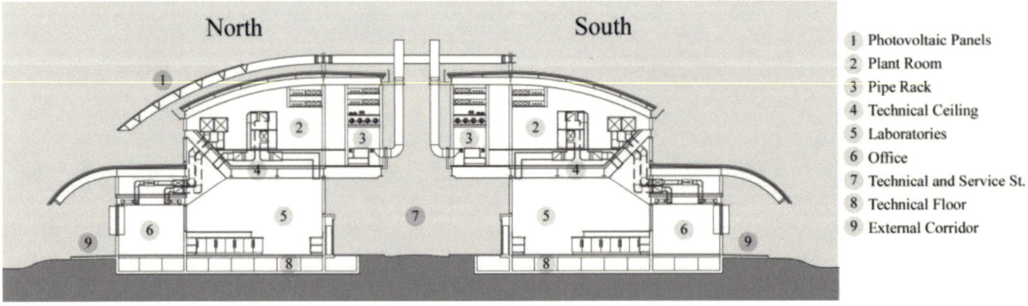

Figure 7. The final design of the Laboratories with the space between the two roofs taken by the technical systems.

3. Thermal comfort in open spaces

Considering the attractive scenery of the Guanabara Bay, in Rio de Janeiro, and the intention to promote an environment for encounters and enjoyment in outdoor spaces, the architectural premises brought a horizontal composition, with buildings connected by open

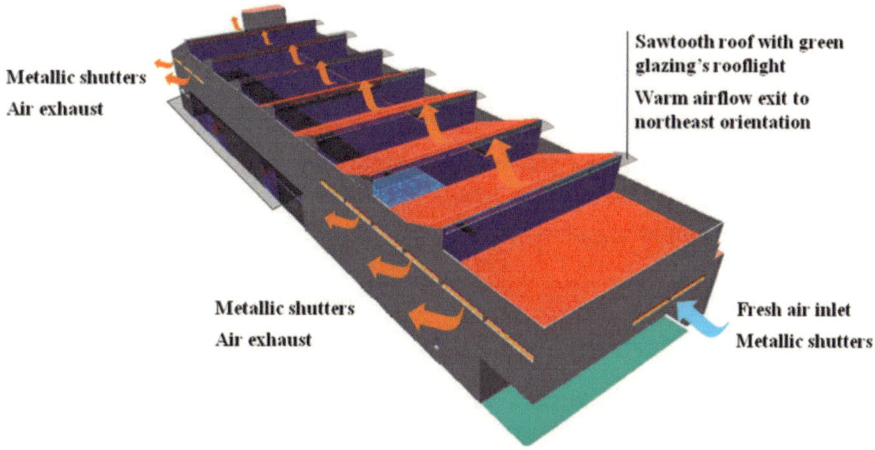

Figure 8. Design concept of the typical naturally ventilated factory- type building for services and utilities.

spaces and transition spaces between outdoors and indoors. Given the warm-humid characteristics of the local climate, the transitional spaces followed the principle of protection against the impact of solar radiation, but exposed to wind. In this respect, as shown in figure 9, ssimulations of air flow around and between the buildings has shown lower velocities particularly near the eastern edge of the laboratory buildings, but better conditions in centre of the patios as well as in the central void of the conventional centre.

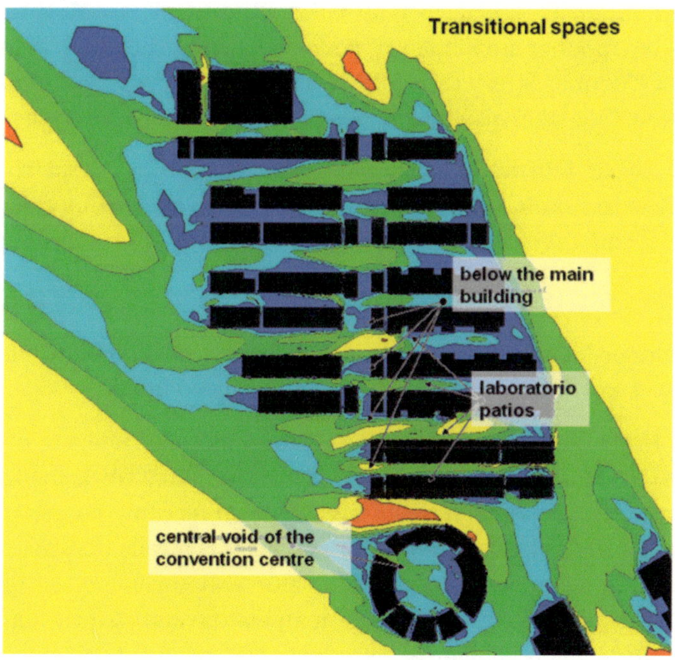

Figure 9. Simulation of air flow around and between the buildings of the new masterplan. Data about air speed was one of the fundamental variables to the prediction of thermal comfort in the open spaces of the masterplan.

The predictions of thermal comfort in outdoor spaces were established using the Outdoor Neutral Temperature (Tne), presented by Aroztegui [6], which considers as reference the concept of Neutral Temperature (Tn) introduced by Humphreys [7], who defined Neutral Temperature (tn) as the room temperature considered thermally neutral to a given population, observing the local conditions. The author presents a linear ratio between mean monthly temperature (tmm) and Neutral Temperature (tn), valid indoors in situations with low air speeds and mean radiant temperature close to air temperature.

$$Tn = 17.6 + 0.31 \bullet tmm \quad (1)$$

Where: tn = Neutral Temperature [ºC]; tmm = mean monthly temperature [ºC]

It is important to notice that the equation for the calculation of the Neutral Temperature is valid for the value range 18.5 ºC - 30.5 ºC, considering individuals in sedentary activity and wearing light clothing. For different human activities, the following corrections can be applied: light work (M=210W), -2.0ºC; moderate work (M=300W), -4.5ºC; heavy work (M=400W), -7.0ºC.

Aroztegui [6] proposed the Outdoor Neutral Temperature based on the same variables of the Neutral Temperature for internal spaces previously defined by Humphreys, to which variables related to sun irradiance and wind speed were incorporated. With respect to sun irradiation, the direct component should not be the only factor, but also diffuse irradiance and surrounding reflections. Regarding wind, the author highlights the need for simplifications, as variables associated with wind are difficult to value, as it is affected in space and time by random accidents at pedestrian level. Looking at other references, Givoni's Index of Thermal Stress (ITS) [8], is based on an empirical equation for indoor neutral temperature, that takes also into account variables that are characteristics of outdoor.

To establish the sweat rate in sedentary activity, considering mean conditions for the individual and the surrounding characteristics (with relative humidity ranging from 35% to 65%), the following equation for outdoor neutral temperature was established:

$$tne = 3.6 + 0.31\ tmm + \{100 + 0.1\ Rdn\ [1 - 0.52\ (v\ 0.2 - 0.88)]\} / 11.6\ v\ 0.3 \quad (2)$$

Where: tne = Outdoor Neutral Temperature [ºC]; tmm = mean monthly temperature [ºC]; Rdn = normal direct solar irradiance [W/m2]; v = wind speed [m/s].

Outdoor neutral temperature is estimated for a mean monthly temperature, corrected to a 50% relative humidity situation. Therefore, in the context of this work, the formulation of the new effective temperature was used to correct the mean monthly temperature values to an equivalent value of a 50% relative humidity situation. This means that, instead of using only the air mean monthly temperature value, this value was considered in terms of the New Effective Temperature, which considers the air temperature and also air humidity, providing equivalent temperature values, having as reference a 50% relative humidity situation.

According to ASRHAE [9], the New Effective Temperature (TE*) is the operative temperature of an enclosure at 50% relative humidity that would cause the same sensible plus latent heat exchange from a person at the actual environment, and it can be calculated by the following equation.

$$TE^* = to + w \bullet Im \bullet LR \bullet (pa - 0.5 \bullet psTE^*) \quad (3)$$

Where: to = operative temperature [ºC]; w = skin wetness [dimensionless]; Im = index of clothing permeability [dimensionless]; LR = Lewis relation; pa = vapour pressure [kPa]; psTE* = saturation pressure of the new effective temperature [kPa].

The Operative Temperature (to) is the uniform temperature of an imaginary black enclosure in which an occupant would exchange the same amount of heat by radiation plus convection as in the actual non uniform environment. It is numerically the average of bulb temperature (tbs) and mean radiant temperature (trm), weighted by their respective heat transfer coefficients (hc and hr). ASHRAE defines the equation for the Operative Temperature as follows [9]:

$$to = hr \bullet trm + hc \bullet tbs / (hr + hc) \quad (4)$$

Where: trm = mean radiant temperature [ºC]; tbs = dry bulb temperature [ºC]; hr = radiant exchange coefficient [W/m2 ºC]; hc = convective exchange coefficient [W/m2 ºC].

In this work the calculation of TE* adopted the proposed equations by Szokolay [10], according to which the new effective temperature is given by lines in the psychometric chart, crossing the curve of relative humidity of 50% for the given temperature [11]. These lines inclination equal to 0.023 • (TE*-14), if TE*<30, and 0.028 • (TE*-14), if TE*>30. Knowing the operative temperature and absolute humidity of a specific location, the new effective temperature was calculated through iterative process.

In this process, the New Effective Temperature is the mean temperature of all the hours from the previews thirty days. Assuming a tolerance range of ± 2.5°C to the outdoor neutral temperature, at least 90% of the users would be satisfied with the thermal environment conditions. Assuming a tolerance range of ± 3.5°C the satisfaction percentage drops down to 80%. In this research, the more restrictive range was applied, working with a satisfaction index superior to 90% of all users.

Three typologies for outdoor environments were studied configuring nine possible different environmental conditions, as shown in table 1 [11], in order to quantify the impact of different degrees of exposure to sun and wind in the overall thermal comfort in open spaces.

	rv	r*v	rv*	r*v*	r	r*	v	v*	-
cold	7,6%	14,5%	0,1%	0,1%	0,0%	0,0%	28,8%	0,4%	0,0%
comfort	57,0%	75,4%	42,4%	54,1%	0,3%	1,1%	68,8%	78,3%	1,7%
hot	35,4%	10,1%	57,5%	45,8%	99,8%	98,9%	2,4%	13,2%	98,3%

Table 1. Comparative results of the considered configurations

The key outdoor environments studied in this analysis were: the open central atrium of the convention centre, the patios of semi-enclosed gardens between the laboratories and the terraces at the rooftop of the main office building, at 10 meters high, (see figures10 to 12) being these three areas for mid and long-term permanence, also being some of the major circulation routes between buildings[1].

Figure 10. View of the open atrium of the Convention Centre.

Figure 11. The gardens between the two laboratory wings. Images of the still "immature" landscape.

Technical studies proved that in the central atrium of the convention centre, when exposed to the sun, one will be in comfort condition approximately for half of the time. When shaded, this figure goes up to nearly 85% of the time, showing a major improvement of outdoors thermal comfort, as seen in table 2 (see figure 10). Aiming for even better results, more and wider apertures between the multiuse rooms, which connect the central atrium to the immediate surroundings of the building, would increase the air flow in the open centre.

[1] All the analytical assessments were done considering the period of occupancy set by Petrobras (from 7am to 6pm), during weekdays of the whole year.

Figure 12. View of the terraces at the roof top of the Central Building, from the west to the east terrace.

Regarding the patios between the laboratories, considering total exposure to direct solar irradiation when the wind speed is lower, the predicted comfort conditions are identified only for 13% of the time. This percentage increases to 23% when 90% of incident solar irradiance is blocked by tree shading. On the other hand, when the wind speed is higher, the hours in thermal comfort raise to 67.5% of the time in the spots with direct solar irradiance, increasing to 98% of the time when the sun is blocked by the trees, as shown in tables 3 and 4 (see figure 11).

Considering the terraces at the rooftop of the main office building (the Central Building), the insulated sandwich metal roof, specified in the final design, provides a high percentage of time in comfort conditions (77%), against the metallic screen solution (64%), proposed in the

Month	cold	comfort	hot
January	0,0%	74,7%	25,3%
February	0,0%	82,3%	17,7%
March	0,0%	93,4%	6,6%
April	0,0%	91,5%	8,5%
May	0,0%	90,5%	9,5%
June	0,0%	100,0%	0,0%
July	0,0%	91,8%	8,2%
August	0,0%	96,3%	3,7%
September	0,0%	83,6%	16,4%
October	0,0%	82,2%	17,8%
November	0,0%	72,7%	27,3%
December	0,0%	83,5%	16,5%
Year	0,0%	86,9%	13,1%

Table 2. Thermal comfort at the open central area of the convention centre (shading)

January	0,0%	7,5%	92,5%
February	0,0%	13,6%	86,4%
March	0,0%	21,5%	78,5%
April	0,0%	21,1%	78,9%
May	0,0%	22,5%	77,5%
June	0,0%	46,6%	53,4%
July	0,0%	28,1%	71,9%
August	0,0%	39,7%	60,3%
September	0,0%	22,7%	77,3%
October	0,0%	18,6%	81,4%
November	0,0%	17,4%	82,6%
December	0,0%	19,5%	80,5%
Year	0,0%	23,2%	76,8%

Table 3. Thermal comfort at the open areas between the laboratories (low mean wind speed and shading)

Month	cold	comfort	hot
January	0,0%	100,0%	0,0%
February	0,0%	100,0%	0,0%
March	0,4%	99,6%	0,0%
April	0,6%	99,4%	0,0%
May	1,3%	98,7%	0,0%
June	5,9%	94,1%	0,0%
July	3,0%	97,0%	0,0%
August	7,9%	92,1%	0,0%
September	0,5%	99,5%	0,0%
October	0,0%	100,0%	0,0%
November	0,0%	100,0%	0,0%
December	0,0%	100,0%	0,0%
Year	1,8%	98,2%	0,0%

Table 4. Thermal comfort at the open areas between the laboratories (high mean wind speed and shading)

conceptual stage, as shown in tables 5 and 6. However, it should be noticed that the thermal performance of the metal screen roof, coupled with local shading strategies (such as small trees, green wired structures, umbrellas, etc), results in a percentage of hours in comfort in the terraces which is verified to be very close to the predicted results for the case of the sandwich metal roof (75%), as seen in table 7 (see figure 12).

Interestingly enough, although the percentage of hours in comfort is fairly close in both cases (insulated sandwich metal roof versus screen roof), the conditions related to discomfort show a significantly different performance. The discomfort in the case of the metal screen with local shading devices are associated with feeling "cold" for 15% of the time, due to slightly high wind speeds, and feeling "hot" for 10%. In the case of the insulated sandwich metal roof, the total hours of discomfort are practically due to feeling "hot". Despite the fact that the sandwich metal roof provides less time in discomfort, it has a distribution curve of thermal conditions tending to a more extreme part of the "hot" zone, i. e. its values are closer to the limit of hot discomfort. On the other hand, the metal screen solution with shading devices presents values closer to better comfort situations, i.e. they are located in the central part of the comfort zone.

Month	cold	comfort	hot
January	1,6%	71,9%	26,5%
February	1,4%	66,8%	31,8%
March	0,0%	78,9%	21,1%
April	0,0%	77,3%	22,7%
May	0,0%	67,1%	32,9%
June	0,0%	85,0%	15,0%
July	0,0%	65,8%	34,2%
August	0,0%	76,4%	23,6%
September	0,9%	89,1%	10,0%
October	1,6%	79,4%	19,0%
November	0,4%	84,3%	15,3%
December	0,9%	83,1%	16,0%
Year	0,6%	77,1%	22,3%

(insulated sandwich metal roof)

Table 5. Thermal comfort at the rooftop area of the central building

Month	cold	comfort	hot
January	7,1%	64,8%	28,1%
February	9,5%	58,2%	32,3%
March	9,9%	55,8%	34,3%
April	10,0%	54,7%	35,3%
May	10,0%	64,9%	25,1%
June	12,3%	66,4%	21,3%
July	5,6%	68,8%	25,6%
August	7,4%	59,9%	32,7%
September	17,7%	61,8%	20,5%
October	19,0%	60,1%	20,9%
November	14,5%	67,4%	18,1%
December	10,4%	64,5%	25,1%
Year	11,1%	62,3%	26,6%

Table 6. Thermal comfort at the rooftop area of the central building (metal screen)

Month	cold	comfort	hot
January	12,6%	72,7%	14,7%
February	13,6%	69,5%	16,9%
March	12,8%	73,6%	13,6%
April	12,9%	72,1%	14,9%
May	15,2%	72,7%	12,1%
June	16,2%	80,2%	3,6%
July	6,9%	80,5%	12,6%
August	12,8%	75,2%	12,0%
September	23,2%	70,9%	5,9%
October	24,1%	70,0%	5,9%
November	16,5%	78,1%	5,4%
December	18,2%	71,4%	10,4%
Year	15,4%	73,9%	10,7%

(metal screen with local shading)

Table 7. Thermal comfort at the rooftop area of the central building

Without barriers in the rooftop area, and since the wind speed at 10m is higher than at the ground, the wind speed in rooftop areas is more significant. Nevertheless, the cold discomfort periods caused by the wind could be reduced by adopting local wind breaks in the more affected areas of the rooftop. Moreover, given the hot-humid climatic conditions throughout the year in Rio de Janeiro, theoretical discomfort for "cold" can be challenged by real-life practice and be simply solved considering occupants adaption through clothing.

At first sight, comparing the insulated sandwich metal roof solution and the metal screen one, without other designing strategies, both present similar thermal comfort performance. But the metal screen solution has a higher potential to increase thermal comfort conditions, since local shading devices and wind brakes (if required) proved to increase comfort hours significantly. On the other hand, in the case of the insulated sandwich metal roof, in which almost all the situations of discomfort are hot ones, local solutions are not possible, since all solar irradiance is already blocked and it is difficult to induce, yet locally, an increment in the air speed.

Due to the possibilities of local solutions, the choice of metal screen provides a great diversity of environmental conditions, with different local figures of thermal radiation and air speed, which affect the thermal comfort sensation in open spaces. Providing environmental diversity, with areas of thermal sensation varying between slightly hotter to slightly colder, it is known that people can satisfy their comfort needs by choosing wherever they want to be, increasing even more the percentage of thermal comfort satisfaction. Nevertheless, in real practice, the possibility of creating a semi-open space protected from the rain led to the adoption of the sandwich metal roof (see figures 7 and 8). Technical assessment led to a final solution composed by metallic screen at the edges of the roof (shading the windows), sandwich metal panels in the main part of the roof, with strips of green glass along the roof surfaces and a central opening for air exhaustion placed along the entire length of the roof, In order to improve daylight and ventilation on the terraces.

Besides the results about the thermal comfort conditions in open spaces of the three buildings at the core of the expansion of the Research Centre, the study of insolation and wind speed in such areas have also informed the landscape design as a whole. Regarding the creation of shadows, the areas less shaded by buildings (or permanently exposed to the sun) have received a higher protection by the landscape than those already shaded, in order to create favourable conditions to thermal comfort in open spaces between and around buildings, taking into consideration the routes of people as well as the adequate conditions for short and long permanence.

Furthermore, the location of the different *restingas* implanted in the open gardens between the laboratories, extending until sea shore of the site, was reviewed based on the diagnosis of air movement around the buildings. The *restinga* of sand, initially proposed for the southern garden, the most exposed to wind, was replanted to the northern garden, thus avoiding sand grains displacements to other open spaces. In the same token, the analysis of insolation and daylight availability in the terraces of the Central were important for the choice of species that would best adapt to the specific conditions of daylight and heat exposure.

The analytical studies of thermal comfort in open and semi-open spaces of the Research Centre verified the possibility of satisfactory comfort scenarios, especially due to the well-planned design of shading and access to air movement in such areas, in the warm-humid tropical climate of Rio de Janeiro.

4. Daylighting

In response to the design brief, daylighting should be prioritized and maximized in all interior spaces where there is no functional restrictions, in order to provide visual comfort with energy efficiency [1]. Given the specific warm-humid climate of Rio de Janeiro and the major impact of solar irradiation on buildings thermal performance, coupled with the typical partially cloudy and bright local sky conditions, the major challenge of taking maximum benefit of daylight was related to the need for solar protection and avoidance of glare. For this purpose, the building form, together with roof's and facades' components were designed and sized with precision to shade the direct sun, whilst capturing and redirecting daylight to the deeper parts of the interior spaces (see figures 13 and 14).

The design and assessment processes of daylighting focused on the three main building typologies of the research centre, where buildings' orientation and form had a significant impact on the daylighting performance: the linear north and south rows of laboratories, the

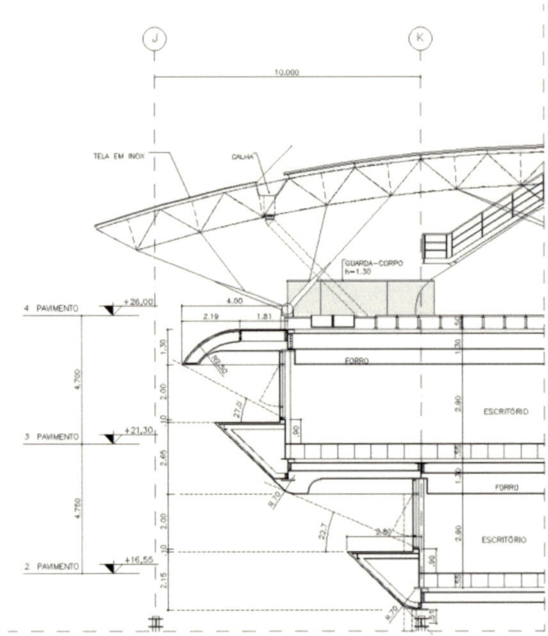

Figure 13. Central Building: design of the building's east facade to provide total solar protection to the working stations.

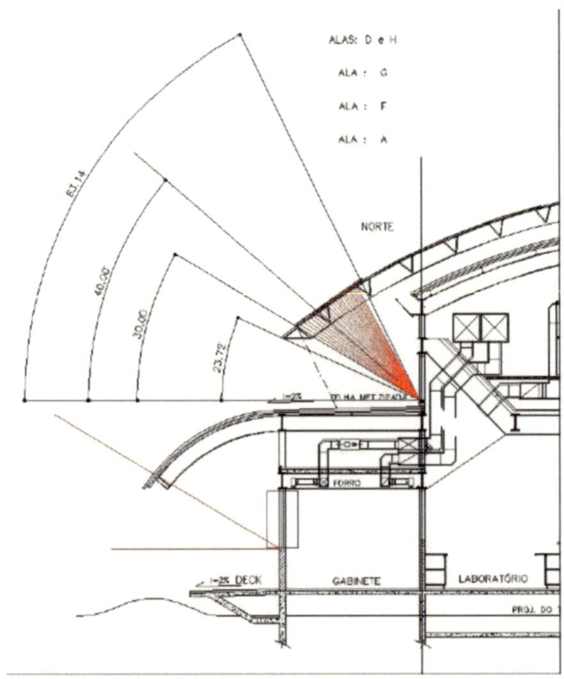

Figure 14. Typical north facing wing of laboratories: design of the shading device for the high-level window aiming to cut the direct solar radiation and redirect it as diffuse light to the deeper parts of the working areas.

multi-storey west and east facing office building, and the typical factory-like typology, in which the conflicts between natural ventilation, solar protection and penetration of daylight had a determining role in the design of the roofs. A number of in-depth simulation parametric studies were used to analyze architectural design possibilities for solar protection and penetration of daylight in all three cases [12].

In order to estimate internal daylight performance, Brazilian Standard NBR 15215 -3 calculation method was adopted [13]. Quantitative reference parameters in lux, from Brazilian Standard NBR 5413 [14], were complemented by Germany Standards DIN 5034, Daylight in interiors [15], DIN 5035, Artificial lighting of interiors [16] and LEED guidelines [17], adopting performance criteria as follows:

a. Uniformity: above 66%;
b. Daylight Factor: above 2%;
c. Illuminance Levels related to work planes: above 265 lx;
d. Daylight Factor and illuminance Levels should be reached on 75% of work plane areas, at least.

Computer simulation parameters for daylight assessment:

a. CIE overcast sky luminance distribution;
b. Unobstructed external horizontal plane illuminance: 13.250 lx[2];
c. Floor, wall and ceiling reflectance: 0.35; 0.60; 0.85;
d. Work plane height: 0.70m (offices); 1.00m (utility buildings);
e. Glass transmittance: 0.89 (general); 0.69(main office building and factory sheds).

In the main office building, the terraces (designed to hold social and working activities) showed areas with daylighting levels higher than 300 lx, nearly corresponding to a Daylight Factor (DF) of 2%. In the working environments close to the east facade facing the bay, the predominant DF identified was over 2%, despite the shading device, allowing appropriate levels up to the point of circulation between working stations, of approximately 2.50 meters deep. Further than this point, DF was below 2%, revealing the need for supplementary artificial lighting. In the west façade, facing the first phase of the Research Centre, the figures for the DF were over 2% in almost the whole area close to the openings, of about 5.5 meters deep. On the other hand, in the deeper plan areas (between 5,50 and 13.00 meters from the facade), the values of DF were predicted to be over 2% in 20% of the floor area (at working height) and 1.5% in 30% (see figures 13, 15 and 16).

---

[2] In parallel to the development of this project, ISO and CIE launched the Standard General Sky, presenting 15 sky relative luminance distribution types, including CIE Standard Clear Sky and CIE Standard Overcast Sky. Unfortunately, there was not enough measurement data to feature Brazilian sky in accordance with ISO/CIE 15 types. Hence, CIE Standard Overcast Sky was adopted to calculate values of DF. However, unobstructed external horizontal illuminance was established taking into account Brazilian climate conditions. The unobstructed external horizontal illuminance parameter (13.250 lx) adopted on computer simulations, reports 80% of annual occurrence. Increasing the unobstructed external horizontal illuminance parameter, the annual occurrence decreases, and the autonomy of daylight can be estimated.

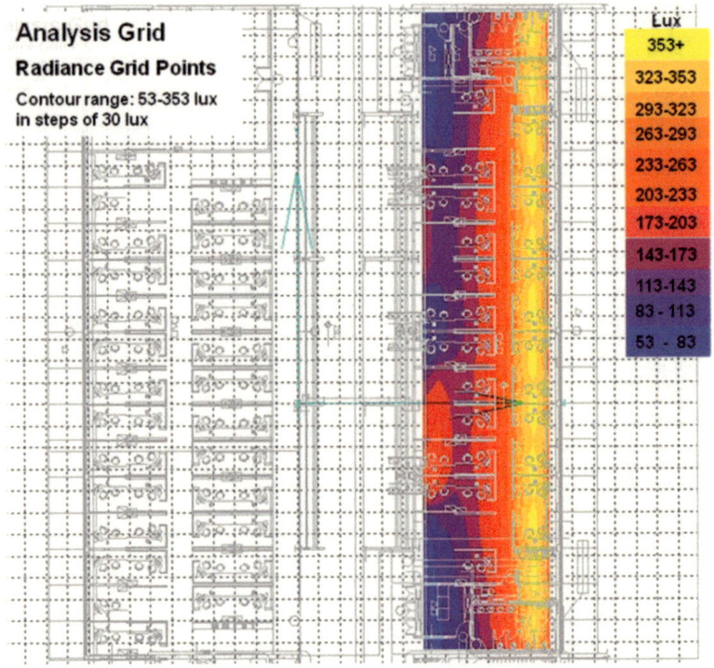

Figure 15. Central Building: daylight distribution on the 2nd floor of the east facing offices (relatively narrower plan in relation to the west facing offices).

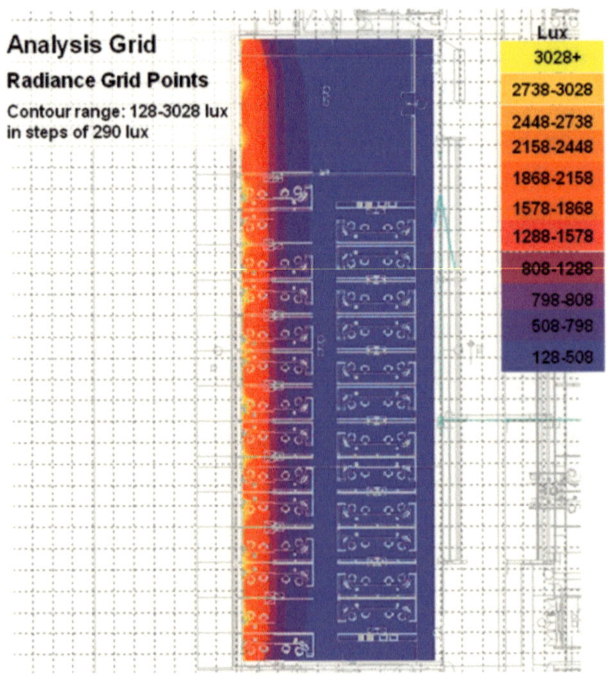

Figure 16. Central Building: daylight distribution on the 2nd floor of the west facing offices (the deepest plan area of the building).

In the laboratory buildings of north and south orientations, the performance of daylight in the two main working areas: the main (and bigger) working space and the contiguous office cell, showed minimum values of DF of approximate 2% in 75% of the main working plan of the south facing laboratories and 70% in the case of the north orientation. In the office cells the result was the same in both cases, being 50% (see figures 14, 17 and 18). The overhangs placed over the high level windows of the north facing laboratory buildings to block the direct solar radiation didn't compromise daylight performance, as it incorporates horizontal louvers which redirect diffuse light into the deeper areas of the main working laboratory room. In the south facing laboratories, the high-level window was kept with the maximum view of the sky to maximize daylight penetration, being shaded only on the side vertical plans against the early morning and late afternoon sun.

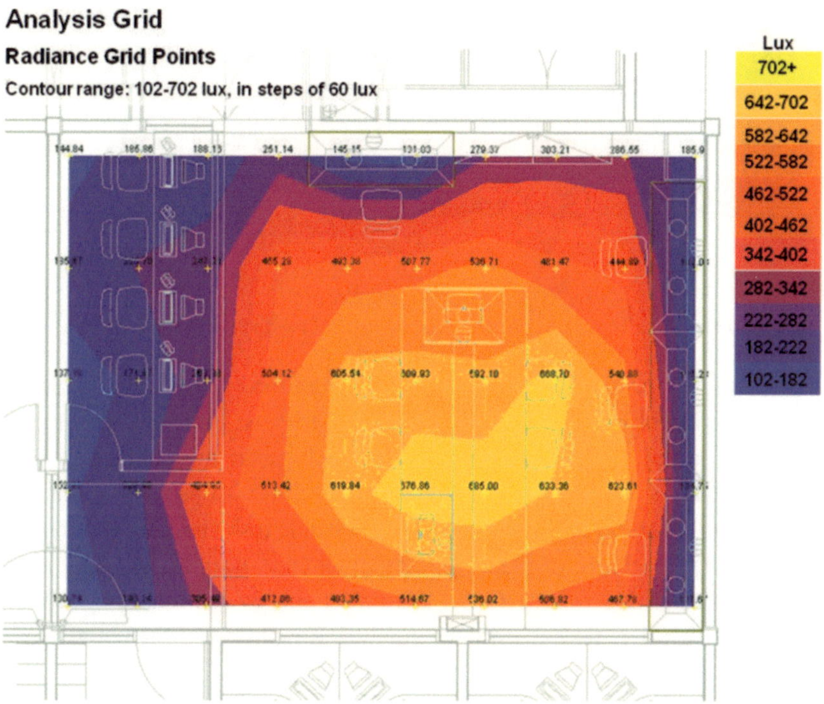

Figure 17. Laboratories: distribution of daylight in the typical south facing rooms.

In the sawtooth roof buildings (the factory type), characterized by top daylight, the minimum illuminance levels established by the national standard [14] were satisfied. However, changes in the design of the shed component in one of the buildings were needed to decrease the initially predicted illuminance levels. For that purpose, a decrease in the height of the shed and the reduction of over 50% of the sheds were tested, not just to reduce daylight (and the risk of glare), but also to cut down critical thermal loads.

Considering the adopted criteria, the illuminance level 265 lx is achieved with the Daylight Factor of 2%, found as a minimum value in most of the working plans in the laboratories and in the main office building. Although this value is close to the recommendation for

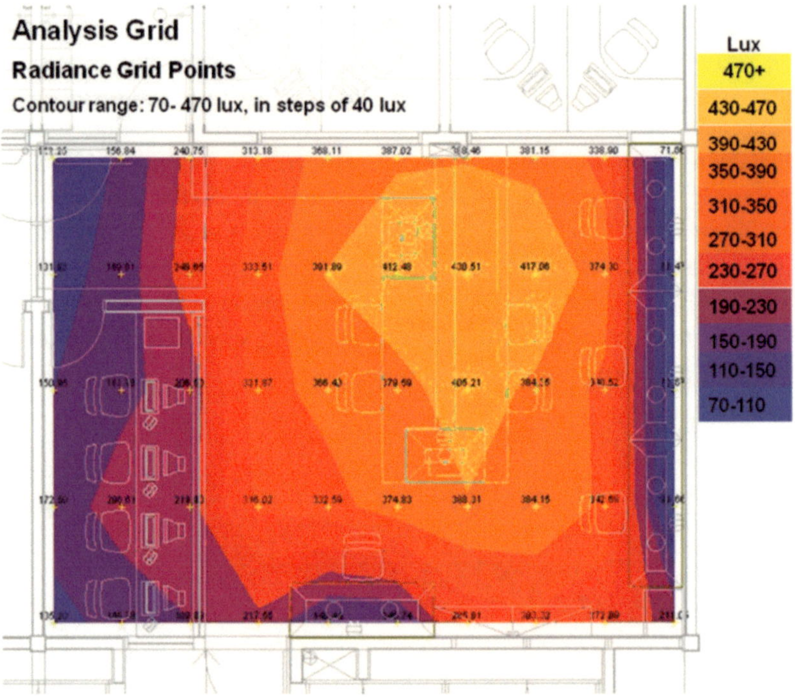

Figure 18. Laboratories: distribution of daylight in the typical north facing rooms.

reading rooms (300 lx), prescribed by the Brazilian Standard NBR 5413, it does not comply with the requirements for office activities, where the advised nominal illuminance is 500 lx [14]. Nevertheless, 500 lx could be considered to be high in the context of contemporary offices and the related work conditions, harming the visibility of computer screens. Adjusting Brazilian Standards guidelines and visual task needs, the complementary artificial lighting was adopted, placing supplementary luminaries over restrict areas controlled by the user. This strategy allows estimated energy savings of approximately 40% to 45% of artificial lighting, taking into account the Research Centre usual working hours between 7 am to 4 pm, during 5 days a week [12]. Almost all main office areas and laboratory rooms presented well balanced daylight distribution. Yet, the uniformity below 0.66 was identified in some places, such as in the office building second floor of the west facing wing of the main office building, where uniformity resulted in 0.35, indicating the need for supplementary artificial lighting, to adjust lighting levels to the task requirements around the room.

In summary, the outcomes of daylight uniformity were:

a. Main office building, 3rd floor rooms, west orientation : 0.82 – 0.88;
b. Main office building, 2nd floor rooms, east orientation: 0.82 – 0.88;
c. Main office building, 2nd floor rooms, west orientation: 0.35 – 0.75;
d. North oriented laboratories: 0.94;
e. South oriented laboratories: 0.94.

Daylight Factor and Uniformity calculations provided artificial lighting design optimization. By means of isolines plotted on work planes, one could map each room and identify natural lighting availability inside the building, detecting zones and establishing the proper strategy for each one [12]. In this sense, artificial light keeps up with natural light variations, adjusting lighting levels and distribution in order to improve luminous environment quantity and quality, intending good visibility and visual comfort.

The daylighting analysis made use of concept of Daylight Autonomy [12], to predict periods only on account of natural light providing the proper visual task lighting level. Although the minimal value of 2% for the Daylight Factor is adopted by some international standards to consider a well-daylit space, regarding tropical countries, daylight availability can be increased due to high sky luminance conditions, consequently, the minimal Daylight Factor value can be decreased. On this basis, minimal value of 1.5% for DF was considered accepted in the context of this work, whereas the minimal of 2% remained as an initial reference. Being so, by maintaining the same criteria, external illuminance parameter was increased to reach the same internal illuminance value related to DF 1.5% and 2%. In essence, beyond the performance indicators, quality of daylight in most of the working areas, including daylight distribution, views, visual communication and absence of contrast and glare, was achieved with the design of the façade's components in relation to the orientation and building form.

5. Naturally ventilated buildings

The main natural ventilated buildings in the extension of the Research Centre were two: *Operational Support Building* and *Utilities Centre.* In addition, both the main office building and the laboratory buildings have spaces optimized to be naturally ventilated for a percentage of the year (see topic: *Air conditioned buildings, in the sequence).* The assessment of the naturally ventilated internal spaces involved a set of analytical work developed with the support of advanced computer simulations of thermal and computer fluid dynamics, according to the following procedures [19]:

a. elaboration of a 5 year weather file (1998 - 2004) with hourly data from the Tom Jobim International Airport station [4];
b. identification of the critical summer month (February 2003) and a summer reference design day for in-depth assessments of heat flow through the building envelope;
c. establishment of comfort parameters for naturally ventilated internal spaces;
d. definition of base cases including building's thermal zoning, occupation profile, construction materials and internal thermal loads;
e. advanced computer simulations of fluid dynamics (carried out with the software CFX 5.7) to determine air flow around buildings, pressure coefficients, air velocity on the openings of buildings' envelope and convection exchange coefficients;
f. thermal dynamic simulations (carried out with the software TAS v.9.0.5) in order to quantify buildings' thermal performance and, consequently, estimate the potential number of hours in comfort;

g. quantification of thermal loads for air-conditioned spaces;
h. elaboration of design guidelines for a better environmental and energy performance of buildings.

In the typical warm-humid climate of Rio de Janeiro, the thermal performance of naturally ventilated buildings is dependent on the provision of efficient shadowing strategies, alongside enhanced cross ventilation and possibly external thermal insulation (to be tested on a case by case scenario). However, the ultimate success of naturally ventilated spaces will depend on the definition of the comfort band and its acceptable conditions.

At the time of the project (2004-2005) there were two national buildings' standards related to the general internal environmental conditions of working spaces: NR- 15 [20] and NR 17 [21]. The first one refers mainly to extremely hot environments, such as factories, and recommends resting periods related to people's exposure to above certain indoor temperatures, whereas the second one says that the environment must be adequate to the psycho-physiological characteristics of the workers and to the nature of the task. For working spaces in which intellectual demand and constant attention are required, NR-17 recommends the following design parameters: effective temperature between 20°C and 23°C; air relative humidity not inferior to 40%; air speed not superior than 0.75 m/s. Even though NR-17 applied theoretically to all environments of interest in this work, its parameters were impossible to be attained in a free running mode, in any time of the reference year, given its narrow limits.

Aiming to maximize natural ventilation, the adaptive comfort model adopted by ASHRAE [22] and created by De Dear [23] was adopted here. It brings an empirical model which includes acclimatization, clothing options, behavior patterns and tolerance to climatic variability (based on the New Effective Temperature (TE*), seen in *Thermal comfort in open spaces)* (see figure 19).

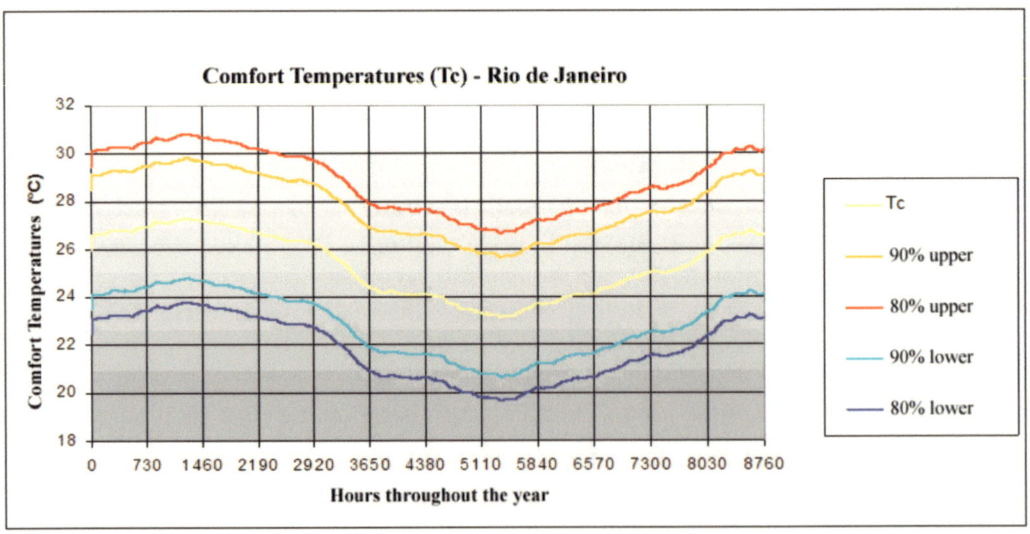

Figure 19. Comfort zone for the climate of Rio de Janeiro, created after De Dear's adaptive model [23].

For naturally ventilated internal spaces, dry bulb temperature, relative humidity and mean radiant temperature extracted from the simulations' results are used to calculate TE* for each hour, which is assessed in relation to the comfort zone. Comparing the TE* figures calculated for each hour of the reference year with the limits of the adaptive comfort zone created for the climate of Rio de Janeiro, one can observe that 50% of occupation time are in accordance with the acceptable thermal comfort conditions [19]. In other words, in a hypothetical scenario where the internal conditions are equivalent to the calculated TE*(for the external climatic conditions), comfort would be obtained for 50% of the occupied hours, which was then considered as a reference for the performance of the naturally ventilated environments.

The thermal performance of buildings and the potential for natural ventilation were determined by advanced thermal dynamics' computer simulations with the software TAS. The simulated environment was divided in a number of thermal zones, in which the influence of neighboring rooms is taken into account. In that way, several thermal phenomena can be evaluated in terms of simultaneity, location and interaction. Building modeling was carried out in two stages: geometry characteristics, and materials specification and occupation schedules.

Preliminary analytical studies showed that the sawtooth roofs of the factory-type buildings facing southeast in order to get the minimum direct sun to the maximum diffuse light, incurred in negative impacts on the natural ventilation of the internal environments, as the southeast coincides with the direction of prevailing winds. For this reason, the shed elements of the sawtooth roofs were re-designed changing the position of the opening for ventilation and including a solar and wind protection device (see figures 20 and 21). Without changing the original orientation of the roof (the best one to minimize solar gains), the new solution prevented the unfavourable effect of the prevailing wind on the air outlet of the stack effect, simultaneously taking advantage of the negative pressure generated to maximize the exhaustion of the internal air flow. In addition, thermal insulation was proved to be beneficial in the buildings' envelop and external solar protection was included on the windows and roof openings of all main buildings. The specification of materials and building components used in the thermal simulations are shown in Table 8.

The factory-type building form played a decisive role in the success of the passive strategies, since the double and triple floor to ceiling heights and the sawtooth roof favoured natural ventilation through stack effect, whilst bringing diffuse daylight into the deeper parts of the floor plan with protection against solar radiation. Shading and ventilation were fundamental strategies for the achievement of satisfactory thermal comfort conditions in the naturally ventilated buildings, followed by external insulation coupled with internal thermal inertia of the buildings' envelop. In the case of interior environments with higher internal heat gains, mechanical ventilation had to be introduced, especially in the hot periods of the year (from December to February). This was the case of the changing rooms in the*Operational Support Building*, where the mechanical ventilation had to provide up to 15 air changes per hour. The mechanical ventilation proved to have a positive impact also in the distribution of air flow within the environments, reaching with more homogeneity the occupied zone.

Figure 20. The Operational Support Building, one of the factory type buildings featuring triple floor to ceiling heights and shed structures in the roof for daylight and stack ventilation.

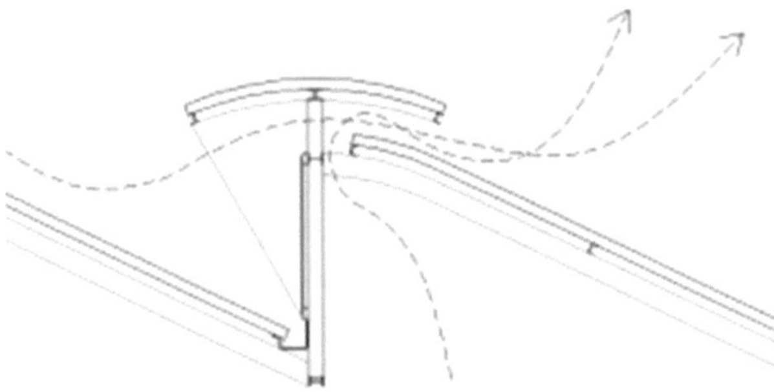

Figure 21. Design of shed showing the ventilation aperture on the back of the shaded glazed area.

Preliminary solution	Final solution
EXTERNAL WALL	
150mm concrete (d=1,200kg/m^3) U = 1,77 W/m^{2}ºC	ceramics + cement + 25mm gypsum +200mm air gap + 120mm concrete (d=2.500kg/m^3) U = 1,63 W/m^{2}ºC
FLOOR	
Ceramics + cement + 150mm concrete (d=2.200kg/m3) + steel +1000mm air gap + 12,5mm gypsum U = 0,82 W/m^{2}ºC	Ceramics + cement + 100mm concrete (d=2.500kg/m3) + steel +1000mm air gap + 12,5mm gypsum U = 0,85 W/m^{2}ºC
GLAZING	
6mm clear single glazing U=5,73W/m^{2}ºC	8mm clear single glazing U=5,66 W/m^{2}ºC
CEILING/ROOF	
aluminium + 50mm rock wool + aluminium (metallic sandwich panel) U=0,58 W/m^{2}ºC	

Table 8. Materials and building components applied in the thermal dynamic simulations of the naturally ventilated factory type buildings

Looking at the yearly percentage of comfort hours in the naturally ventilated environments, the highest performance was found in the changing rooms in the Operational Support Building, featuring approximately 60% of the yearly occupation hours within comfort conditions (considering the hours found in "cold" conditions are insignificant) (see table 9). In addition, it was estimated that the addition of the mechanical system would increment the number of air changes per hour could raise the hours in comfort to 75% (an increase of 65% comparatively to its performance before the implementation of the initial architectural modifications) [19]. On the other hand, the lowest performance was in the kitchen of the same building marking 51% of hours in comfort (see table 10). In conclusion, it can be said that the results of the thermal performance of the naturally ventilated internal environments (seen in number of hours within the comfort zone) shows the adequacy of the architectural response to the specific warm-humid conditions of the climate of Rio de Janeiro.

	Daily occupation 5am-7pm					
Month	**PDD = 20%**			**PDD = 10%**		
	Cold	**Comfort**	**Hot**	**Cold**	**Comfort**	**Hot**
January	0.0%	32.8%	67.2%	0.0%	20.6%	79.4%
February	0.0%	32.3%	67.7%	0.0%	24.7%	75.3%
March	0.0%	47.9%	52.1%	0.0%	34.2%	65.8%
April	0.0%	47.9%	52.1%	0.0%	34.2%	65.8%
May	0.0%	49.5%	50.5%	0.0%	38.1%	61.9%
June	2.0%	87.0%	11.0%	8.1%	65.8%	26.1%
July	1.6%	76.2%	22.2%	3.8%	59.4%	36.8%
August	7.3%	76.7%	16.1%	15.5%	56.4%	28.2%
September	0.3%	76.3%	23.3%	4.0%	57.0%	39.0%
October	0.0%	60.3%	39.7%	0.3%	42.9%	56.8%
November	0.0%	47.6%	52.4%	0.6%	31.5%	67.9%
December	0.0%	44.4%	55.6%	0.0%	29.5%	70.5%
Year	**1.1%**	**58.5%**	**40.4%**	**3.2%**	**42.6%**	**54.2%**

Table 9. Operational Support Building: Percentage of predicted yearly hours in comfort in changing rooms.

	Daily occupation 5am-7pm					
Month	**PDD = 20%**			**PDD = 10%**		
	Cold	**Comfort**	**Hot**	**Cold**	**Comfort**	**Hot**
January	0.0%	21.4%	78.6%	0.0%	8.4%	91.6%
February	0.0%	25.0%	75.0%	0.0%	9.3%	90.7%
March	0.0%	38.8%	61.2%	0.0%	19.1%	80.9%
April	0.0%	38.8%	61.2%	0.0%	19.1%	80.9%
May	0.0%	45.1%	54.9%	0.0%	21.3%	78.7%
June	0.0%	93.3%	6.7%	0.0%	66.4%	33.6%
July	0.0%	70.8%	29.2%	0.0%	41.6%	58.4%
August	0.0%	81.8%	18.2%	0.0%	56.7%	43.3%
September	0.0%	63.3%	36.7%	0.0%	39.3%	60.7%
October	0.0%	46.1%	53.9%	0.0%	23.5%	76.5%
November	0.0%	32.4%	67.6%	0.0%	16.7%	83.3%
December	0.0%	34.6%	65.4%	0.0%	16.8%	83.2%
Year	**0.0%**	**51.3%**	**48.7%**	**0.0%**	**29.8%**	**70.2%**

Table 10. Operational Support Building: Percentage of predicted yearly hours in comfort in kitchen of the Operational Support Building.

6. Air conditioned spaces

This work looks at the thermal performance of the specific areas within the air conditioned buildings which have the potential to be naturally ventilated for parts of the year: the office cells of the laboratories and the working spaces of the main office building. The assessment of the air conditioned and mixed-mode spaces followed the main procedures described for the analysis of the naturally ventilated environments, with the inclusion of some specific studies:

a. in addition to the annual performance assessment, specific studies were developed for the critical summer month of the weather file (February 2003), aiming to analyze the maximum thermal loads and the heat flows through the building envelop;
b. comfort parameters different from the ones used for the natural ventilated buildings were established for the air conditioned spaces;
c. thermal dynamic simulations: were first carried out with the software TAS without the air conditioning mode, in order to determine the potential number of hours in comfort;
d. subsequently, the loads for the air conditioning and mixed-mode strategies were simulated;
e. finally, the results were analysed and changes were proposed in the architectural design, materials and criteria of operation of buildings in order to increase the naturally ventilated hours and improve comfort conditions, where possible, and reduce cooling loads where air conditioning was necessary either for part or the total time of the year.

With respects to the thermal comfort parameters for air conditioned spaces, air conditioning settings were established based on the international standard ISO 7730 [24] and the correlated regulations ISO 7726 [25], ISO 8996 [26] and ISO 9920 [27], and compared with the Brazilian national regulations.

ISO 7730 estimates the predicted percentage of dissatisfied people (PPD) in a given thermal environment and recommends a PPD value inferior to 10%. However, such design parameters are limited to the following conditions: dbt between 10 °C and 30 °C, rh between 30% and 70% and air velocity under 1 m/s. Within the context of the national standards, as previously presented for the naturally ventilated buildings at the time of the project there were two buildings' regulations concerning the internal environmental conditions in general working spaces, NR- 15 [20] and NR 17 [21]. For the air conditioned spaces, there were two other national standards: NBR 6401 [28] and the *Orientação Técnica sobre Padrões Referenciais de Qualidade do Ar Interior* (Technical Orientation on Air Quality Standards) by the National Sanitary Supervision Agency, ANVISA [29]. In both cases, the recommended ranges for dry bulb temperature (dbt) and relative humidity (rh) are: dbt = 23–26°C and rh = 40–65%; dbt_{max} = 26.5–27 °C and rh_{max} = 65%; dbt_{max} = 28 °C and rh = 70% (for access areas); and considering air velocity at 1.5 m ($v_{a.1,5m}$) = 0,025- 0,25m/s. Based on the previous references, exploratory studies considered combinations of dbt, rh and v_a for a PPD inferior to 10%, assuming:

a. Metabolic rate (M) for sedentary activity (ISO 8996) M = 70 W/m^2 (1.2 met);
b. Clothing thermal resistance I_{clo} (ISO 9920) of 0.5 clo (shirt with short sleeves, light trousers, underwear, socks and shoes);

c. mrt (mean radiant temperature) = dbt, since the envelopes are shadowed and/or insulated. Therefore (ISO 7726) to = tbs; and
d. v_a < 0.25 m/s for light or sedentary activities during summer, if to < 26 °C (ISO 7730; ASHRAE 55). Above 26 °C, air velocity should be under 0.8 m/s.

In summary, the recommended air-conditioning settings considered dbt = 26 °C, rh = 65% and v_a = 0,1m/s, complying with ISO 7730 and the Brazilian national regulations, what resulted in higher air temperatures than those frequently adopted by the current Brazilian practice.

A mixed-mode strategy, alternating natural ventilation and active cooling was suggested for the office cells of the laboratory buildings and the specific areas of the main office building, where a minimum of 30%[3] of the occupation time was found to be within thermal comfort conditions in the free-running mode. The parameters adopted in this assessment for the mixed-mode strategy were:

a. window opening at tbs=20 °C, with a gradual increase in the natural ventilation rate;
b. window closing and air-conditioning activation at tbs>26 °C, keeping an indoor tbs of 26 °C ou 24 °C (according to the comfort criteria established for each occupation area); and
c. window closing when external wind velocity $v_a \geq$ 5.0m/s, as found in the ventilation assessments as the threshold for acceptable internal air velocities.

A comparative analysis of thermal performance and energy efficiency of the various solutions was carried out according to the following criteria:

a. the periods of comfort during the free running mode were determined by the adaptive comfort model [22, 23], for a dissatisfaction index of 10% and 20%; the percentage of hours of "comfort", "warm" or "cold" was calculated for each thermal zone during the occupied time;
b. the maximum thermal load was established by the highest load of the design reference day, and it was provided to inform the sizing of the air-conditioning system; only the internal loads were considered (regardless of the air exchange loads in the coil). It is worth to notice that the thermal loads' values of the annual assessments which exceed the simulations of the reference day are never superior to 5% of the occupation period; therefore, the recommendations for the sizing of the air-conditioning system follow the reference day recommendations;
c. a profile of annual loads created based on the frequency of occurrence (both accumulated and by intervals) of thermal loads during the year; and
d. a profile of cooling loads of the selected spaces was also created for the reference day.

[3] This value is based on a general cost–benefit analysis for European projects, since there are no benchmarks for tropical climates or for the Brazilian economic reality [30]. Although the figure of 30% was taken as a reference, it could not be a benchmark, since economic criteria including air-conditioning running costs as well as the associated costs of operable façades have to be considered in the definition of such a target.

Preliminary parametric investigations were carried out for a typical office cell of one of the laboratory buildings to provide an overview of the possibilities and limitations of a mixed-mode approach in such spaces (see figure 22). The test-room was in the most exposed side of the laboratory buildings, with one exposed façade towards the west, however with openings shaded, opaque surfaces with light colour finishing and with a maximum occupancy of two people. Simulation parameters included north and south orientated offices, and the possibility of users' control of both windows for natural ventilation and the air-conditioning system. The specification of materials applied in these simulations are described in table 11.

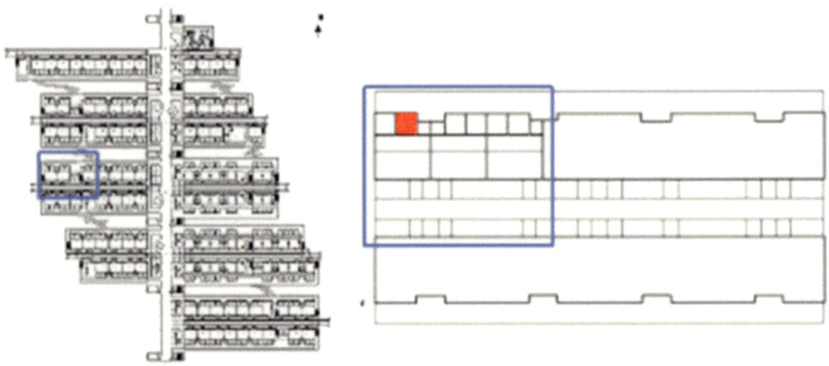

Figure 22. Location of the laboratory used as the first base-case for the thermal assessment.

Occupancy parameters	Elements	Materials	U value (W/m^{2}ºC)
2 people + 15 W/m^2 equipment load + 12 W/m^2 lighting	Windows	Aluminum frame (5%) + transparent float glass (one sheet -6 mm)	5,73
	Door	Wood (3cm)	
	Walls	Cellular concrete panel (15cm), no coat	0,90
	Floor	Concrete 2000kg/m^3 (15cm) + air cavity (50cm) + concrete 400 kg/m^3 (15cm) in contact with the soil	1,10
	Roof	Plasterboard (2cm) + air cavity (50cm) + concrete 400 kg/m^3 (15cm) + air cavity (100cm) + sandwich panel (15cm)	0,25

Table 11. Materials and building components applied in the thermal dynamic simulations of the typical office space of the Laboratories' office cells.

The results of the preliminary assessment indicated a potential annual period of natural ventilation ranging from 13% to 30%, higher values were found with the inclusion of thermal mass, adding thermal inertia to the internal spaces and considering the entire window as a ventilation aperture, opened for 24hrs during the days of thermal comfort conditions [3]. Excluding the hottest period of the year (from December to February), the performance of natural ventilation increases to 50% of the occupied hours. Based on such results, the mixed-mode strategy has proved to be a positive approach, especially for environments with low internal gains. Furthermore, regarding the set points of the air-conditioning system, the exploratory simulations showed a total reduction in thermal loads of 22% for set point conditions of 26 °C and 60% rh, compared to the scenario of 24 °C and 60% rh, therefore showing the significant impact of comfort standards on building energy efficiency, in the specific case of the Petrobras Research Centre [3].

More in-depth spaces of comparing different environmental control settings of the air conditioning system showed a even more significant impact in the final cooling demand of the spaces, apart from allow a longer period of natural ventilation. Looking at the the case of the east-facing working areas in the main office building, a change in the operation parameters from 26 °C and 65% rh to 24 °C and 50% rh reflected in 28% increase in the room's total annual loads. A higher impact was found for the laboratories' offices, marking 42%, and resulting in a decrease of one month in the natural ventilation potential period. However, the final simulations for annual loads considered the environmental scenario of 24 °C and 50%, as determined in the engineering design as an operation parameter for the cooling system, whereas the settings for the sizing of the cooling systems were 22 °C and 50%.

As a consequence of introducing hours of natural ventilation, in the case of the south-facing laboratories, the mixed-mode strategy led to a reduction of 10% in the total annual loads in comparison with the full air-conditioning mode, and a 50% decrease was found considering the maximum thermal loads. For the working areas in the main office building, the mixed-mode strategy was only not possible in the west-wing offices natural ventilation was not recommended due to acoustic issues related to the motorway that separates phase one and phase two of the Research Center[4].

Regarding the specification of windows, transparent laminated double glazing windows with air filling (10+28+6 mm) were tested against transparent laminated 8mm single glazing. The double glazing has not presented significant benefits for the thermal performance of the offices of the laboratories, due to the periods of natural ventilation. Similar findings occurred for the east-facing rooms of the main office building. As opposed to that, in the

[4] Natural ventilation in such spaces was only possible through the openings on the internal façade, adjacent to the central corridor, which proved to be insufficient, so that the west-wing offices were designed from the outset to operate on the air-conditioning mode [19].

case of the west-wing offices, which present fixed façades, as already mentioned, transparent double glazing was recommended for both façades in order to increase the efficiency of the air-conditioning system.

Regarding the specification of windows, transparent laminated double glazing windows with air filling (10+28+6 mm) were tested against transparent laminated single glazing (8mm). The double glazing has not presented significant benefits for the thermal performance of the offices of the laboratories, due to the periods of natural ventilation. Similar findings occurred for the east-facing rooms of the main office building (see table 12). As opposed to that, in the case of the west-wing offices, which ended up with fixed windows, transparent double glazing was recommended for both façades in order to increase the efficiency of the air-conditioning system.

	total annual load (MWh)		
office / room	clear single glazing	clear + green single glazing	
	26°C / 65%	26°C / 65%	24°C / 50%
east (mixed-mode)	58,8	57,0	79,1
west (full a/c)	62,9	61,8	81,3

Table 12. Total annual loads for the east and west office rooms, with different glazing types and air-conditioning settings.

The thermal performance of the working spaces of the main office building was strongly influenced by the big roof, a fundamental element of the design of such building, covering the terraces and the office floors. Different design solutions were tested for the roof, aiming to find a balance between daylighting and the thermal conditions of both the offices and the open spaces of the terraces (as seen in the *Thermal comfort in open spaces*). In comparison to the original design based on a metallic screen, permeable to air and light, the final solution of the insulated sandwich metal panels showed reduction of 20% in the cooling loads of the office areas located at the top floors o, mainly due to a reduction in the solar heat gains through the open corridor between the two office wings (see Table 13).

With special concerns to the main office building, in spite of the environmental and energy assessment carried out, some architectural aspects, such as raised floors and insulated ceilings, resulted in internal environments similar to those from the commercial international standards (see Table 13). Alternatively, the working spaces of the laboratories, with reduced dimensions and relatively low thermal loads, additionally to internal thermal mass use, presented a comparatively higher performance, mainly due to the natural ventilation (see Table 14).

Preliminary solution	Final solution
EXTERNAL WALL	
25mm gypsum +50mm glass wool +25mm concrete, d=2.200kg/m3 U = 0,66 W/m²ºC	25mm gypsum + 50mm rock wool +200mm air gap + aluminium U = 0,43 W/m²ºC
FLOOR	
raised carpet floor (metal+120mm concrete d=2.200kg/m³ + 375mm air gap + metal + carpet) U = 0,85 W/m²ºC	
GLAZING	
6mm clear single glazing U=5,73W/m²ºC	- 8mm clear single glazing, U=5,66 W/m²ºC - 8mm green single glazing, U=5,66 W/m²ºC
CANOPY	
51mm metallic sandwich panel (EPS filling) U=4,76 W/m²ºC	Composition: perforated metal panel (40% void)+ metalic sandwich panel + 8mm green single glazing+ shed

Table 13. Materials and building components applied in the thermal dynamic simulations of the main office building: preliminary vs. final solution

Preliminary solution	Final solution
EXTERNAL WALL	
25mm gypsum +50mm glass wool +25mm concrete, d=2.200kg/m³ U = 0,66 W/m²ºC	120mm celular concrete d=2.500kg/m³ + plasterboard U = 1,07 W/m²ºC
FLOOR	
110mm concrete d=2.200kg/m³ + 700mm air gap + 150mm concrete, d=2.800kg/m³ U = 1,38 W/m²ºC	200mm aereted concrete d=2.500kg/ m³ + impermeabilization + cement + ceramics U = 1,52 W/m²ºC
GLAZING	
6mm clear single glazing U=5,73W/m²ºC	8mm clear single glazing U=5,66 W/m²ºC
CEILING/ROOF	
gypsum +400mm air gap +110mm concrete, d=2.200kg/m³ U=1,82 W/m²ºC	12,5mm gypsum +1530mm air gap + metallic sandwich panel U=0,39 W/m²ºC

Table 14. Materials and building components tested in the thermal dynamic simulations of the office cells of the Laboratories: preliminary vs. final solution

Regarding the thermal performance of building components, it was verified that, in general, U values close to 1,00 W/m2 ºC and higher for the walls of both naturally ventilated and air conditioned spaces and modes, coupled with light external colours, give a good balance between heat losses and heat gains in the specific climatic conditions of Rio de Janeiro. On the other hand, the roofs had to be more insulated, given the intensity of solar radiation throughout the year, with U values between 0,25 W/m2 ºC and close to 0,50 W/m2 ºC in the great majority of the buildings, reinforcing the validity of the initial hypothesis of double roofs with extra shading and insulation for a better thermal performance of the internal spaces.

It must be noticed that the mixed-mode approach to the control of the internal thermal environments is a new concept in Brazil, where most air-conditioned buildings are sealed boxes with raised floors and insulated ceilings. Nevertheless, in the context of this project it is considered that the mixed-mode approach can bring economic and environmental benefits, introducing periods of natural ventilation and allowing for fluctuations in the internal environmental conditions within the limits of a given comfort zone, whilst also reducing the thermal loads of the air-conditioning periods. It also increases the user interaction with the exterior, fact which has proved to have positive psychological effects in the occupation of buildings [31].

Despite the technical reasoning that would support the use of a mixed-mode approach to internal environmental control, in the final design of buildings and systems, all environments in which air-conditioning should be necessary for any period of the year became fully air-conditioned main due to a established design culture. Notwithstanding, the specification of operable windows, coupled with an air distribution zoning in the office environments in both in the laboratories and in the working areas of the main office building (one fan-coil for each unity), granted the occupant with the possibility of choosing for periods of natural ventilation, if technically possible and desirable.

7. Final considerations

The challenge of bringing together design considerations about outdoors comfort, good daylight, the need and possibility of natural ventilation and energy efficient buildings' envelop was geared towards quality spaces for the occupants and visitors, rather than focusing only on energy efficiency issues. In this respect, the core objective of the environmental design approach was to create inviting open and semi-open spaces as well as comfortable internal spaces where the occupants are intuitively led to interact with means of controlling their own environmental conditions, ultimately achieving levels of environmental and energy performance beyond what was shown by the analytical predictions.

The potential environmental performance of the buildings that encompass the extension of the Petrobras Research Centre in Rio de Janeiro, which was firstly developed at the design level, has been put to test by its real occupation which took place in June 2010. An Initial and informal feedback has confirmed the satisfaction of the occupants with the environmental quality of both naturally ventilated and air conditioned buildings, translated into internal visual communication and access to daylight, views towards the outside from all working spaces, physiological cooling through natural ventilation and inviting shaded transitional spaces.

Possibly, the positive response of the occupants about their working environments will maximize the use of natural ventilation in the working spaces where environmental quality can induce a better performance than the predicted in the simulations, since a more flexible and adaptable notion of comfort could be put in practice by the occupants as a consequence of the overall quality of the spaces. This could be the case of the office cells in the laboratories, for instance, where the relatively small internal loads, coupled with the internal thermal mass and operable windows facing the gardens could spontaneously lead to the preference for natural ventilation rather than the air conditioning system.

Overall, the environmental quality and performance potential achieved in the design of the extension of the Petrobras Research Centre in Rio de Janeiro define a new reference for future environmentally responsive buildings in Rio de Janeiro, exploring environmental attributes from outside the boundaries (and often limited) of energy performance standards.

Author details

Joana Carla Soares Gonçalves[1,2,*]
[1]Laboratório de Conforto Ambiental e Eficiência Energética (LABAUT), Departamento de Tecnologia da Arquitetura (AUT), Faculdade de Arquitetura e Urbanismo, Universidade de São Paulo (FAUUSP), São Paulo, Brasil
[2]Environment & Energy Studies Programme Architectural Association School of Architecture, London, Great Britain

Denise Duarte, Leonardo Marques Monteiro and Mônica Pereira Marcondes
Laboratório de Conforto Ambiental e Eficiência Energética (LABAUT), Departamento de Tecnologia da Arquitetura (AUT), Faculdade de Arquitetura e Urbanismo, Universidade de São Paulo (FAUUSP), São Paulo, Brasil

Norberto Corrêa da Silva Moura
Departamento de Tecnologia da Arquitetura (AUT), Faculdade de Arquitetura e Urbanismo, Universidade de São Paulo (FAUUSP), São Paulo, Brasil

Acknowledgement

Thanks to: Anésia Frota, Anna Miana, Alessandra Shimomura, Bruna Luz, Cecília Muller, Gisele de Benedetto, Jose Cremonesi, Jose Ovidio, Luciana Ferreira, Marcia Alucci, Rafael

---

* Corresponding Author

Brandão and Rodrigo Cavalcante, for the support in the development of the environmental analytical work. The authors are also thankful to the architects from Zanettini Arquitetura S.A., especially to Siegbert Zanettini and Érika Bataglia for their collaborative work throughout the environmental assessments and, finally, to Petrobras S.A. for the support provided during the design project and authorizing the publication of the information presented in this text.

8. References

[1] Dias, Maria Angela (2010). O Concurso e os Projetos Finalistas. In: Arquiteturas em Contextos de Inovação: Centro de Pesquisa e Desenvolvimento na Cidade Universitária da UFRJ. Maria Angela Dias (Org.). Rio de Janeiro: Petrobras Cenpes, p. 59-81.

[2] Bernardes, S. (1970). Memória justificativa do 1º projeto Cenpes. Rio de Janeiro, p. 7.

[3] Brandão, R. S.; Marcondes, M. P.; Benedetto, G. S.; Gonçalves, J. C. S.; Duarte, D. H. S.; Ramos, J. O. (2008). The new research centre of the Brazilian Petroleum Company in Rio de Janeiro, Brazil: the achievements in the thermal performance of air-conditioned buildings in the tropics. Energy and Buildings, v. 40, p. 1917-1930, 2008.

[4] Metar (2005). Dados Climáticos do Aeroporto do Galeão 1998-2004. São Paulo: Instituto de Astronomia, Geofísica e Ciências Atmosféricas da Universidade de São Paulo.

[5] Gonçalves, Joana Carla Soares; Duarte, Denise; Moura, Norberto (2010). Arquitetura, clima e desempenho ambiental. In: Arquiteturas em Contextos de Inovação: Centro de Pesquisa e Desenvolvimento na Cidade Universitária da UFRJ. Maria Angela Dias (Org.). Rio de Janeiro: Petrobras Cenpes, p. 188-221.

[6] Aroztegui, José Miguel (1995). Cuantificacion del impacto de las sombras de los edificios, in: Encontro Nacional Sobre Conforto no Ambiente Construído (ENCAC), 3, 1995, Gramado. Anais...Gramado, ENCAC.

[7] Humphreys, Michael A. (1978). Outdoor Temperatures and Comfort Indoors. Garston. Watford. Building Research and Practice, v. 6, p. 92-105, Mar/Apr.

[8] Givoni, Baruch (1981), Climate considerations in building and urban design, New York, John Wiley & Sons, 1981.

[9] ASHRAE, American Society of Heating, Refrigerating and Air Conditioning Engineers (2005), Handbook of Fundamentals, Atlanta, ASHRAE.

[10] Szokolay, S. (2001). Use of the new effective temperature: ET* in practice. In: PLEA 18, Florianopolis: PLEA, P. 1003 – 1008.

[11] Monteiro, L. M.; Duarte, D.; Gonçalves, J.; Alucci, M. P. (2008). Conforto térmico como condicionante do projeto arquitetônico-paisagístico: o caso dos espaços abertos do novo centro de pesquisas da Petrobras no Rio de Janeiro, Cenpes II. Ambiente Construído, Porto Alegre, v. 8, n. 4, p. 61-86, out./dez. Available at: <http://www.seer.ufrgs.br/index.php/ambienteconstruido/article/view/4807/4722>

[12] Moura, Norberto Corrêa da Silva; Miana, Anna Christina; Gonçalves, Joana Carla Soares; Duarte, Denise Helena Silva. Arquitetura e desempenho luminoso: Cenpes

II, o novo centro de pesquisas da Petrobras, no Rio de Janeiro, Brasil. Ambiente Construído, Porto Alegre, v. 9, n. 2, p. 151-172, jan./mar. 2009. Available at: <http://www.seer.ufrgs.br/index.php/ambienteconstruido/article/view/7434/5477>

[13] ABNT, Associação Brasileira de Normas Técnicas (2005). NBR 15215-3: Iluminação natural, Parte 3 - Procedimento de cálculo para a determinação da iluminação natural em ambientes internos. Rio de Janeiro.

[14] ABNT, Associação Brasileira de Normas Técnicas (1991). NBR 5413: iluminância de interiores. Rio de Janeiro.

[15] Deutsches Institut Für Normung (1985a) DIN 5034: daylight in interiors: part 1 u. 2. Berlin.

[16] Deutsches Institut Für Normung (1985b). DIN 5035: artificial light of interiors: part 1 u. 2. Berlin.

[17] United States Green Building Council (2002). Green building rating system: for new construction and major renovations (LEED-NC), version 2.1.

[18] Mardaljevic, J. (1999). Daylight Simulation: Validation, Sky Models And Daylight Coefficients (PhD thesis). De Montfort University Leicester. p. 210-215

[19] Marcondes, M. P., Mueller, Cecília Mattos, Brandão, Rafael Silva, Benedetto, Gisele Severiano, Shimomura, A. R. P., Frota, Anésia Barros, Brunelli, Gustavo, Gonçalves, Joana Carla Soares, Duarte, Denise Helena Silva (2010). Conforto e Desempenho Térmico nas Edificações do Novo Centro de Pesquisas da Petrobras, no Rio de Janeiro. Ambiente Construído (Online), Porto Alegre, v.10, p.7 - 29. Available at: <http://www.seer.ufrgs.br/index.php/ambienteconstruido/article/view/4807/4722>

[20] Ministério do Trabalho (1978). NR 15: atividades e operações insalubres, anexo 3: limites e tolerância para exposição ao calor. Brasília.

[21] Ministério do Trabalho (1973). NR 17: ergonomia e segurança do trabalho. Brasília.

[22] ASHRAE, American Society of Heating, Refrigerating and Air Conditioning Engineers (2004). ASHRAE 55-2004: thermal environmental conditions for human occupancy. Atlanta.

[23] De Dear, Richard; Brager, G.; Cooper, D. (1997). Developing an Adaptive Model of Thermal Comfort and preference. Final Report, ASHRAE RP-884, Macquire University.

[24] ISO, International Organization Standardization (1994). ISO 7730: moderate thermal environments: determination of the PMV and PPD indices and specification of the conditions for thermal comfort. Geneve.

[25] ISO, International Organization Standardization (1998). ISO 7726: ergonomics of the thermal environment: instruments for measuring physical quantities. Geneve.

[26] ISO, International Organization Standardization (1990). ISO 8996: ergonomics: determination of metabolic heat production. Geneve, 1990.

[27] ISO, International Organization Standardization (1995). ISO 9920: ergonomics of the thermal environment: estimation of the thermal insulation and evaporative resistance of a clothing ensemble. Geneve.

[28] ABNT, Associação Brasileira de Normas Técnicas (1980). NBR 6401: instalações centrais de ar-condicionado para conforto: parâmetros básicos de projeto. Rio de Janeiro.

[29] Agência Nacional de Vigilância Sanitária (2003). Orientação Técnica sobre Padrões Referenciais de Qualidade do Ar Interior, em Ambientes Climatizados Artificialmente de uso Público e Coletivo. Brasília.

[30] Gonçalves, Joana Carla Soares; Umakoshi, Erica Metie (2010). The Environmental Performance of Tall Buildings. Earthscan, London.

[31] Hoppe, Peter (2002). Different Aspects of Assessing Indoor and Outdoor Thermal Comfort. In: Energy and Buildings, Munique, v. 34, n. 6, p. 661-665.

Chapter 3

Energy Efficient Mobility Management for the Macrocell – Femtocell LTE Network

Dionysis Xenakis, Nikos Passas, Ayman Radwan,
Jonathan Rodriguez and Christos Verikoukis

Additional information is available at the end of the chapter

http://dx.doi.org/10.5772/48251

1. Introduction

The demand for higher data rates and improved energy-efficiency have motivated the deployment of short-range, low-cost, consumer-deployed cellular access points, referred to as femtocells [1]. Femtocells are consumer-deployed cellular access points, which interconnect standard user equipment (UE) to the mobile operator network via the end user's broadband access backhaul. Although femtocells typically support up to a few users, e.g., up to four users [2], they embody the functionality of a regular base station which operates in the mobile operator's licensed band. From the mobile operator perspective, the deployment of femtocells reduces the capital and operational costs, i.e., femtocells are deployed and managed by the end user, improves the licensed spectrum spatial reuse, and decongests nearby macrocell base stations. On the other hand, the end users perceive enhanced indoor coverage, improved Quality of Service (QoS), and significant User Equipment (UE) energy savings.

The deployment of femtocells is one of the most promising energy efficiency enablers for future networks [3-5, 23]. The study in [3] indicates that compared to a standard macrocell deployment, femtocell deployments may reduce the energy consumption on both the access network and the mobile terminals from four to eight orders of magnitude. Analogous results are derived in terms of system capacity per energy unit, although the performance degradation due to increased RF interference between the macro – femto and the femto – femto systems is not investigated. The latter effect is incorporated in [4], where it is shown that in-band macro – femto coexistence results in non-negligible performance degradation on the macrocell network layer. Nevertheless, improved QoS and significantly reduced energy consumption per bit are simultaneously achieved in the UE, with respect to the femtocell deployment density. To further reduce the energy consumption on the femtocell access point (FAP), the authors in [5]

INTECH
open science | open minds

© 2012 The Author(s). Licensee InTech. This chapter is distributed under the terms of the Creative Commons Attribution License http://creativecommons.org/licenses/by/3.0), which permits unrestricted use, distribution, and reproduction in any medium, provided the original work is properly cited.

propose an idle mode procedure according to which the pilot transmissions are disabled in the absence of nearby cellular user activity. Compared to static pilot transmission, the proposed procedure is shown to significantly reduce the overall signaling overhead due to mobility.

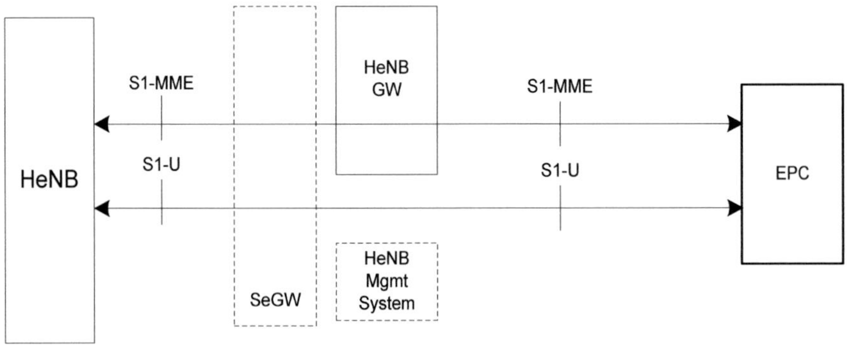

Figure 1. E-UTRAN HeNB Logical Architecture [6]

The Release 9 series of standards for the 3rd Generation Partnership Project (3GPP) the Long Term Evolution (LTE) system [6] is one of the first standards to provision the deployment of femtocells. In the context of LTE, a macrocell is referred to as evolved Node B (eNB), while a femtocell is referred to as Home eNB (HeNB). An LTE user is member of a Closed Subscriber Group (CSG) either if it is permitted to utilize a particular set of closed access femtocells or if it receives prioritized service on a particular set of hybrid access femtocells [7]. The standard describes the cell identification and access control procedures in the presence of LTE femtocells, along with the mobility management procedure for CSG femtocells. Fig. 1 depicts the logical architecture to support femtocells in the LTE system.

As shown in Fig. 2, two of the evolved packet core (EPC) network entities are directly involved in the support of HeNBs, i.e., the Mobility Management Entity (MME) and the Serving Gateway (S-GW). The MME implements the functions of core network (CN) signaling for MM support between 3GPP access networks, idle state mobility handling (e.g. paging), tracking area list management, roaming, bearer control, security, and authentication. On the other hand, the S-GW hosts the functions of lawful interception, charging, accounting, packet routing and forwarding, as well as mobility anchoring for intra and inter-3GPP MM. In the presence of femtocells, the evolved UMTS terrestrial radio access (E-UTRA) air interface architecture consists of eNBs, HeNBs, and HeNB gateways (HeNB GW). The eNBs provide user and control plane protocol terminations towards the UE, and support the functions of radio resource management, admission control, scheduling and transmission of paging messages and broadcast information, measurement configuration for mobility and scheduling, as well as routing of user plane data towards the S-GW. The functions supported by the HeNBs are the same as those supported by the eNBs, while the same implies for the procedures run between the HeNBs and the EPC. The HeNB GW acts as a concentrator for the control plane aiming to support of a large number of HeNBs in a scalable manner. The deployment of HeNB GW is optional; however, if present, it appears to the HeNBs as an MME and to the EPC as an eNB.

The eNBs interconnect with each other through the X2 interface, while they connect to the EPC through the S1 interface [3]. The same implies for the connection between the HeNBs and the EPC, whereas different from the eNB case, the X2 interface between HeNBs is not supported. Fig. 2 illustrates the overall LTE network architecture in the presence of HeNBs.

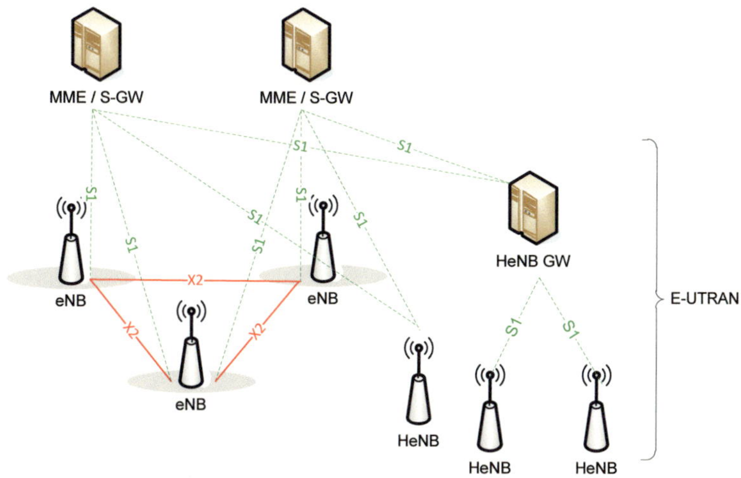

Figure 2. Support of femtocells in the LTE network architecture

In a cellular environment, MM typically consists of three phases [8] a) serving cell monitoring and evaluation, b) cell search and measurement reporting, and c) mobility decision/execution. The serving cell quality is monitored and evaluated on a periodic basis to sustain the service quality over an acceptable threshold. If the service quality falls below a policy-defined threshold, e.g. received signal strength or energy consumption, cell search and measurement reporting is triggered. The cell search and measuring procedure (which bands to sense, in what order, what measurement period and sampling rate to adopt, etc) can be either network-configured or user equipment (UE) based depending on the radio interface standard, the current UE state (e.g. idle or connected), the UE capabilities, and so on. In the former approach, the serving cell exploits its awareness on the surrounding cellular environment to configure the UE to derive and report back signal quality measurements on a predefined set of frequency bands or cells, e.g. provides the UE with a neighbor cell list (NCL) [8]. On the contrary, the UE-based approach is built on the UE capability to autonomously determine when and where to search for neighbor cells without any network intervention. In both cases, a handover (HO) decision entity incorporates the derived signal quality measurements to decide on whether the UE should move to another cell. This entity can reside either on the network (network-controlled approach) or the UE side (mobile-controlled approach) while the decision criteria can incorporate various performance measures such as a) signal quality measures, e.g. received signal strength and SINR, b) user mobility measures, e.g. speed, direction, and c) energy consumption at the UE side, e.g. Joule or Joule/bit. The mobility procedure where the user has no active connections (idle mode) is referred to as cell selection if the user is not camped on a cell or as cell reselection if the user is already camped on a cell. On the other hand, cell HO refers to the mobility procedure performed to seamlessly transfer ongoing user connections from the serving to the target cell (connected mode).

MM in the macrocell – femtocell network comprises many technical challenges in all three phases. Given the femtocell sensitiveness on user mobility and ambient radio frequency (RF) interference, serving cell monitoring and evaluation should be performed in a more frequent basis to sustain an acceptable service quality when connected to a femtocell. Considering the relatively small number of physical cell identifiers in prominent radio air-interfaces, more complicated yet backwards compatible cell identification procedures are required to facilitate cell searching and identification. Furthermore, maintaining and broadcasting a comprehensive Neighbor Cell List (NCL) to facilitate cell search and measurement reporting is not scalable in an integrated femtocell – macrocell network. To this end, novel UE-based cell search procedures are required to fully exploit the underlying femtocell infrastructure. The effectiveness of these procedures will have a great impact on the UE energy autonomicity and perceived QoS as explained in the following.

In the presence of ongoing user connections, cell quality measurements are usually performed during downlink (DL) and uplink (UL) idle periods provided either by Discontinuous Reception (DRX) or by packet scheduling (i.e. gap assisted measurements) [6]. However, the DRX periods are typically utilized for UE energy conservation while the measurement gaps can be utilized to extend the user service time. Taking this into account and considering that a) the short femtocell range results in more frequent cell search and measurement report triggering even under low to medium mobility scenarios, and b) the large number of neighboring cells will substantially increase the aggregated measurement time in dense femtocell deployments, it follows that cell search and measurement reporting may severely deteriorate the user-perceived QoS and deplete the UE battery lifetime. Moreover, searching for and deriving measurements on nearby yet non accessible femtocells should also be avoided, e.g. when a nearby femtocell belongs to a closed access group where the user is not subscribed. In prominent cellular standards, the mobility decision is typically based on signal quality, coverage or load balancing criteria [8, 9]. Given their preferential QoS and significantly reduced energy consumption on the UE side, femtocells are expected to be prioritized over macrocells during the mobility decision phase. However, the mobility decision and execution in an integrated femtocell environment is a non-trivial issue. Femtocell identification introduces non-negligible delay overhead while the limited femtocell capacity in terms of supported users may substantially increase the HO failure probability. The tagged user access status on the candidate femtocells should also be taken into account both to avoid unnecessary signaling overhead and minimize the HO failure probability due to HO rejection [9]. The femtocell sensitiveness on user mobility can substantially increase the number of mobility decision and execution events, increasing thus the network signaling overhead due to mobility management and compromising the UE service continuity when in connected mode.

HO decision affects various aspects of the overall network performance, which mainly include the Signal to Interference plus Noise Ratio (SINR) performance, the interference performance as well as the energy-efficiency at the access network nodes. Current literature includes various HO decision algorithms for the macrocell – femtocell network [10-12], which primarily focus on prioritizing femtocells over macrocells with respect to user mobility criteria. Emphasis is given in reducing the number of the network-wide HO execution events, owing to the short femtocell radius and the ping-pong effect [9]. Nevertheless, the strongest cell HO decision policy [8] is considered for both macro-macro and femto-femto HO scenarios. According to it, the serving cell proceeds to a HO execution whenever the Reference Signal Received Power (RSRP) [6] of a neighbor cell exceeds over the respective RSRP status of the serving cell plus a policy-defined HHM, for a policy-defined time period namely the Time To Trigger (TTT). The HHM is typically introduced to mitigate UE measurement inconsistencies, encompass frequency-related propagation divergences and minimize the ping-pong effect [9], i.e. consecutive HOs originating from the user movement across the cell boundaries. If comparable downlink Reference Signal (RS) power transmissions are assumed amongst the LTE cells, the strongest cell HO policy facilitates mobility towards a LTE cell with preferential propagation characteristics. However, this is not the case of the macrocell – femtocell LTE network where femtocells are expected to radiate comparably lower downlink RS power for interference mitigation on the macrocell layer [1]. Divergent RS power transmissions are expected even amongst the femtocell layer, in accordance with the adopted self-optimization procedure [5]. Apart from RS power transmission divergences, substantial RF interference divergences are also expected amongst the LTE cells. RF interference is an inevitable product of the unplanned femtocell deployment, both in terms of location and operating frequency, even if advanced interference cancellation and avoidance techniques are adopted [1-2, 14-16]. The RF interference divergences amongst the LTE cells may severely deteriorate the user-perceived QoS due to service outage and substantially increase the network signaling due to mobility, if the interference-agnostic strongest cell HO decision policy is adopted.

In conclusion, apart from improved indoor coverage and enhanced user-perceived QoS, femtocells natively achieve significant energy savings at both the access network and the UE side. To this end, more sophisticated HO decision algorithms are required in the presence of LTE femtocells to fully exploit the native femtocell superiority both in terms of enhanced QoS and reduced energy consumption. The remainder of this chapter discusses an energy-efficient HO decision policy for the macrocell - femtocell LTE network which aims at reducing transmit power at the mobile terminals [17]. The employment of the proposed policy is based on adapting the HO Hysteresis Margin (HHM) with respect to a mean SINR target and standard LTE measurements of the candidate cells' status. The incorporation of the SINR target guarantees QoS, while the utilization of standard LTE measurements provides an accurate estimation of the required UE transmit power per candidate cell. The benefit for employing the proposed HO decision policy is three-fold; improved energy-efficiency at the LTE UEs, lower RF interference, and guaranteed QoS for the ongoing user links. Another important feature of the proposed HO decision policy is that even though it is fundamentally different from the predominant strongest cell HO policy, it is employed in an LTE backwards-compatible manner by suitably adapting the HHM.

The remainder of this chapter is organized as follows. Section 2 models the macrocell – femtocell LTE in network under the viewpoint of MM and discusses the predominant strongest cell handover decision policy. Section 3 describes the proposed HO decision policy, while section 4 discusses the network signaling procedure required to employ it. Section 5 includes selected simulation results to illustrate its performance in terms of energy consumption per bit, UE power consumption, cell power consumption, and number of HO execution events. Finally, Section 5 concludes the chapter.

2. System model and strongest cell handover decision policy

2.1. System description

A two-tier LTE network is considered, operating within the LTE band set $\boldsymbol{N} := \{1, \ldots, N\}$. A macrocell station is referred to as evolved Node B (eNB), while a femtocell station as Home eNB (HeNB). To resourcefully sustain its ongoing services, user u is assumed to have a mean SINR target, denoted by $\overline{\gamma}_{target}^{u}$. Let $\boldsymbol{C_n}$ denote the set of LTE cells operating in band $n \in \boldsymbol{N}$, including both eNBs and HeNBs, and $\boldsymbol{U_n}$ the set of users receiving service from an LTE cell within $\boldsymbol{C_n}$. Assuming that user $u \in \boldsymbol{U_n}$ is connected to cell $s \in \boldsymbol{C_n}$, the respective mean uplink SINR for a tagged time interval T is given as follows:

$$\overline{\gamma}_{u\to s}^{T} = \frac{\overline{P}_{u}^{T} \cdot \overline{h}_{u\to s}^{T}}{\sum_{c\in \boldsymbol{C_n}-\{s\}} \overline{P}_{c}^{T} \cdot \overline{h}_{c\to s}^{T} + \sum_{u'\in \boldsymbol{U_n}-\{u\}} \overline{P}_{u'}^{T} \cdot \overline{h}_{u'\to s}^{T} + \left(\overline{\sigma}_{s}^{T}\right)^{2}} \quad (1)$$

where $\overline{P}_{u}^{T}$ denotes the power transmission of user u, $\overline{h}_{u\to s}^{T}$ the channel gain from user u to cell s, $\overline{P}_{c}^{T}$ the power transmission of cell c, $\overline{h}_{c\to s}^{T}$ the channel gain between cells c and s, and $\overline{\sigma}_{s}^{2}$ the noise power at cell s, all averaged within the time interval T. Accordingly, the mean downlink SINR is given as follows:

$$\overline{\gamma}_{s\to u}^{T} = \frac{\overline{P}_{s\to u}^{T} \cdot \overline{h}_{s\to u}^{T}}{\sum_{c\in \boldsymbol{C_n}-\{s\}} \overline{P}_{c}^{T} \cdot \overline{h}_{c\to u}^{T} + \sum_{u'\in \boldsymbol{U_n}-\{u\}} \overline{P}_{u'}^{T} \cdot \overline{h}_{u'\to u}^{T} + \left(\overline{\sigma}_{u}^{T}\right)^{2}} \quad (2)$$

where $\overline{P}_{s\to u}^{T}$ denotes the power transmission of cell s to user u, $\overline{h}_{s\to u}^{T}$ the channel gain from cell s to user u , $\overline{h}_{u'\to u}^{T}$ the channel gain from user $u'$to user u, and $\overline{\sigma}_{u}^{2}$ the noise power at user u, all averaged within the time interval T.

Let us now focus on the expected UE transmit power for maintaining a link between a tagged user u and cell c. Let $\boldsymbol{L_u} \subseteq \bigcup_{n\in N} \boldsymbol{C_n}$ indicate the candidate cell set for user u, which consists of accessible LTE cells and has been identified during the network discovery phase. Using Eq. (1) for the mean SINR target $\overline{\gamma}_{target}^{u}$, it can be readily shown that the mean UE power transmissions for maintaining a link between user u and cell $c \in \boldsymbol{L_u}$ can be estimated as follows:

$$\overline{P}_{u\to c}^{T} = \frac{\overline{\gamma}_{target}^{u} \cdot \left(\sum_{c'\in \boldsymbol{C_n}-\{c\}} \overline{P}_{c'}^{T} \cdot \overline{h}_{c'\to c}^{T} + \sum_{u'\in \boldsymbol{U_n}-\{u\}} \overline{P}_{u'}^{T} \cdot \overline{h}_{u'\to c}^{T} + \left(\overline{\sigma}_{c}^{T}\right)^{2}\right)}{\overline{h}_{u\to c}^{T}} \quad (3)$$

Note that Eq. (3) includes the impact of handing over to cell $c \in \boldsymbol{L_u}$, given that the RF interference caused by the ongoing user link, i.e., $\overline{P}_{u}^{T} \cdot \overline{h}_{u\to s}^{T}$, is not included. Eq. (3) also corresponds to the UE power consumption, owing to transmit power, for maintaining a link between user u and cell c. The LTE standard describes a wide set of network and UE link quality measurements [18], which can be utilized to estimate the expected SINR in Eq. (1) and (2), and the average UE power transmission in Eq. (3). Table I summarizes standard LTE measurements, and includes the notation followed in this paper for a tagged user u, cell c, and measurement interval T.

Measurement	**Definition**	**Performed by**	**Notation**
Reference signal received power (RSRP)	The linear average over the power contributions (in [W]) of the resource elements that carry cell-specific reference signals within the considered measurement frequency bandwidth. For RSRP determination the cell-specific reference signals R0 shall be used while if the UE may use R1 in addition to R0 if it is reliably detected. The reference point for the RSRP shall be the antenna connector of the UE.	UE	$RSRP_{c\to u}^{T}$
E-UTRA Carrier Received Signal Strength Indicator (RSSI)	The linear average of the total received power (in [W]) observed only in OFDM symbols containing reference symbols for antenna port 0, over $R_{c,DL}$ number of RBs by the UE from all sources, including co-channel serving and non-serving cells, adjacent channel interference, thermal noise etc. RSSI is not reported as a stand-alone measurement rather it is utilized for deriving RSRQ.	UE	$RSSI_{c\to u}^{T}$
Reference Signal Received Quality (RSRQ)	The ratio $R_{c,DL}$×RSRP/(E-UTRA carrier RSSI) where $R_{c,DL}$ is the number of RB's of the E-UTRA carrier RSSI measurement bandwidth. The measurements in the numerator and denominator shall be made over the same set of RBs. The reference point for the RSRQ shall be the antenna connector of the UE.	UE	$RSRQ_{c\to u}^{T}$
Downlink Reference Signal Transmitted Power (DL RS Tx)	The linear average over the power contributions (in [W]) of the resource elements that carry cell-specific reference signals which are transmitted by a tagged cell within its operating system bandwidth. For DL RS TX power determination the cell-specific reference signals R0 and if available R1 can be used. The reference point for the DL RS TX power measurement shall be the TX antenna connector.	E-UTRAN	$P_{c,RS}^{T}$
Received Interference Power	The uplink received interference power, including thermal noise, within the physical RB's bandwidth of N_{sc}^{RB} resource elements. The reported value is averaged over uplink physical RB. The reference point for the measurement shall be the RX antenna connector.	E-UTRAN	$\bar{I}_c^{T}$

Table 1. Basic UE and LTE cell measurement capabilities

Note that the $\bar{I}_c^{T}$ measurement in Table I corresponds to the linear average of the RIP measurements performed within the tagged cell's operating bandwidth, i.e., the utilized Resource Blocks [19]. To the remainder of this paper, we focus on the HO decision phase, which is performed in the serving LTE cells. The network discovery procedure is outside the scope of this paper, and it is assumed that for all UEs connected to it, each serving LTE cell has a consistent candidate cell set, and link quality measurements describing its status.

2.2. Strongest cell handover decision policy

In the context of LTE, the strongest cell HO decision policy results in a HO execution whenever the RSRP of an accessible cell exceeds over the RSRP of the serving cell plus a policy-defined HHM, for a policy-defined time period namely the Time To Trigger (TTT) [9]. The HHM is utilized to mitigate frequency-related propagation divergences, and the ping-pong effect. Based on our system model, the strongest cell HO policy for the LTE system is described as follows:

$$\arg\max_{c\in \boldsymbol{L_u}} RSRP^{TTT}_{c\to u,(dB)} \coloneqq \{c \mid RSRP^{TTT}_{c\to u,(dB)} > RSRP^{TTT}_{s\to u,(dB)} + HHM_{c,(dB)}\} \tag{4}$$

where $HHM_{c,(dB)}$ corresponds to the HHM for cell $c \in \boldsymbol{L_u}$, and the notation $X_{(dB)}$ to the value of X in decibels (dB). Taking into account the definition of the RSRP in [15], it follows that:

$$RSRP^{T}_{c\to u} = P^{T}_{c,RS} \cdot \overline{h}^{T}_{c\to u} \tag{5}$$

Substituting Eq. (5) to Eq. (4), it follows that the strongest cell policy facilitates mobility towards cells with higher RS power transmissions or improved channel gain. As a result, even though the strongest cell policy may improve the channel gain for the tagged LTE user (Eq. 5), it does not necessarily improves the SINR performance (Eq. 1, 2), given that neither the RF interference, nor the actual RS power transmissions of the target cells, are taken into account. The same implies for the UE power transmissions, which are not necessarily being reduced (Eq. 3) having a negative impact on both the UE power consumption and the RF interference network-wide.

3. The proposed handover decision policy

The proposed HO decision policy, referred to as UE Transmit Power Reduction (UTPR) policy in the following, consists of handing over to the cell with the minimum required UE transmit power, while maintaining the prescribed mean SINR target. The following analysis is pursued to derive the HHM required for minimizing the UE transmit power, based on the available set of standard LTE measurements in Table I. It is assumed that user u receives service from cell s, which has consistent LTE measurements describing the status of every candidate cell $c \in \boldsymbol{L_u}$ for user u, for the time interval $T = TTT$.

Using (5) under the assumption of a symmetric channel gain, the following estimation can be made:

$$\overline{h}^{T}_{u\to c} \cong \overline{h}^{T}_{c\to u} = \frac{RSRP^{T}_{c\to u}}{P^{T}_{c,RS}} \tag{6}$$

By the RIP measurement definition in [18], it follows that:

$$\overline{I}_c^T = \left(\sum_{c' \in C_n - \{c\}} \overline{P}_{c'}^T \cdot \overline{h}_{c' \to c}^T + \sum_{u' \in U_n} \overline{P}_{u'}^T \cdot \overline{h}_{u' \to c}^T + \left(\overline{\sigma}_c^T\right)^2\right) \quad (7)$$

Using Eq. (3), (6), and (7), it can be shown that the UE power transmission on the serving cell s is given by (8).

$$\overline{P}_u^T \triangleq \overline{P}_{u \to s}^T = \frac{\overline{\gamma}_{target}^u \cdot P_{s,RS}^T \cdot \overline{I}_s^T}{RSRP_{s \to u}^T} \quad (8)$$

Following a similar approach, the UE transmit power on the candidate cell c can be estimated as follows:

$$\overline{P}_{u \to c}^T = \frac{\overline{\gamma}_{target}^u \cdot P_{c,RS}^T \cdot \left(\overline{I}_c^T - \overline{P}_u^T \cdot \overline{h}_{u \to c}^T\right)}{RSRP_{c \to u}^T} \quad (9)$$

where the term $\overline{P}_u^T \cdot \overline{h}_{u \to c}^T$ is introduced to include the positive impact of handing over to cell $c \in \boldsymbol{L_u}$, if cells c and s operate in the same LTE band (if not, it is omitted), i.e, if $c, s \in \boldsymbol{C_n}$. Accordingly, handing over to the candidate cell c, is expected to result in reduced UE transmit power compared to the one used in the current serving cell s, if the following are in effect:

$$\overline{P}_{u \to s}^T > \overline{P}_{u \to c}^T \Rightarrow \quad (10)$$

$$\frac{\overline{\gamma}_{target}^u \cdot P_{s,RS}^T \cdot \overline{I}_s^T}{RSRP_{s \to u}^T} > \frac{\overline{\gamma}_{target}^u \cdot P_{c,RS}^T \cdot \left(\overline{I}_c^T - \overline{P}_u^T \cdot \overline{h}_{u \to c}^T\right)}{RSRP_{c \to u}^T} \Rightarrow \quad (11)$$

$$RSRP_{c \to u}^T > RSRP_{s \to u}^T \cdot \frac{P_{c,RS}^T \cdot \left(\overline{I}_c^T - \overline{P}_u^T \cdot \overline{h}_{u \to c}^T\right)}{P_{s,RS}^T \cdot \overline{I}_s^T} \quad (12)$$

where Eq. (11) is derived by using Eq. (8), and (9), and Eq. (12) by rearranging (11). Note that the parameter $\overline{P}_u^T$ is given by Eq. (8). By taking the respective parameter values in dB, Eq. (12) can be rearranged as follows:

$$RSRP_{c \to u,(dB)}^{TTT} > RSRP_{s \to u,(dB)}^{TTT} + HHM_{c,(dB)}^{UTPR} \quad (13)$$

where the parameter $HHM_{c,(dB)}^{UTPR}$ is given by (14).

$$HHM_{c,(dB)}^{UTPR} = \begin{cases} 10 \log \dfrac{P_{c,RS}^{TTT} \cdot \left(\overline{I}_c^{TTT} - \overline{P}_u^{TTT} \cdot \overline{h}_{u \to c}^{TTT}\right)}{P_{s,RS}^{TTT} \cdot \overline{I}_s^{TTT}} & c, s \in \boldsymbol{C_n} \\ 10 \log \dfrac{P_{c,RS}^{TTT} \cdot \overline{I}_c^{TTT}}{P_{s,RS}^{TTT} \cdot \overline{I}_s^{TTT}} & otherwise \end{cases} \quad (14)$$

It can be seen that Eq. (13) can be utilized as a HO decision criterion for minimizing the UE power transmissions in the two-tier LTE network. To achieve this, Eq. (14) can be incorporated in the standard LTE HO procedure, as an adaptive HHM. Given that a HHM for mitigating the side-effects of user mobility is still required, the $HHM_{c,(dB)}^{UTPR}$ parameter should be incorporated as an additional HHM in the strongest cell HO decision policy. Taking this into account, the proposed UTPR HO decision policy can be described as follows:

$$\arg\max_{c\in \boldsymbol{L}_u} RSRP^{TTT}_{c\to u,(dB)} \coloneqq \left\{c \mid RSRP^{TTT}_{c\to u,(dB)} > RSRP^{TTT}_{s\to u,(dB)} + HHM_{c,(dB)} + HHM^{UTPR}_{c,(dB)}\right\} \quad (15)$$

Summarizing, the proposed UTPR policy is based on standard LTE measurements, while it is employed by introducing an adaptive HHM to the standard LTE HO procedure. The employment of the UTPR policy does not require any enhancements for the LTE UEs, however, an enhanced network signaling procedure is necessitated. Next section provides some insights on how the proposed policy could be employed in the context of the macrocell – femtocell LTE network.

4. Network signaling to employ the proposed handover decision policy

To identify and ultimately utilize CSG femtocells within its proximity, each LTE UE maintains a CSG whitelist. The respective CSG whitelist per LTE user is also maintained in the Mobility Management Entity (MME), residing in the LTE Core Network (CN), in order to perform access control during the mobility execution phase. The closed and hybrid access LTE femtocells broadcast their CSG identity (CSG ID) along with a CSG indicator set to 'TRUE' or 'FALSE', respectively. Both these fields along with the E-UTRAN Cell Global Identifier (ECGI), used for global LTE cell identification, are signaled within the System Information Block Type 1 (SIB1) in the LTE downlink [6]. Although this information is not required during the LTE cell search and measurement phase, it is considered prerequisite during the LTE mobility decision and execution phase. To this end, a cell identification procedure is performed, where the UE is reconfigured to obtain the ECGI of the target LTE cell [6]. In the following, we identify and discuss two different LTE network signaling approaches to facilitate the employment of the proposed UTPR-based HO decision policy.

The employment of the proposed UTPR policy necessitates the incorporation of standardized LTE cell measurements on the tagged user's neighbor cell set, i.e. the downlink RS transmitted power P^{c}_{RS} and Received Interference Power $\boldsymbol{I}_c$, $\forall c \in \boldsymbol{L}_u$. These measurements can be commuted through the S1 interface [6] to the serving LTE cell. The entire HO decision parameter set will be referred to as HO context in the following. Depending on whether the required HO context is reported and maintained in a LTE CN entity or not, e.g. the MME, two different network signaling approaches are identified i.e. the reactive and the proactive [24] In the reactive approach the HO context is obtained on request towards the target LTE cell, while in the proactive approach it is directly obtained on request to the MME. To employ the latter, the LTE cells are required to report their HO context status to the MME on a periodic basis. The reporting periodicity should be MME-configured and adapted according to the HO context request history, the LTE CN status and so on. Assuming that the serving eNB can be either a regular eNB or a HeNB, Fig. 3 and 4 illustrate the detailed network signaling [6] required in the reactive and the proactive HO context derivation approaches, respectively. Without loss of generality, it is considered that the serving and the target cell are connected to the same MME.

The cell search and measurement signaling steps for both approaches, i.e. steps 1-7 in the reactive and steps 5-11 in the proactive, are in accordance with [6]. During these steps, the serving eNB configures the UE to identify an appropriate neighbor cell set and derive consistent RSRP and RSRQ measurements. Notice that the measurement configuration and reporting phase in LTE is triggered on critical events [20], e.g. when the serving cell RSRP is below a network-configured threshold for a network-configured time period TTT. To facilitate subsequent parameter acquisition, each measurement report includes a measurement timestamp. The proximity configuration and indication signaling in Fig.3 and 4 is utilized for UE-based autonomous HeNB discovery, while the System Information (SI) acquisition and report signaling is required for HeNB identification and access control validation [6]. The serving eNB utilizes the reported UE measurements, sent on critical LTE events, for HO decision triggering (steps 8 in the reactive and 12 in the proactive approach) [21, 22].

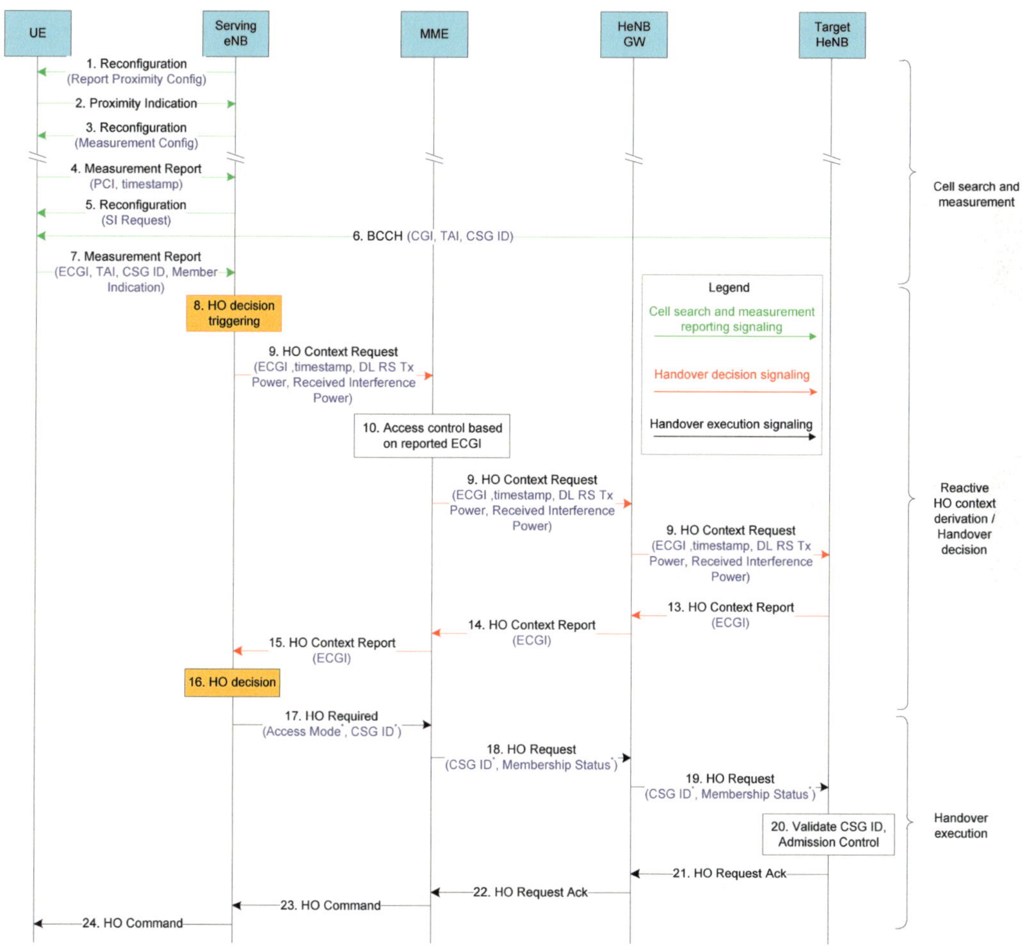

Figure 3. Network signaling procedure for the reactive handover approach

Upon HO decision triggering, the serving eNB initiates a HO context request towards the MME including the corresponding measurement timestamp and target ECGI, i.e. steps 9 in Fig. 3 and 13 in Fig. 4. To minimize unnecessary network signaling, the MME verifies the access status of the tagged UE on the target ECGI in steps 10 and 14, respectively. If the tagged user is not allowed to access the target eNB, the MME notifies the serving LTE cell accordingly.

The key difference between the reactive and the proactive approaches is that in the former the MME forwards the HO context request towards the target eNB (steps 11-15), while in the latter the MME may directly provide the required HO context by utilizing the reports derived in steps 1-4 (Fig.4). It should be noted that the proactive context derivation signaling phase is indicatively located in steps 1-4, since it can be performed asynchronously with respect to the rest HO signaling procedure. In the absence of HO context close to the required measurement timestamp, the MME may decide to forward the HO context request towards the target eNB as in the reactive approach. Upon HO context acquisition, the HO decision algorithm in the serving eNB proceeds to a HO execution whenever necessary. In that case, a common HO execution signaling follows for both approaches (steps 17-24) [6].

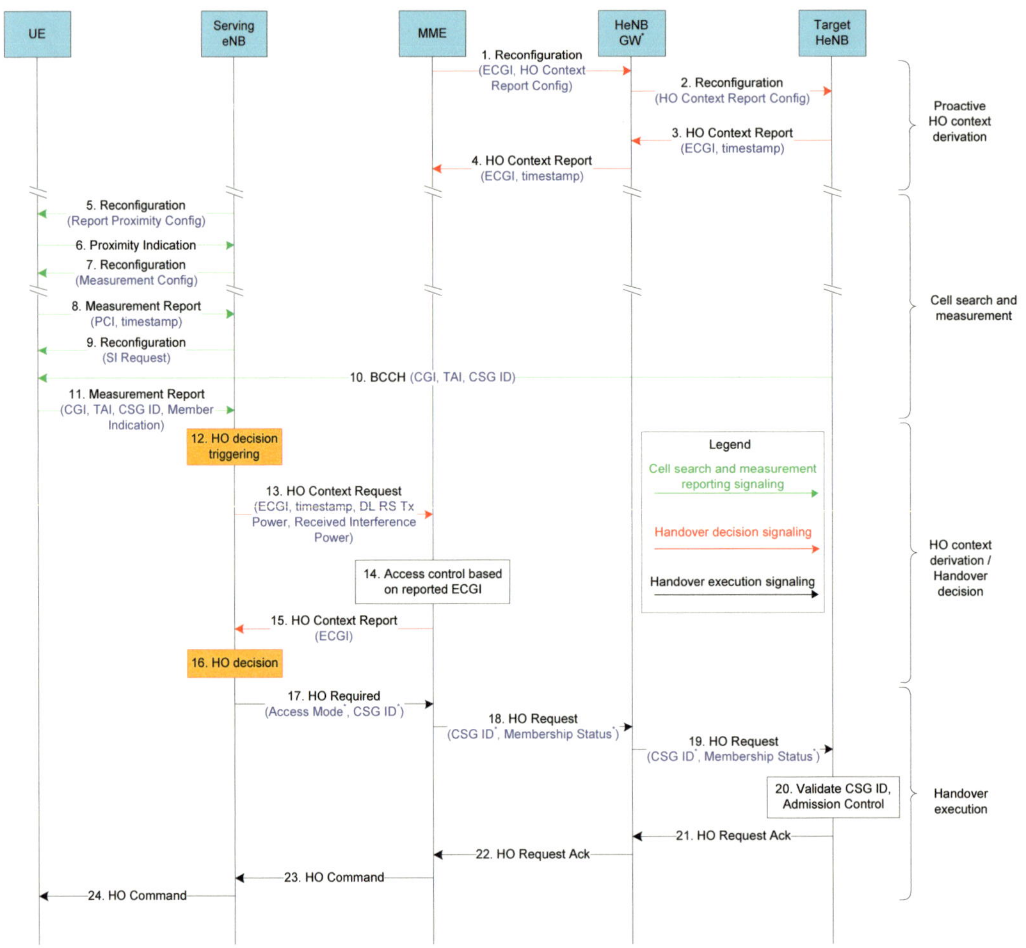

Figure 4. Network signaling procedure for the proactive handover approach

The HO context requests and reports can be signaled in an aggregated manner in both the access (eNB, HeNB) and the core LTE network (MME, HeNB GW). For example, on multiple HO context requests towards a tagged eNB, the MME may send an aggregated HO context request including all the required measurement timestamps. A similar approach can be applied for the HO context report in the reverse direction. Although the reactive approach minimizes the required signaling between the MME and the target LTE cell, the overall network signaling will be highly correlated to the occurrence rate of HO triggering events. On the other hand, more frequent yet more deterministic signaling overhead is expected in the proactive approach, provided that the MME configures the HO context reporting periodicity on the eNBs. In addition to that, the proactive approach may significantly reduce the resulting HO decision delay compared to the reactive approach, provided that the HO context resides on the context-aware MME rather than the target LTE cell. However, certain operational enhancements are required in the MME to resourcefully support the proactive approach, in contrast with the reactive approach where no further LTE CN enhancements are needed.

5. Numerical results

This section includes selected numerical results to evaluate the performance of the proposed UTPR HO decision policy in the macrocell – femtocell LTE network. The simulation scenario is based on the evaluation methodology described in [22], while the proposed HO decision policy is compared against a strongest cell based policy, referred to as SCB policy in the following.

A conventional hexagonal LTE network is considered, including a main LTE cluster composed of 7 LTE cells, where each LTE cell consists of 3 hexagonal sectors. The wrap-around technique is used to extend the LTE network, by copying the main LTE cluster symmetrically on each of the 6 sides. A set of blocks of apartments, referred to as femtoblocks, are uniformly dropped within the main LTE cluster according to the parameter d_{FB}, which indicates the femtoblock deployment density within the main LTE cluster, i.e., the percentage of the main LTE cluster area covered with femtoblocks. Each femtoblock is modeled according to the dual stripe model for dense urban environments in [22]. According to it, each femtoblock consists of two stripes of apartments separated by a 10 m wide street, while each stripe has two rows of $A = 5$ apartments of size 10×10 m. For a tagged femtoblock, femtocells are deployed with a femtocell deployment ratio parameter r_{fc}, which indicates the percentage of apartments with a femtocell [22]. Each femtocell initially serves one associated user, while in general, it can serve up to 4 users. Femtocells and femtocell users are uniformly dropped inside the apartments. Each LTE user is member of up to one CSG, where the CSG ID per user and femtocell is uniformly picked from the set $\{1,2,3\}$. Each LTE sector initially serves ten macrocell users, which are uniformly distributed within it. Unless differently stated, it is assumed that $\bar{v} = 3$ km/h and $s_u = 1$ km/h.

The macrocell stations operate in a LTE band centered at 2000MHz, divided into R RBs of width 180 KHz and utilizing a 5MHz bandwidth. The macrocell inter-site distance is set to

500m, while the operating band for each femtocell is uniformly picked from a band set including the macrocell operating band and its two adjacent frequency bands of 5MHz bandwidth. The adopted Modulation and Coding Schemes (MCS) are in accordance with [21], while the Exponential Effective SINR Mapping method is used to obtain the effective SINR per RB and the consequential UE throughput [22]. The minimum required SINR per UE is set to $\overline{\gamma}_t^u = 3$ dB, while the communications are carried out in full buffer as in [22]. The shadowing standard deviation for the macro and femto systems are 8 and 4 dB respectively, and the macrocell and femtocell noise figures are set to 5 and 8 dB in that order. The macrocell downlink RS power transmissions are normally distributed with a mean value of 23 dBm and a standard deviation of 3dB, while the respective femtocell downlink RS power transmissions are uniformly distributed within the [0,10] dBm interval. The UE power class is set to 23dBm and the maximum transmission powers for the macrocell and femtocell stations are set to 43 and 10dBm [22], respectively. The adopted path loss models are depicted in Table II, where d and d_{indoor} are the total and indoor distances between the tagged cell and the tagged user in meters, respectively. The term $0.7d_{indoor}$ takes into account the penetration losses due to indoor walls, w corresponds to the number of walls separating the UE and the target cell, while $L_{ow} = 15$ dB and $L_{iw} = 5$ dB correspond to the penetration losses of the building external and internal walls, respectively. The frequency-selective fading is considered to follow the Rayleigh distribution [8]. Finally, the overall simulation time is set to 200 sec and the simulation unit is set to 1 sec. The key simulation parameters are summarized in Table II.

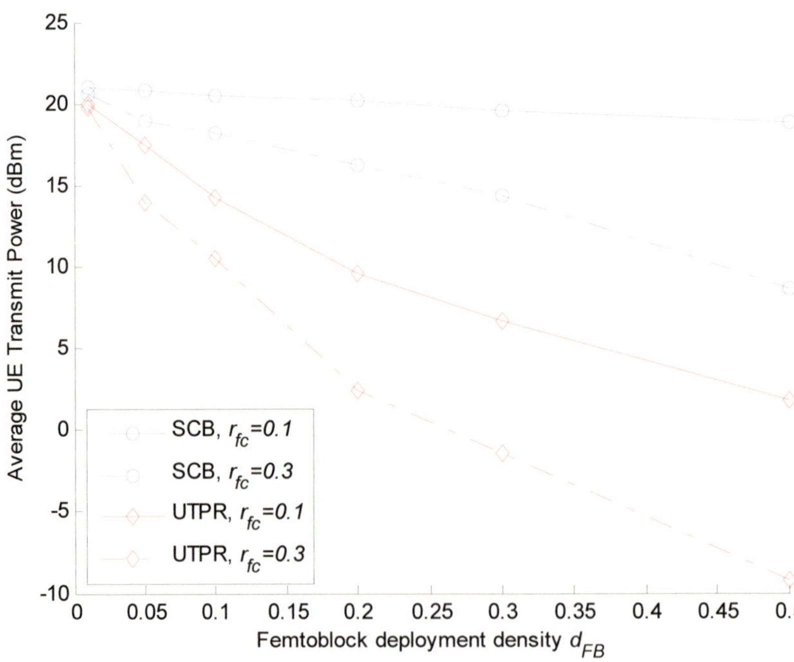

Figure 5. Average UE transmit power versus the d_{FB}

Network layout			
Macrocell layout			7 clusters, 7 sites per cluster, 3 sectors per site, freq. reuse 1
Macrocell inter-site distance			500 m
Initial number of UEs per macrocell sector			10 UEs
Macrocell UE distribution			Uniform within each sector
Femtoblock layout			Dual stripe model for dense urban environments [22]
Femtoblock distribution in the main LTE cluster			Uniform
Femtocell station and UE distribution within an apartment			Uniform
Initial number of UEs per femtocell station			1 UE
Maximum number of supported UE per femtocell			4 UEs
System operating parameters			
Parameter	**Macrocell**		**Femtocell**
Carrier frequency	2000 MHz		Uniformly picked from the set {1990, 2000, 2010} MHz
Channel bandwidth	10 MHz		10 MHz
Maximum Tx Power	$\overline{P}_{max}^{c,T} = 46$ dBm		$\overline{P}_{max}^{c,T} = 20$ dBm
Antenna gain	14 dBi		0 dBi
Noise figure	5 dB		8 dB
Shadowing standard deviation	8 dB		4 dB
RS transmit power (DL RS Tx)	Normally distributed with a mean value of 23 dBm and standard deviation 3dB		Uniformly distributed within the [0,20] dBm interval
CSG ID distribution	Does not apply		Uniform within {1, 2, 3}
Link-to-system mapping	Effective SINR mapping (ESM) [22]		
Path Loss Models			
UE to Macrocell	UE outdoors	$PL(dB) = 15.3 + 37.6log_{10}d$	
	UE indoors	$PL(dB) = 15.3 + 37.6log_{10}d + L_{ow}$	
UE to Femtocell	UE in the same apartment stripe	$PL(dB) = 38.46 + 20log_{10}d + 0.7d_{indoor} + w \cdot L_{iw}$	
	UE outside the apartment stripe	$PL(dB) = \max(15.3 + 37.6log_{10}d, 38.46 + 20log_{10}d) + 0.7d_{indoor} + w \cdot L_{iw} + L_{ow}$	
	UE inside a different apartment stripe	$PL(dB) = \max(15.3 + 37.6log_{10}d, 38.46 + 20log_{10}d) + 0.7d_{indoor} + w \cdot L_{iw} + 2 \cdot L_{ow}$	
Interior / Exterior wall penetration loss (indoor UEs)			5 / 15 dB
UE parameters			
UE power class	$\overline{P}_{max}^{u,T} = 23$ dBm		
UE antenna gain	0 dBi		
Mean UL SINR target	$\overline{\gamma}_t^u = 3$ dB		
CSG ID distribution	Uniformly picked from {1, 2, 3}		
Traffic model	Full buffer similar to [8]		
Mobility model [13]	User speed	$v_t = N(\bar{v}, s_u)$ m/s	
		Mean user speed	$\bar{v} = 3$ km/h
		User speed standard deviation	$s_u = 1$ km/h
	User direction	$\varphi_t = N\left(\varphi_{t-1}, 2\pi - \varphi_{t-1}\tan(\frac{\sqrt{v_t}}{2})\Delta t\right)$	
	where Δt is the time period between two updates of the model, and $N(a, b)$ the Gaussian distribution of mean a and standard deviation b		
Other simulation parameters			
Overall simulation time			200 sec
Simulation time unit			$\Delta t = 1$ sec

Table 2. System-level simulation model and parameters

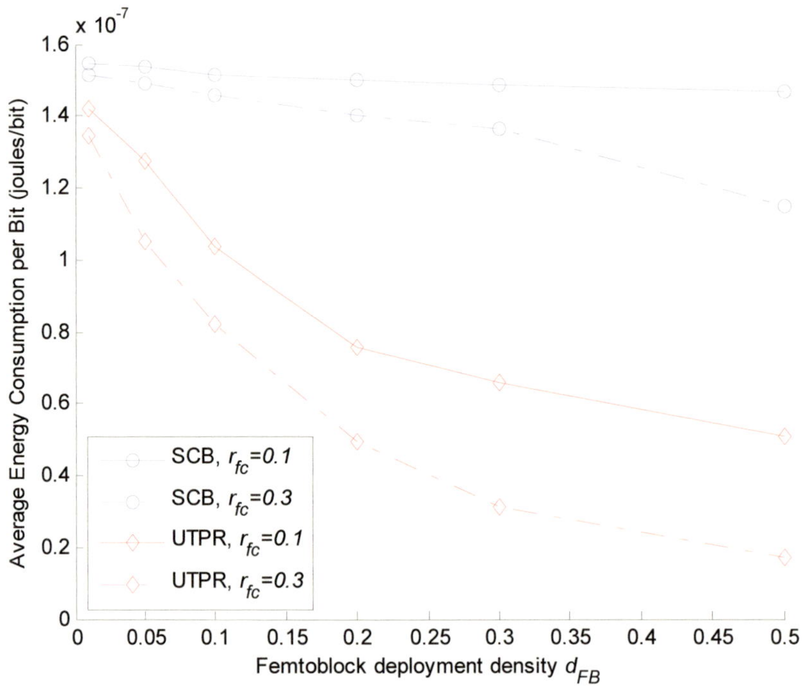

Figure 6. Average UE energy consumption per bit versus the d_{FB}

Fig. 5 and 6 depict the performance of the SCB and UTPR decision policies in terms of UE average transmit power and average energy consumption per bit, owing to transmit power, respectively. Notice that an increased femtoblock deployment density d_{FB} corresponds to an increased number of femtocells and UEs within the main LTE cluster. The same implies for an increased femtocell deployment ratio r_{fc}, which corresponds to an increased femtocell and UE density within each femtoblock. As expected, an increasing femtoblock deployment density d_{FB} or femtocell deployment ratio r_{fc} results in lower UE power and energy consumption per bit for both approaches. However, a higher femtocell deployment ratio r_{fc} is required in order for the SCB policy to benefit from the LTE femtocell presence, both in terms of UE power and energy consumption per bit. On the contrary, the UTPR policy's awareness on the downlink RS and received interference power enables mobility towards LTE cells with lower UE power consumption, while maintaining the tagged user's SINR target. In more detail, for $r_{fc} = 0.1$ and $r_{fc} = 0.3$ the proposed policy results in significantly lower UE power consumption compared to the SCB policy, varying from 1 to 16 dB and 1 to 20 dB respectively. Significantly lower UE energy consumption per bit is also achieved, varying from 10 to 85% compared to the SCB policy, in accordance with the femtoblock deployment density and the femtocell deployment ratio.

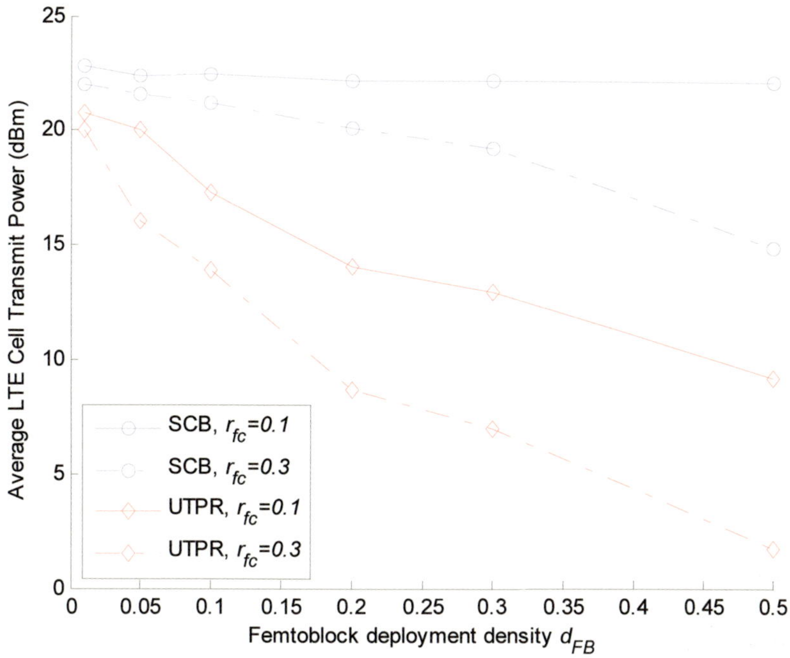

Figure 7. Average LTE cell transmit power versus the d_{FB}

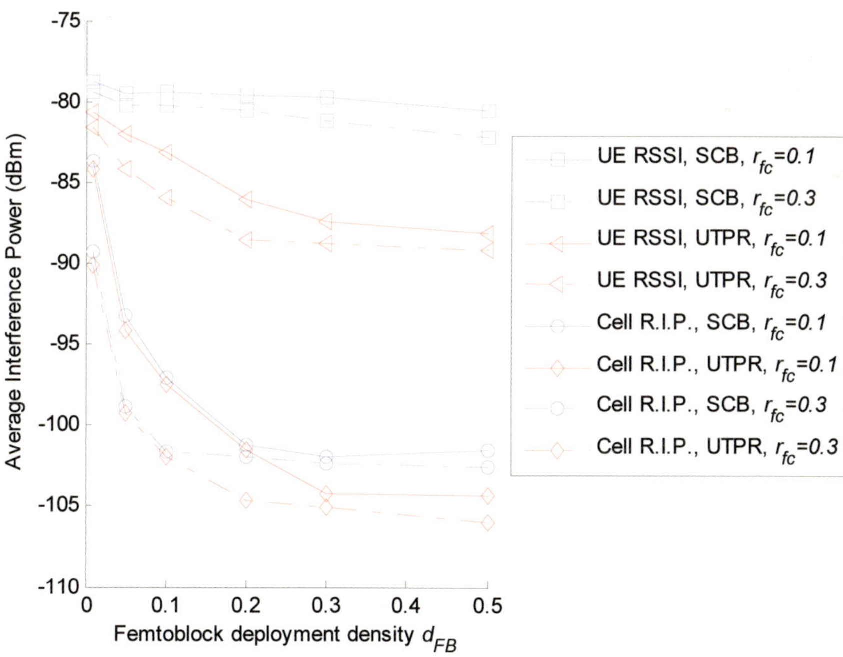

Figure 8. Average UE RSSI and Cell Received Interference Power versus the d_{FB}

The UTPR policy reduces the average transmit power in the LTE cells as well (Fig. 7), as a result of the substantial interference mitigation achieved in the LTE downlink in terms of RSSI and in the LTE uplink in terms of Received Interference Power at the LTE cells (Fig. 8). These are a direct outcome of the proposed policy's tendency to facilitate mobility towards cells which utilize bands with lower Received Interference Power. The latter reduces the number of UE interferers in congested LTE bands and condenses the overall UE power transmissions per band. Although the incorporation of the proposed UTPR policy achieves substantial energy consumption and interference mitigation gains, an increased HO probability is observed compared to the SCB policy (Fig. 9). This follows from the proposed policy's tendency to extend the femtocell utilization time, resulting to an increased sensitiveness on user mobility. To this end, the HO execution events are even more frequent when the femtocell deployment ratio per femtoblock increases. As in the SCB case, standard mobility-centric HO margin $HHM_{c,(dB)}^{UTPR}$ adaptation techniques can be utilized [10-12] to moderate the network-wide number of HO execution events (Fig. 10). The following results are derived for $d_{FB} = 0.05$ and $r_{fc} = 0.2$, while three different mean user speed values are considered i.e. 3, 60 and 125 km/h.

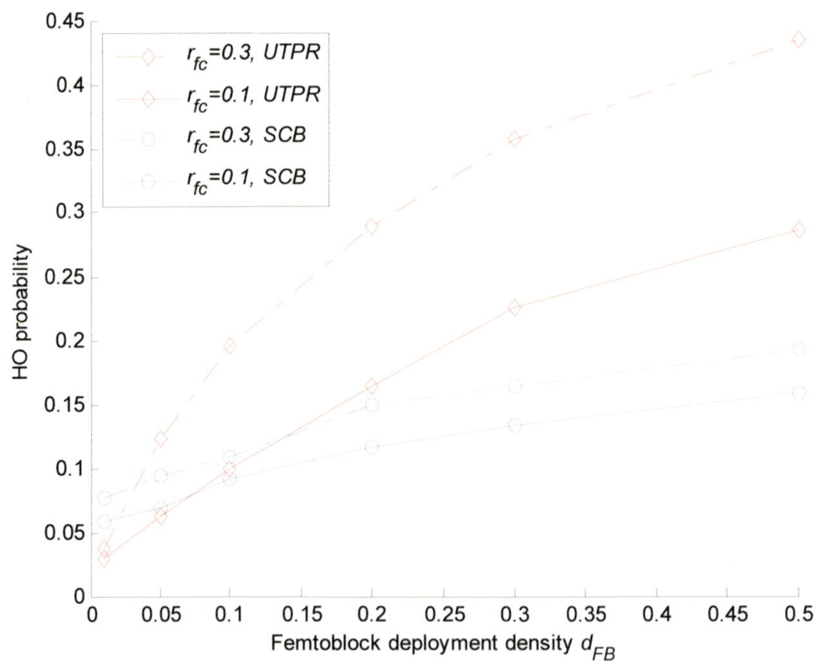

Figure 9. HO probability versus the d_{FB}

Fig. 10 illustrates the HO probability versus the $HHM_{c,(dB)}^{UTPR}$ value. As expected, an increasing user speed raises the HO probability for both policies. However, it can be seen that for a suitable $HHM_{c,(dB)}^{UTPR}$ parameter adaptation, the HO execution events for the UTPR policy are moderated and converge to the number of HO execution events corresponding to the SCB policy with lower $HHM_{c,(dB)}^{UTPR}$ values.

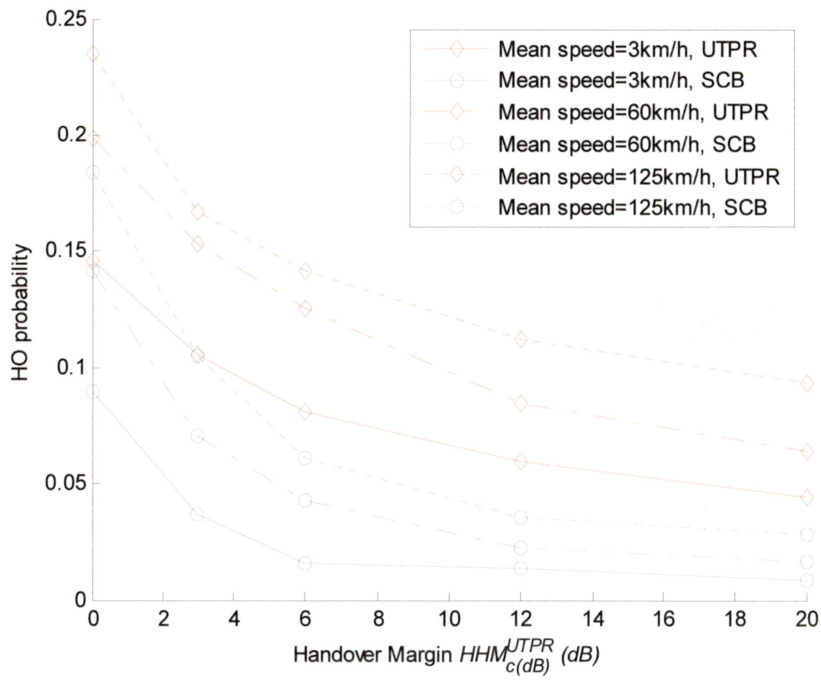

Figure 10. HO probability versus the Handover Margin

6. Conclusion

The random femtocell deployment may result in degraded SINR performance, increased outage probability, and enlarged network signaling, if the interference-agnostic strongest cell policy is employed during the HO decision phase. This chapter discussed the key feature of femtocell deployment and presented a novel HO decision policy for reducing the UE transmit power in the macrocell – femtocell LTE network given a prescribed mean SINR target for the LTE users. This policy is fundamentally different from the strongest cell HO policy, as it takes into account the RS power transmissions and the RF interference at the LTE cell sites. The proposed policy is backwards compatible with the standard LTE handover decision procedure, as it is employed by adapting the HHM with respect to the user's mean SINR target and standard link quality measurements describing the status of the candidate cells. Even though employing the proposed policy necessitates an enhanced network signaling between cells, numerical results demonstrate greatly lower network-wide RF interference, and reduced UE power consumption owing to transmit power, compared to the strongest cell HO policy. The impact of using an increased HHM for mobility mitigation has also been investigated in terms of HO probability.

Author details

Dionysis Xenakis and Nikos Passas
Dept. of Informatics and Telecommunications, University of Athens, Greece

Ayman Radwan and Jonathan Rodriguez
Instituto de Telecomunicações – Aveiro, Campus Universitário de Santiago, Portugal

Christos Verikoukis
Telecommunications Technological Centre of Catalonia, Barcelona, Spain

Acknowledgement

This paper has been partially funded by the C2POWER (FP7-ICT- 248577) and the GREENET (FP7-PEOPLE 264759), projects, and co-financed by the EU (European Social Fund – ESF) and Greek national funds through the Operational Program "Education and Lifelong Learning" of the National Strategic Reference Framework (NSRF) - Research Funding Program: Heracleitus II. Investing in knowledge society through the European Social Fund.

7. References

[1] V. Chandrasekhar, J. Andrews, A. Gatherer, "Femtocell networks: a survey", IEEE Communications Magazine, vol.46, no.9, pp.59-67, Sept. 2008.
[2] A. Galindo-Serrano, L. Giupponi, M. Dohler, "Cognition and Docition in OFDMA-Based Femtocell Networks," 2010 IEEE Global Telecommunications Conference, pp.1-6, Dec. 2010.
[3] H. Leem, S. Y. Baek, D. K. Sung, "The Effects of Cell Size on Energy Saving, System Capacity, and Per-Energy Capacity," IEEE Wireless Communications and Networking Conference, pp.1-6, Apr. 2010.
[4] F. Cao, Z. Fan, "The tradeoff between energy efficiency and system performance of femtocell deployment," The 7th International Symposium on Wireless Communication Systems (ISWCS), pp.315-319, Sept. 2010.

[5] I. Ashraf, L.T.W. Ho, H. Claussen, "Improving Energy Efficiency of Femtocell Base Stations Via User Activity Detection," IEEE Wireless Communications and Networking Conference 2010, pp.1-5, Apr. 2010

[6] 3GPP, "E-UTRA and E-UTRAN Overall Description.", TS 36.300 V10.1.0 (2010-10)

[7] A. Golaup, M. Mustapha, L.B. Patanapongpibul, "Femtocell access control strategy in UMTS and LTE," IEEE Communications Magazine, vol.47, no.9, pp.117-123, Sept. 2009.

[8] S. Sesia, I. Toufik, M. Baker, "LTE – The UMTS Long Term Evolution: From Theory to Practice", John Wiley & Sons, ISBN: 978-0-470-69716-0, 2009.

[9] K. Dimou, M. Wang, Y. Yang, M. Kazmi, A. Larmo, J. Pettersson, W. Muller, Y. Timner, "Handover within 3GPP LTE: Design Principles and Performance," IEEE 70th Vehicular Technology Conference Fall, pp.1-5, Sept. 2009.

[10] W. Shaohong, Z. Xin, Z. Ruiming, Y. Zhiwei, F. Yinglong, Y. Dacheng, "Handover Study Concerning Mobility in the Two-Hierarchy Network," IEEE 69th Vehicular Technology Conference, pp.1-5, Apr. 2009.

[11] A. Ulvan, R. Bestak, M. Ulvan, "Handover Scenario and Procedure in LTE-based Femtocell Networks", The 4th International Conference on Mobile Ubiquitous Computing, Systems, Services and Technologies, Oct. 2010.

[12] Z. Becvar, P.Mach, "Adaptive Hysteresis Margin for Handover in Femtocell Networks", 6th International Conference on Wireless and Mobile Communications, pp.256-261, Sept. 2010.

[13] J. Zhang, G. de la Roche, "Femtocells : technologies and deployment", John Wiley & Sons Ltd, ISBN 978-0-470-74298-3, 2010.

[14] G. Boudreau, J. Panicker, G. Ning, R. Chang, N. Wang, S. Vrzic, "Interference coordination and cancellation for 4G networks," IEEE Communications Magazine, vol.47, no.4, pp.74-81, Apr. 2009

[15] M. Yavuz, F. Meshkati, S. Nanda, A. Pokhariyal, N. Johnson, B. Raghothaman, A. Richardson, "Interference management and performance analysis of UMTS/HSPA+ femtocells," IEEE Communications Magazine, vol.47, no.9, pp.102-109, Sept. 2009.

[16] O. Simeone, E. Erkip, S. Shamai Shitz, "Robust Transmission and Interference Management For Femtocells with Unreliable Network Access," IEEE Journal on Selected Areas in Communications, vol.28, no.9, pp.1469-1478, Dec. 2010.

[17] D. Xenakis, N. Passas, and C. Verikoukis, "A Novel Handover Decision Policy for Reducing Power Transmissions in the two-tier LTE network", IEEE International Communications Conference (IEEE ICC) 2012, Ottawa, Canada, June 2012.

[18] 3GPP, "Physical layer; Measurements", TS 36.214 V10.0.0 (2010-12)

[19] 3GPP, "LTE physical layer; General description", TS 36.201 V10.1.0 (2010-12)

[20] 3GPP, "Radio Resource Control (RRC); Protocol specification", TS 36.331 V10.0.0.

[21] 3GPP, "User Equipment (UE) radio transmission and reception", TS 36.101 V10.1.0.

[22] Femto Forum, "Interference Management in OFDMA Femtocells", Femto Forum, Mar. 2010.

[23] A. Radwan, J. Rodriguez, "Energy Saving in Multi-standard Mobile Terminals through Short-range Cooperation", EURASIP Journal on Wireless Communications and Networking, to appear.

[24] D. Xenakis, N. Passas, and C. Verikoukis, "An energy-centric handover decision algorithm for the integrated LTE macrocell–femtocell Network", Computer Communications, Elsevier, 2012, to appear. http://dx.doi.org/10.1016/j.comcom.2012.04.024

Chapter 4

Tools and Solution for Energy Management

Soib Taib and Anwar Al-Mofleh

Additional information is available at the end of the chapter

http://dx.doi.org/10.5772/48401

1. Introduction

Energy efficiency can be defined as utilizing minimum amounts of energy for heating, cooling, lighting and the equipment that is required to maintain conducive conditions in a building [1,2]. An important factor impacting energy efficiency is not only the building envelope but also the management of energy within the premises. The amount of energy consumed varies depending on the design of the building, the available electrical systems and how they operate. The heating and cooling systems consume the most energy in a building; however control system such as programmable thermostats and building energy management systems can significantly reduce the energy use of these systems. Some buildings also use zone heating and cooling systems, which can reduce heating and cooling in the unused areas of the building. In commercial buildings, integrated space and water cooling/ heating systems can provide the best approach to energy-efficient heating [3]. Energy audits can be conducted as a useful way of determining how energy efficient the building is and what improvements can be made to enhance its efficiency. Tests should be undertaken to ensure that the heating and cooling systems as well as equipments and lightings work effectively and efficiently.

Building cooling and heating also produces Carbon Dioxide (CO_2) emissions, but this sector receives less attention compared to other pollution contributors such as the transportation and industry sectors. In addition to energy conservation and energy efficiency approach, a strategic plan to introduce renewable energy resources would be an advantage to any sector as it will reduce the carbon dioxide emissions as well they could be used for heating, cooling, ventilation and lighting systems [4]. As illustrated in Figure 1, the percentage of energy consumed by various type of building is presented. It was shown that the rental and service buildings utilize the highest energy consumption. It is easier to design energy efficient features for energy management of new buildings using available tools; however existing buildings comprise approximately 99% of the building stock. Although energy efficiency initiatives for existing buildings can be demonstrated to be cost effective, there has

INTECH
open science | open minds

© 2012 The Author(s). Licensee InTech. This chapter is distributed under the terms of the Creative Commons Attribution License http://creativecommons.org/licenses/by/3.0), which permits unrestricted use, distribution, and reproduction in any medium, provided the original work is properly cited.

been limited success in convincing large organizations and building owners to undertake energy efficiency projects such as retrofits, and retro commissions [5].

An important factor that rises while doing a comparison is the use of benchmarks as representative standards against the buildings to be compared and monitored. For example, the comparison of energy consumption with a square meter of floor area to the benchmark will allow the decision maker to observe and assess the amount of energy consumed. Therefore determination for the improvements can be made to minimize the consumption within that specific area. An effective energy management of a buildings do not necessarily cost more to build as compares to normal buildings, provided they are well maintained and manages effectively. Due to the use of energy efficient tools they are set to be very reliable, comfortable and as productive as a normal building. Numerous studies have focused on improving energy efficiency in commercial buildings. As stated by Escriva [6] engineers and researchers have developed complex methods to improve energy efficiency, but buildings are often managed by non-specialized technicians who need understandable and cost-effective actions to implement in their buildings. Therefore basic actions for the base improvements in energy efficiency for commercial buildings had to be stated and implemented.

2. Energy efficiency concept

The simple concept of energy flow can be shown in Figure 2. The whole idea is to minimize the energy loss either from the primary to secondary or from secondary to end use. Implementation of energy practices is part of the solution in reducing the loss from the infrastructure requirement to the energy demand. Energy is essential in increasing productivity and ensuring a high quality of life, thus the relation between energy and economic growth is crucial. However, the proportionality of economic growth and energy demand constitutes the depletion of energy resources. One essential and effective way to manage around the depletion of resources while at the same time fostering economic growth is by applying an energy efficiency concept [7, 8]. Since the late 1970s, a considerable increase in energy efficiency has been achieved in response to energy price hikes, supply uncertainties, government policies, and independent technology improvements [9,10]. Nowadays, the focus of international incentives for people to practice energy efficiency is to inject considerable amounts of awareness concerning climate change into the society. Energy efficiency is a generic term which refers to the usage of less energy to produce the same amount of services or useful output [11, 12].

However, the term energy efficiency depends heavily on its application. Thus there is no defined and clear qualitative measure of it. The input, output, analysis and monitoring parameters are very crucial in executing energy efficiency to the maximum in order to achieve the desired result. In general, the energy efficiency indicators are of the form of

$$Energy\ \ Efficiency = \frac{output\ power}{input\ power} \tag{1}$$

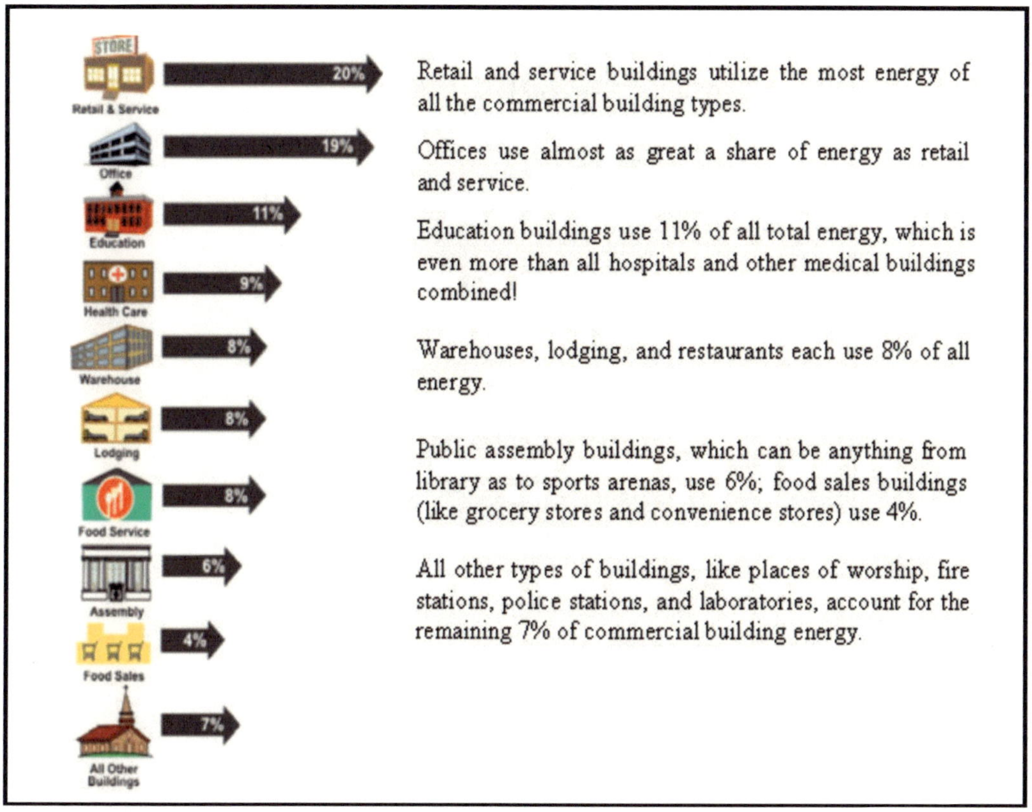

Figure 1. Building energy efficiency (Electric Power Research Institute)

The opposite of energy efficiency is referred to as energy intensity. The output of energy efficiency results can be physically determined, or enumerated in monetary units. The only plausible evaluation of energy efficiency is by looking at its indicators and then evaluating them by observing the results achieved targets, and relative situations among other groups. Energy efficiency number normally gives different interpretations relates to energy processes, programmers, investments, conservation properties, as well as system performances. Comparisons can be made to the past and to the projected future. This is important to ensure that the distribution of energy efficiency technologies and procedures can be systematically promoted. Evaluation and monitoring of energy efficiency practices give beneficial motivation mainly on the financial side. In many factories and buildings, overhead accounts duly increase the energy cost; therefore it saves a lot of money when total energy management is being practiced. Improvements can be adjust and implemented through accounting systems and at the end more accurately allocate energy costs within plants could be shown [8]. Diakaki et all in [9] investigate the feasibility of the application of multi objective optimization techniques to the problem of the improvement of the energy efficiency in buildings, so that the maximum possible number of alternative solutions and energy efficiency measures may be considered

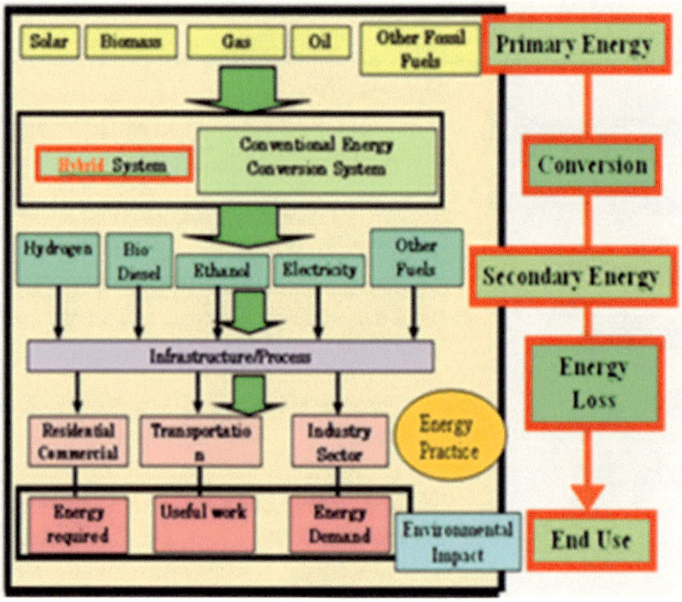

Figure 2. General layout of energy flow

3. Energy efficiency indicators

Analyses of energy efficiency usually include taking several placements of systems indicators and later grouping and evaluating them together. However, their effectiveness is always subject to the stipulation, particularly regarding, data quality and reliability, as well as availability [13]. Placement of indicators shows that there are 4 groups of energy efficiency indicators.

i. Thermodynamic: This group of indicators relies on the measurement of data by thermodynamic science applications while simple ratios and more complicated measures are used to measure the actual energy usage of an ideal process.
ii. Physical-thermodynamic: A much-improved version of thermodynamic units. However, the output is measured as a physical quantity. The purpose of using a physical approach is to measure severing parameters in terms of passenger miles or tones of product.
iii. Economic-thermodynamic: Another hybrid version of energy efficiency indicators where the input is still being measured in terms of thermodynamic units. The output meanwhile is measured according to market prices.
iv. Economic: A measure of this indicator defines changes in energy efficiency by the market values. Both input and output are in terms of the market prices.

Specific energy consumption (SEC) reduction is defined as the improvement of energy efficiency (EE) by industrial players. SEC is one of the EE indicators as it gives a ratio of energy consumption to the beneficial output (physical) of a process. SEC also can serve as an energy intensity indicator, especially for single processes that generate one single product.

The fifth group with addition to Patterson's definition is environmental EE indicators. They are special for measuring energy related specific emission which is direct to environmental issues. However these indicators only allow for the comparison of the efficiency of processes which require the same end use service. It is a very evident that the energy quality problem is a fundamental problem across all energy efficiency indicators when trying to compare process with different quality inputs and outputs [8, 14].

3.1. Management of energy efficiency indicators

Energy efficiency indicators, while functioning to provide information about EE consumption and its end results, also function to compare and provide benchmarks for present and future technologies. Benchmarking is quite a tedious process, where external influences that affect economic, financial and other non-includable parameters have to be excluded from the judgment [15]. The influence of external factors tends to dynamically increase which can easily frustrate assessments [13]. According to [8] the main problems with EE indicators are.

i. Inhomogeneous data,
ii. Geographical differences that make the ratios and indicators vary continuously, and
iii. Interpretations of ratios that diverge accordingly

The evaluation of EE indicators heavily depend on the transparency of the data collected and calculation of indicators. In order to reach this 3 targets are identified as follows [9].

i. Progressive harmonization of data and regularity of updated database for data management.
ii. Defining the status of common technology for EE assessment. This can also be applied on the consequences on Carbon dioxide (CO_2) emissions and the related ratios.
iii. Acquiring necessary mechanisms in order to regularize the findings to the real time experiences. Harmonization of interpretations is a necessity to ensure the reliability of indicators produced from common database.

3.2. Energy efficiency labels and standards

Energy efficiency standards and labels usually come together. Standards are technical settings of energy efficiency, while labels provide guidelines to consumers to select more efficient appliances when they make a purchase [16].

3.2.1. Labels

Energy efficiency labels are educational labels that are affixed to explain the energy performance of manufactured products, and to give the consumer the necessary data for making knowledgeable purchases [16]. According to [17] there are three kinds of labels.

i. Endorsement Labels: Fundamentally given according to products that meet specified criteria.

ii. Comparative labels: Allows a customer to evaluate product performance against similar products using discrete categories of performance or a continuous scale.
iii. Information-only labels: Provides data on a product's presentation.

Energy labels can stand alone or balance energy standards [18, 19]. They provide information that allows consumers to select efficient products. The effectiveness of energy labels is very dependent on how they present information to the consumer. The sample of European label is shown in Figure 3a where it will tells you about the energy efficiency of electrical appliances. Grade A++ is now the most efficient, and Grade G is the least efficient. While Figure 3b shows energy label used in Malaysia where the number of star reflect the most efficient.

(a) (b)

Figure 3. Energy Efficiency label (a) European Energy Level and (b) Malaysia energy level

3.2.2. Standards

There are several definitions of energy efficiency standards. According to Greg at all [20] an energy standard is defined as a minimal requirement for efficiency, or the measured energy consumption for the household appliance. Duffy [21] stated that the energy efficiency standards are government mandated standards that define minimum levels of efficiency or maximum levels of energy consumption and that must be met by all products sold in the particular authority.

However definition given by McMahon and Turiel [22] which points out the energy efficiency standard as the prescribed energy performance of a manufactured product, sometimes keeping out the manufacture of products with less energy efficiency than the minimum standards. There are three types of energy-efficiency standards.

i. Prescriptive standards: Requiring that a particular feature or device be installed (e.g., insulation) or not installed (e.g., pilot lights) in all new products;
ii. Minimum energy-performance standards (MEPS) : Prescribing minimum efficiencies (or maximum energy consumption usually as a function of size or capacity) that

manufacturers must achieve in each and every product, specifying the energy performance but not the technology or design details of the product; and

iii. Class-average standards: Specifying the average efficiency of a manufactured product, allowing each manufacturer to select the level of efficiency for each model so that the overall average is achieved.

3.2.3. Products covered by energy efficiency standards

Products covered under National Commission on Energy Policy (NCEP) as shown in Table 1. In normal condition heating and cooling consumed more energy utilization compare to others and in some case it contributed to nearly 80% of energy bill [23]. The standard which was implemented includes those set by the legislation as well as standard adopted by DOE through rulemaking. The impact cover primary energy savings and water saving, net present value of customer benefits and estimated reduction in CO_2.[24]

Heating and Cooling	Residential Central Air Conditioners and Heat Pumps Room Air Conditioners Commercial Unitary Air Conditioners and Heat Pumps Direct Heating Equipment Residential Furnaces, Boilers Mobile Home Furnaces Swimming Pool Heaters
Cleaning & Water	Clothes Washers Clothes Dryers, Dishwashers Water Heaters Faucets, Showerheads Toilets/Water Closets, Urinals
Lighting	Fluorescent Lamps and Ballasts Incandescent Reflector Lamps High-Intensity Discharge Lamps
Food Preservation and Cooking	Refrigerator Freezers Kitchen Ranges and Ovens
Other Products	Distribution Transformers Small Electric Motors Televisions

Table 1. Products covered by energy efficiency standard

3.2.4. Developing labels and standards programs

The steps to developing energy efficiency labels and standards are shown in Figure 4. These steps are described in the following paragraph.

i. First step is to decide whether and how to implement energy labels and standards. A government's decision whether or not to develop an energy-efficiency labeling or standards-setting program is complex and difficult. Many factors determine whether such a program is beneficial in any particular country.

ii. Second step: Develop a testing capability. Testing by manufacturers and private laboratories need to be accredited and recognized. Government costs are reduced and product-marketing delays are avoided if governments rely mainly on private testing and conduct audits themselves.

iii. Third and fourth steps: Design and implement a labeling program, and analyze set standards: label requirements can be established in a variety of ways, usually involving consumer research as important part of the process. A label can provide a single rating or large number of data, and energy performance measurement of competing products.

iv. Standard setting: a standard can be set to eliminate the less efficient models currently on the market. And to encourage importers and local manufacturers to develop the most economically efficient products, several types of analyses should be conducted to ensure that a standard is achieved.

v. Fifth step: Maintain and enforce compliance: After the label design process is mandated or a standard is set, those responsible for the labeling and standard setting must certify, monitor, and enforce compliance.

vi. Sixth step: Evaluate the labeling or standards setting program: If a government is to maintain an energy efficiency labels and standards program over the long run it will have to monitor the programs performance to gather guidance for adapting the program changing circumstance and to clearly demonstrate to funding agencies and the public so that expected benefits are actually being achieved.

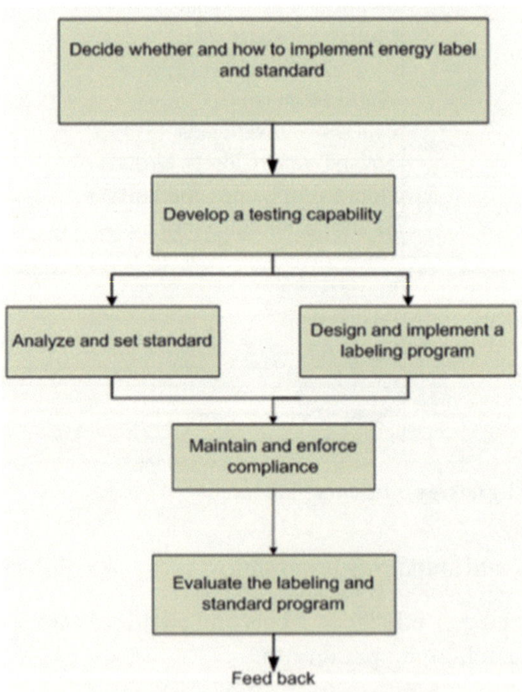

Figure 4. Steps in developing energy efficiency labels and standards

For developing policy-relevant indicators in the residential sector, the following data are required [25]:

i. Energy consumption by major end-uses and by energy sources;
ii. Main activity variables for the sector, including number of households and residential floor area;
iii. Information on the stock and efficiency not only of large appliances, but also of small appliances given their growing importance; and
iv. Information on heating and cooling degree-days to adjust for weather conditions

3.2.5. Labels and standards relationship to another energy program

Energy efficiency labels and standards work best in combination with other energy policies designed to shift the market toward better energy efficiency. It is important that consumers receive a consistent message on energy efficiency. Collective efforts from everyone, including government policy makes an energy efficient economy together with an array of policy instruments that can influence manufacturing, supply, distribution product purchases, and operation maintenance of energy consuming products in our society. When working effectively, these policy instruments accelerate the penetration of energy efficiency technology throughout the market. Energy efficiency labels and standards are considered by many researchers such as Mahlia [17] to become the main of a country's energy efficiency portfolio.

3.2.6. Labels and standards of energy efficiency scope

Energy efficiency label and standards can be applied to any product that consumes energy. The national benefits of labels and standards applied to the most prevalent and energy intensive appliances. The benefits from labels and standards for less common or less energy intensive products are often too small to justify the cost [27].

The first mandatory minimum energy efficiency standards were introduced as early as 1962 in Poland, for a range of industrial appliances. The French government set standards in 1966 and 1978 and other European countries introduced legislation mandating efficiency information labels and standards through the 1960s and 1970s. Mandatory labeling programs have developed in parallel with standards. In 1976 France introduced mandatory labeling followed by Japan, Canada, and USA. A report by [23] show that the ENERGY STAR program of the USA Department of Energy was able to verify product performance and also could identify product that do not meet these standard. In China when the Ministry of Constructions issued a revised energy design standard for new heating JGJ 26-95, an increase of 50% of energy saving was achieved [26]. The labels and standard are being updated continuously worldwide [27-29]

4. Constraints in implementation the energy management practice

The administrative personnel may feel that finding previous energy bills and equipment manuals is extra work. If the future data (bill) on electricity and oil is obtained from normal administrative work (reporting), this would not be the case. The need for energy efficiency, both for economic and environmental reasons, has never been greater. The International Energy Outlook 2000 [30], predicts that energy consumption will increase by 60% over period 1997 to 2020. Energy use worldwide will continue to grow at an average annual rate of 1.1%, and by 2020 the world consumption will rise from (380) quadrillion British thermal units (Btu) in 1997 to (608) quadrillion Btu in 2020, as shown in Figure 5.

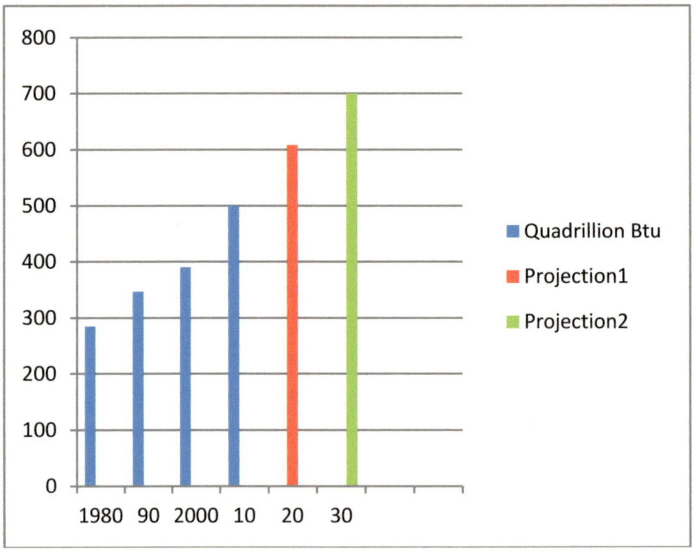

Figure 5. Energy use world wide

The main factor responsible for the increase in energy demand especially in the developing countries is the economic growth. Confronting the growing world energy demand raises two questions. Will there be enough energy available, and in what forms. Reassuringly, there appears to be no prospect of absolute shortage, at least for the next twenty or thirty years and probably beyond. Fossil energy resources are still abundant: for example, the cumulative production of natural gas has used only one sixth of the 325 trillion cubic meters of known reserves. Coal reserves are even larger, providing a basis for continuous production for hundreds of years. Known uranium resources, even on the basis of present knowledge, could meet the current level of demand for a period of 8000 years if advanced nuclear technology is developed [31].

Consideration of energy in relation to the building environment throughout the world's developed countries reveals that 20 – 55% of all delivered energy can be directly associated with buildings and industry. Consequently, new technologies applied to the built environment and industry may be expected to make a significant contribution to a reduction in energy consumption. By raising the efficiency of energy utilization, it is possible to reduce energy consumption by 10 – 30%, representing a savings of around 3Mtce per year. Some progress has been made in recent years as building energy management (sensors, HVAC control equipment) and information technology (IT) systems have evolved to a point where they can support and integrate the activities involved in energy management. With respect to environmental impact and economics, the ability to make well founded decisions regarding energy consumption and supply is of the utmost importance. This requires some means to assess the current performance and agreed targets against which to judge performance [32-34].

5. Energy efficiency targets

Energy efficiency practices could the solution to ensure energy is available to satisfy all demands, to ensure energy is used and supplied with minimal cost/ environmental impact and to ensure energy is not wasted. The challenge of achieving these targets is the driving force for the development of computer-based energy management tools that enable users to understand how energy is consumed within their properties and how they can improve the use of energy resources for effective task processing [34-40]. The Energy Efficiency Initiative, a report published by the IEA, Danish Energy Agency and the Energy Charter 1999, identified four most essential elements of a normal framework for effective energy policies.

i. Focus market interest on energy efficiency Actions include: Fostering voluntary agreements, establishing and enforcing building codes, minimum energy performance standards, integrating energy efficiency in procurement practices and using government purchasing to stimulate the market for advanced technology.
ii. Ensure access to good technology Actions include: Encouraging the development, adaptation and diffusion of energy efficient technology, improving district heating systems and expanding the use of combined heat and power.
iii. Develop and maintain a supportive institutional framework: Actions include: Integrating energy efficiency in sectoral policies and ensuring the availability of impartial expertise.
iv. Act to ensure continuity: Actions include: Establishing policy clarity, demonstrating leadership, implementing effective evaluation, monitoring techniques and strengthening international collaboration

Other outcomes of the implementation of energy efficiency in industries can be given as [41]:

a. Industries become aware of the actual and rational energy utilization performance, as well as Energy Efficiency and Energy Conservation measures that can be applied to improve energy utilization efficiency through the establishment of energy use norms for industrial sub-sectors and processes.
b. Industries comply with regulations / guidelines designed to encourage the use of energy efficient equipment and practices.
c. Awareness about, and attitude towards, energy efficiency and environmental improvement by industries widespread.
d. Industries are using and benefiting the local energy support services (ESCOs) in the implementation of their energy efficiency projects.

e. Industries are implementing proven and cost-effective energy efficiency technology projects.
f. Industries utilize locally manufactured equipment with comparable efficiencies to imported quality industrial equipment.
g. The energy authorities is able to increase its capacity and capability in providing energy advisory services to the public and the private sectors

A proposal by Escriva to impose seven actions of the base improvement in energy efficiency in commercial building had shown a considerable financial saving is being made. These actions include an accurate operational data measurement, a proper schedule, automatically monitored of the consumption of electricity, individual responsibility for energy use in each building, proactive action to increase energy efficiency, facilities modification to enable easier energy management and an excellent communication between user and the building managers. Even many national governments and international organization have developed new regulations, what is lacking today is the universal energy efficiency index for building. In order to address this issue Gonzalez et all [42] had proposed an energy efficiency index for buildings that relates the energy consumption within a building to reference consumption. The proposed energy index can be obtained in a simple manner by combination standard measurement, simulation and public data base; furthermore the index is upgradable whenever new data are available.

6. Benefit of energy efficiency

The benefits resulting from the implementation of energy efficiency in management systems are numerous and can be grouped as direct and indirect. Direct benefits are the current benefits, while indirect benefits are the expected short or long term benefits. Some of the direct and indirect benefits are given in the following section [30, 31].

6.1. Direct or short term benefit

Some of the direct benefit can be states as follow.

a. Reduction of fuel and electricity bills
b. Identification of energy saving opportunities and environmental compatibility concerning the building
c. Energetic retrofitting of the buildings
d. Reduction of operation and maintenance costs
e. Improved public image

6.2. Indirect or long term benefits

6.2.1. Environmental benefits

Greenhouse gases include carbon dioxide which is produced every time we use energy from fossil fuels - oil, coal and natural gas. With businesses producing almost half of the world's carbon emissions their impact is huge [32]. The increase in our planet's temperature has already caused sea levels to rise, making floods more frequent and severe. And as the temperature gets hotter, it's predicted that we shall see more extreme weather. It is found that the businesses are currently responsible for about half of all the world's carbon emissions. Even one small office can produce 3-5 tons of carbon dioxide (CO_2) in a year. Unless they reduce their carbon emissions, businesses will start paying the price of climate change through more expensive energy supplies and higher insurance premiums [32].

6.2.2. *Business benefits*

Many companies think of energy as a fixed overhead but saving energy is actually one of the easiest ways to reduce costs and improve your standing. The following are some of the benefits from the practices discussed [41].

i. Save money simply by switching machines off after use, or turning the heating down in warm weather or switching off the air conditioning system when not occupying a room, the owner can make real savings on energy bills.
ii. You can offer better value to your customers by cutting your overheads as production costs will go down, making your products and services more competitive.
iii. Enhance your reputation. Increasing numbers of consumers and business customers will now only buy from, or invest in, companies with environmentally friendly policies and production methods. By demonstrating a commitment to saving energy, you will increase your appeal in the market and attract a wider customer base.
iv. Encourage more people to come and works for you. People do not just want to buy from socially responsible businesses; they also want to work for them. So you will also increase your appeal in the recruitment market, and stay an employer of choice.
v. Stay ahead of government regulation. With climate change so high on the political agenda, there are likely to be more initiatives like the government levy and scheme, as well as tighter building regulations from the government. Being prepared for these changes will save time and money when they are introduced [36].

6.2.3. *Home benefits*

Home energy is responsible for carbon dioxide emissions which contribute to climate change. By following the best practice standards, new buildings and refurbished housing will be more energy efficient and will reduce these emissions. New buildings or home extensions can provide new, energy efficiency accommodations and also improved the overall efficiency of the houses that are extended [35]. For the homeowner, specifying an energy efficient extension is a cost effective approach because the additional cost will be recovered in reduced fuel costs. Payback period is less than 10 years, but fuel cost is entirely reduced for the entire life of the building.

6.2.4. *Transport benefits*

By running the fleet more energy efficiently the following benefits will be withdrawn.

i. Reducing your maintenance, operation costs and cutting your vehicle emissions.
ii. Improving your social and environmental reputation.
iii. Minimizing traffic and parking problems where you work.

On average, every time you use up to a 400 liter tank of fuel, you produce: "1.04 tons of carbon dioxide, 1.38 kilos of carbon monoxide, 0.67 kilo of hydrocarbon, 6.15 kilo of nitrogen oxide and 0.14 kilo of particulates" (UK Road Transport Emission Projection 1997). Bio fuel use has cut carbon dioxide emissions by 40,000 tons in the first quarter of 2006. Independently evaluated by the Edinburgh Centre for Carbon Management, the reductions were said to be the equivalent of 50,000 family vehicles taken off British roads [36].

6.2.5. Community benefits

Community Action for Energy is one of the energy efficiency programs that are designed to promote and facilitate local community-based energy projects. Community action for energy can help to improve the quality of life in community by providing new opportunities to improve the comfort, health and well-being of people in community, combat fuel poverty and help the local economy. Example of community activities are community heating or cooling, ground source pump, renewable energy, solar PV, wind, and community energy project [37].

6.2.6. School benefits

Energy efficiency for schools promotes energy management and efficiency for the whole school community. It encourages schools to make links between the school curriculum and the management of energy efficiency and learn about how to make a difference to climate change. Schools can save themselves money on their energy bills, reduce their carbon dioxide emissions and improve their working environment.

7. Barriers to energy efficiency

Theoretically it is easy to define and put measures on energy efficiency parameter, although in practice, the same cannot be said, as there are quite a number of challenges to be faced. However for overall success, there is an urgent need to implement the solution of energy efficiency in a large scale in developing and developed countries [38]. Various barriers that may be giving hindrances to implementation of energy efficiency initiatives have been studied. For instance Reddy [39] has investigated a typology for energy efficiency barriers that related to consumers, manufactures, financial institutions and the government. While DeCanio [40] has investigated that there are a few problems regarding energy efficiency barriers in firms. Painuly and Reddly have identified 6 important factors that need to be addressed to effectively deal with those barriers.

i. Technical – availability of reliable knowledge of energy efficiency technology.
ii. Institutional – availability of right technical input and proper execution of programmers.

iii. Financial – convenient mechanism of finance.
iv. Managerial – training and management.
v. Pricing – proper rationalization of pricing on electricity and fuels.
vi. Information – appropriate level of information.

In Malaysia for example there are several barriers which could hamper the smooth implementation of Industrial energy efficiency improvement project [41]. These are as follows:

a. Limited knowledge/awareness about Energy Efficiency and Energy Conservation techniques/technologies in industries and the lifecycle economic benefits.
b. Limited access to information on energy efficiency techniques as well as energy benchmark.
c. Industries are unwilling to incur what are perceived as "high cost, high risk" transactions.
d. Industries generally focus on investments on production-related improvements.
e. Lack of financiers ready to finance energy efficiency investments.
f. Limited/not stringent regulations on energy efficient standards and implementation.
g. Few or limited energy efficiency technology demonstration projects implemented.
h. Weak local energy support service

An energy gap is a term that points to a phenomenon where a firm is not utilizing the feasibility of technically and economically viable efficiency measures. Weber [43] pointed them to its institutional, market, organizational and behavioral barriers. While Sorrel, et.al [44] grouped the barriers of energy efficiency as neo-classical (economical), behavioral and organizational

The need for border framework to tackle barriers to energy efficiency for significant improvement has been intensified. But literature reviews reveal that there are qualitative issues on the barriers. And a creation of suitable topologies is a must to tackle them. There is currently not enough attention when considering relevant influential factors to the qualifications of the barriers levels, especially industry specific barriers related studies that are very scarcely found [44]. Various definitions of energy efficiency barriers that have been introduced by previous researchers include environmental, economical, and technology related aspects. This prevents a whole implementation of an energy efficient based practice blueprint. Governments as well as private sectors need to churn out more effort with the intention to practice energy efficient so that the longevity of a healthy world can be achieved for future generations [45]. A recent work by Escriva [46] could be an excellent approach to overcome part of the problem. This research work proposes the continuous assessment of energy efficiency in building using different energy rating factor for assessing energy performance and identifying wasted energy.

8. Software tools

For the last four decades there have been many energy software had been developed and most of the energy tools can be run using a simple text-based input. The simple algorithms that were used can calculate energy performance for the specific requirement. Such tools that have been developed are capable of predicting the daily, monthly and also annual energy performance of the proposed or existing building. The software had been tested and shows tolerable results compare to normal practical measurement. Most of the developed commercial software provides the users information of building performance such as energy demand, temperature, operating costs and humidity [47]. Additional information regarding energy analytics, policy management, billing system, meter data management, Green House Gas tracking, continuous benchmarking and commissioning also available. Some of the software currently available are Energy-10, EnergyPlus, Energy Manager AutoStartTM , IES VE, Ecotect, HAP, Trace, eQUEST, PowerDomus, eSight Energy and iEnergyIQ. The list of the software developed but not being commercialized was uncountable and normally they were being used for authorized personnel.

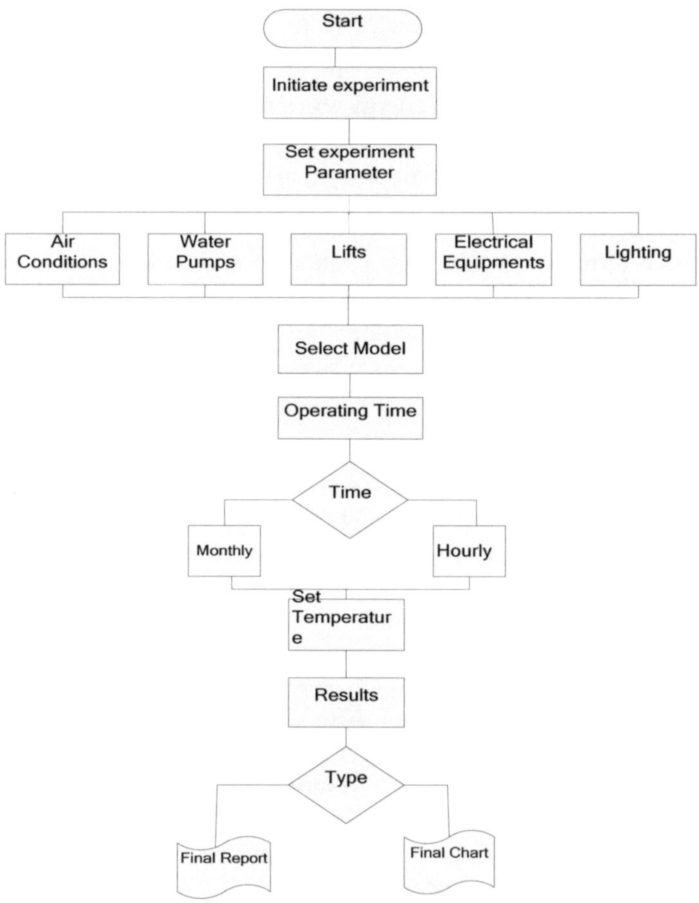

Figure 6. Layout for Smart Energy Evaluation System

One of the software developed at the School of Electrical Engineering, University Science Malaysia was the Smart Energy Evaluation System [48]. Based on Microsoft Visual Basic, a smart interface media is developed to give users ample opportunity to browse and observed energy consumption, bills, temperature, humidity control and its operation hours. The built in intelligent enables the user to obtain analysis regarding energy related issue. The simple layout of the system is shown in Figure 6 and the manager window is shown in Figure 7. Figure 8 shows the results of energy consumption in a particular seminar hall.

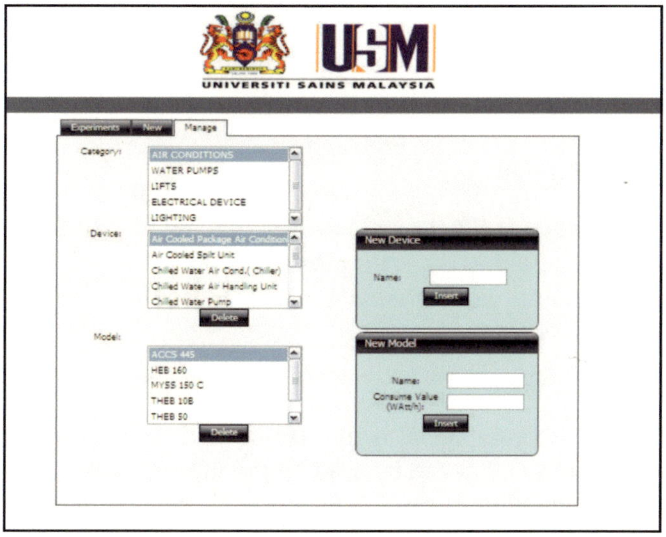

Figure 7. The manager window for Smart Energy Evaluation System

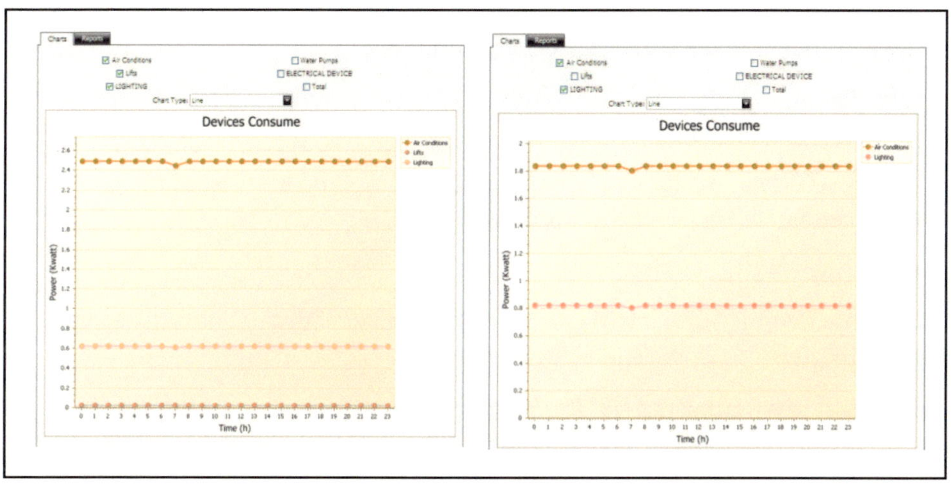

Figure 8. Energy consumption in a seminar room using Energy Evaluation System.

It is quite difficult to select the appropriate software for general purpose because each of them is set to a specific function. Energy software development normally depends on the user's requirements and demands, where the designer had to identify their energy analysis algorithm, model and results presentation. A few of these software solutions are consider expensive, costing more than US $10,000/year for example, but the majority of the commercial product are offered either for a reasonable price or free. The requirement for the designers is to design the software using accurate measurement tools and a proper automated reporting system within an acceptable price. This should include the report of data either hourly or monthly basis so that the verification of the software could be check and in line with historical data. The other important feature that had to be considered in future is the data sharing within the software. Based on the experience it is worth to use the appropriate energy efficient tools as part of energy management for our future sustainability plan. Not only it preserves our environment by reducing the GHG emission but also save unnecessary funding for energy.

9. Case study in Malaysia

The Malaysia Industrial Energy Efficiency Improvement Project (MIEEIP) was developed to remove barriers to efficient industrial energy use. The project is also designed to facilitate the reduction in greenhouse gas emission by the industrial sector [41]. The following programmers were implemented:

a. Energy used bench marking

The overall objective of this component is to establish and develop energy use benchmark for the eight industrial sectors that can be used by the industries as a guide in their energy efficient and energy conservation efforts.

b. Energy Audit

The overall objective of this component is to improve the energy efficiency levels in industries by promoting the practices of energy auditing.

c. Energy Rating

The overall objective of this component is to introduce activities that will inform industries about energy efficient equipment energy rating programmers including cost technical specification, economic and energy performance.

d. Energy Efficiency Promotion

The main objective of this component is to disseminate information on energy efficient practices and technology application to the industries

e. Energy Services Companies (ESCO) Support

The overall objectives of this component are to provide engineering services or consultancies, providing financial solution as well as risk mitigation for energy efficiency activities to industries/client.

Case study 1: Tritex Containers Sdn Bhd (Malaysia)

For more than 20 years the management thought that it was okay to have big boiler for its production process. Little did they realize that oversized boiler was causing inefficient use of energy which eventually meant the company was losing money? Initial finding by the audit team showed that the company had the potential for energy saving as high as

USD50,000 a year. The MIEEIP team discovered that most of the energy losses occurred because of uninsulated pipes in the boiler and leakage in the compressed air system and main distribution loops [41].

Energy Efficiency Activities:

The measures that were recommended involving no-cost and low cost investment included:

a. Fine tuning boiler operations
b. Repairing leaks in distribution pipes and consumption points
c. Recycling leakages in compressed air system
d. Shutting-off compressor and dryer during non-production hours
e. Using smaller compressor during low production period
f. Switching off all nonessential equipment

With an initial investment of only USD15,000 the company put in place a more efficient system that resulted in an annual energy saving of USD30,000. Table 3 shows the estimated annual savings from the measures implemented practices.

No	Energy saving measure	Invesment (USD)	Annual Saving (USD)
1	Fixing air compressor leaks	0	2000
2	Compressor and dyer Shut Off	0	1500
3	Boiler readjustment	0	1200
4	Reduce Boiler Breakdown	0	500
5	Repair Steam Leak	6500	14000
6	Replace Pneumatic Pumps	500	800
7	Insulate Boiler and Condensate	3700	3800
8	Relocat WTP Closer to Boiler	4000	1700
9	Pump Condensate from WTP	6000	4200
	TOTAL	14800	30000

Table 2. Estimated annual saving

Case Study 2: Pan Century Edible Oil (PCEO) Sdn Bhd (Malaysia)

It's total manufacturing capacity of oil refinery product and special product is about one million tones. The factory recorded an annual turnover of USD250 million in 2002. More than 550 tones of steam are generated daily for physical refining, making soap noodles, tank farm heating and fractionation.

During the audit, the team identified significant steam leakages and analyzed the respective losses. With an investment on USD343,000 on the steam optimization programmed, resulted in an annual energy saving of USD229,300. Modification to the factory's cooling towers, replacement of standard high efficient motor, steam conversation scheme, heat recovery scheme and monitoring, which produce another total energy saving of USD100,000. Table 4 shows the estimate annual saving from the implemented measure practices.

	Energy Cost Measure	Investment	Saving GJ/yr	Investment (USD)	Saving (USD
1	Use of low pressure steam	Low cost	2665	21000	27000
2	Pressure reduction valve for distillation	Low cost	6504	9300	40800
3	Insulation improvement	Low cost	10789	16000	68000
4	Steam trap maintenance	Low cost	4395	25000	26000
5	Condensate recovery	Low cost	2943	39000	24000
6	Temperature Control	Low cost	1896	8000	11500
7	Steam ejector replacement	Low cost	5188	15000	32000
	TOTAL		34380	133300	229300

Table 3. Estimate annual energy saving

Case Study 3: JG Container Sdn Bhd

This company had commenced operation in Malaysia since1970 and total energy costs representing 20% of it turnover. By applying energy efficiency practices 57,300 GJ annual energy saves from the new furnace and 2,800 GJ/annum from improvements in two annealing lehr, saving in fuel cost of more than 33% from heat recovery system for the glass furnace. It is found that by total investment of USD2.5 million made a saving of USD0.6 million per year. Table 6 shows the energy efficient measure and the saving affected.

	Energy saving measure	Investment (USD)	Fuel Saving (%)	Payback (year)
1	Rebuilding furnace	2.3 million	33	4
2	Modification of one annealing lehr	17000	42	1
3	Natural gas-powered lehr	133000	67	6.5
4	Water recycle	6000	25	1
	Total	2.5 miilion		

Table 4. Investment and saving for JG Container Sdn Bhd (Malaysia)

Case Study 4: HeveaBoard Bhd (Malaysia)

Saves USD180,000 on annual energy costs, saves 37,000 GJ in fossil fuels and decreases CO2 emissions by nearly 3,000 tonnes each year. Following an energy audit, HeveaBoard, a particleboard manufacturer, replaced its fuel oil fired thermal heater with a wood dust fired thermal oil heater. This allowed the company to capitalize on a cheap source of energy, namely the excessive wood waste left over from production. HeveaBoard also engaged the services of an Energy Service Company (ESCO) to install the system which guarantees results and payment based on energy savings achieved.

Case Study 5: Globalmas Sdn. Bhd. (Malaysia)

A manufacturer of canned food based in Sarawak recorded an annual turnover of USD1.4 million in 2002 and found ways to slash its operating costs. The company identified the best use of energy consumption during various plant processes by reanalyzing electric, gas, and fuel or steam resources. Globalmas discovered that with a total capital expenditure of USD42,000, the factory could reduce its energy costs by nearly half and recover the investment within a period of 18 months.

10. Conclusions

Energy efficiency has become one of the main tasks in energy management. Technical systems and solutions that can save energy and keep the costs down gain an importance decision. At present in buildings, equipment like air-condition, pumps and lighting had consumed high amounts of energy. A small adjustment through energy practices and energy audit can lead to a significant reduced of energy used. It is found that the available commercial tool for analyzing energy consumption and implements energy conserving measures could be a solution to minimize energy utilization. Practices of energy audit and energy efficiency activities in industries had shown a potential saving. An improvement of energy efficiency in building will be among the top priorities in the energy management worldwide together with the implementation of ISO and related standard.

Author details

Soib Taib
School of Electrical & Electronic Engineering, Engineering Campus, Univerisiti Sains Malaysia, Pulau Pinang, Malaysia

Anwar Al-Mofleh
Department of Electrical, Balqa Univeristy, Amman, Jordan

Acknowledgement

The author wish to thank the Ministry of Higher Education (Malaysia) for the Fundamental Research Grant Scheme in supporting this research work.

11. References

[1] Anwar A, Soib T, Hamza, G. Tools for Building Energy Efficiency Estimation, Proceeding of Fifth International Symposium On Mechatronics And Its Applications, Jordan (2008)
[2] Horace H. Energy efficiency a critical view, 2006: Energy (31) 10–20
[3] Matteo C, Paolo S.C, Marco F. Energy demand for space heating through a statistical approach: application to residential buildings 2008 Energy and Buildings (40) 1972–83
[4] Shuichi A, Toshihiko N. Energy-efficiency strategy for CO2 emissions in a residential sector in Japan 2008: Applied Energy (85) 101–114
[5] Johnny W, Heng L, Jenkin L. Evaluating the system intelligence of the intelligent building systems Part 1: Development of key intelligent indicators and conceptual analytical framework 2008: Automation in Construction 284–302

[6] Escriva G. Basic action to improve energy efficiency in commercial buildings in operation 2011: Energy and Building (43) 3106-11.

[7] Soib T, Shanti B, Shuauib L, Seng B, Singh J, Azahari B. Promoting Energy Awareness and Efficiency: USM Strategy 2005: Proceedings of The 8th International Conference on Quality in Research (QIR),EEE1-003

[8] Oikonomou V, Becchis F Stegc L, Russolillo D. Energy saving and energy efficiency concepts for policy making 2009: Energy Policy (37) 4787–96

[9] IEA International Energy Agency. Energy efficiency initiative. Energy policy analysis 1997: (1) p.193-99

[10] Diakaki C, Grigoroudis E, Kolokotsa D. Towards a multi-objective optimization approach for improving energy efficiency in building 2008: Energy and Building (40) 1747-54

[11] Murray G. P. What is energy efficiency? Concept, indicators and methodological issues 1996: Energy Policy (24) No-5 377-390

[12] Kanako T. Assessment of energy efficiency performance measures in industry and their application for policy 2008: Energy Policy (36) 2887– 2902

[13] Bosseboeuf D, Chateau B, Lapillonne B. Cross-country comparison on energy efficiency indicators: The on-going European effort towards a common methodology 1997: Energy Policy (25) 673-82

[14] APERC Energy efficiency indicators, a study of energy efficiency indicators for industry in APEC economies. Asia Pacific Energy Research Centre. Tokyo. 2000: 154-160

[15] Yunchang J.B. Consistent multi-level energy efficiency indicators and their policy implications 2008: Energy Economics (30) 2401-19

[16] Konstantinos P, Haris D, Argyris K, John P. Sustainable energy policy indicators: Review and recommendations 2008: Renewable Energy (33) 966–73

[17] Mahlia T.M.I. Methodology for predicting market transformation due to implementation of energy efficiency standards and labels 2004: Energy Conversion and Management (45) 1785–93

[18] Bertoldi P. European Union Efforts to Promote More Efficient Equipment, European Commission, Directorate General for Energy, Brussels, Belgium 2000

[19] Wiel S, McMahon J. Energy-efficiency labels and standards: A guidebook for appliances, equipment, and lighting. Collaborative Labeling and Appliance Standards Program (CLASP), Washington, DC, February 2001

[20] Greg R, Michael M, Maithili I, Stephen M, James M. Energy efficiency standards for equipment: Additional opportunities in the residential and commercial sectors 2006: Energy Policy (34) 3257–67

[21] Duffy J. Energy Labeling, Standards and Building Codes: A Global Survey and Assessment for Developing Countries. International Institute for Energy Conservation, Washington 1996

[22] McMahon J, Turiel I. Introduction to special issue devoted to appliance and lighting standards 1997: Energy and Buildings (26) 1–4

[23] Rosenquist G, McNeil M, Iyer M, Meyers S, McMahon J. Energy efficiency standards for residential and commercial equipment: Additional opportunities 2004: 4th

International Conference on Energy Efficiency in Domestic Appliances and Lighting, LBNL
[24] Meyers S, Williams A, Chan P. Energy and Economic Impacts of US Federal Energy and Water Conservation Standards Adopted From 1987 Through 2010, Lawrence Berkeley National Laboratory (2011)
[25] ENERGY STAR Appliances Verification Testing – Pilot Program Summary Report 2012
[26] Lang S. Progress in energy efficiency standards for residential buildings in China 2004: Energy and Building (36) 1191-96
[27] Nathalie T, Isabel M. Development of Energy Efficiency Indicators in Russia, Working Paper 2011
[28] Stephen W, McMahon J. Governments Should Implement Energy-Efficiency Standards and Labels Cautiously 2003: Energy Policy (31) 1403–15
[29] Waide P, Benoft L, Hinnells M. Appliance Energy Standards in Europe 1997: Energy and Buildings (26) 45-67.
[30] International Energy Outlook. World energy and economic outlook 2004: Available at: http://www.eia.doe.gov/oiaf/ieo/world.html
[31] IEA International Energy Agency. Indicators of energy use and efficiency. Understanding the link between energy and human activity 1997: 330-35
[32] Hina Z, Devadas V. Energy management in Lucknow City 2007: Energy Policy (35) 4847–68
[33] McKa, C, Khare A. Awareness Development for an Energy Management Program for Social Housing in Canada 2003: Energy and Buildings (36) 237-50.
[34] Anwar A, Soib T, Salah W. Energy Policy and Energy Demand for Malaysian Development Energy Efficiency 2007: Proceeding of Power Engineering and Optimization Conference Shah Alam, Malaysia
[35] Rusell C. Strategic Industrial Energy Efficiency: Reduce Expense, Build Revenues and Control Risk, The Alliance to Save Energy.USA 2003
[36] Fulkerson W, Levine M. D, Sinton J. E, Gadgil A. Sustainable, efficient electricity service for one billion people 2005: Energy for Sustainable Development 4(2)
[37] William A. Benchmarking to trigger cleaner production in small businesses: dry cleaning case study 2007: Journal of Cleaner Production (15) 798-13
[38] Painuly J, Reddy B. Electricity conservation programs—barriers to their implementations 1996: Energy Sources (18) 257–67.
[39] Reddy A. Barriers to improvements in energy efficiency 1991: Energy Policy 19(10) 953–61.
[40] DeCanio S. J. Barriers within firms to energy efficient investments 1993 Energy Policy 21(9) 906–14.
[41] Jan V. D. A. Malaysian Industrial Energy Efficiency Improvement Project 2008
[42] Gonzalez A. B. R, Diaz J. J. V, Caamano A. J, Wilby M R. Towards a universal energy efficiency index for buildings 2011: Energy and Building (43) 980-87
[43] Weber L. Some Reflections on Barriers to Efficient use of Energy 1997: Energy Policy 25(10) 833–5

[44] Sorrel, S, Malley E, Schleich J, Scott S. The Economics of Energy Efficiency—Barriers to Cost Effective Investment. Cheltenham: Edward Elgar, UK 2004

[45] Nagesha N, Balachandra P. Barriers to energy efficiency in small industry clusters: Multi-criteria-based prioritization using the analytic hierarchy process, 2006: Energy (31) 1969–83

[46] Escriva C, Santamaria-Orts O, Mugarra-Llopis F. Continuous assessment of energy efficiency in commercial building using energy rating factors 2012: Energy and Building (article in press)

[47] EIB (Energy Information Bureau), Cases on Energy Savings in Manufacturing. Available at: http://eib.org.my/index.php?page=article&item=98,107,109,110.

[48] Anwar A, Soib T, Salah W. USM Approaches on Energy Efficiency: Benefits and Solution 2007: Proceeding of International Conference on Engineering and ICT, Malaysia

Section 2

Energy Efficiency – Equipment

Chapter 5

High Efficiency Mix Energy System Design with Low Carbon Footprint for Wide-Open Workshops

Tomas Gil-Lopez, Miguel A. Galvez-Huerta,
Juan Castejon-Navas and Paul O'Donohoe

Additional information is available at the end of the chapter

http://dx.doi.org/10.5772/48321

1. Introduction

Existing buildings are a key focus in the EU's strategy to reduce its member countries emissions. As a consequence, the EU Directive for improving energy performance of buildings (Directive 2002/91/EC, 2003; Directive 2010/31/EU, 2010) was brought in to guide member states. They are advised to upgrade the energy performance of the existing buildings with a floor space of more than 1,000 m^2, when they are reformed. This can be achieved through improving the energy performance of its service systems (heating, cooling, ventilation, hot water, and lighting systems) and of the materials that conforms its architectural envelope.

Of the existing buildings, industrial sector has a significant potential for energy savings. Facilities in many cases obsolete, old comfort standards and an out-of-date yet demanding labour legislation (Ley 31/1995, 1995; R.D. 486/1997, 1997; Instituto Nacional de Seguridad e Higiene en el Trabajo, 1997) provide the framework in which refurbishment must be accomplished. Especially critical is the case of partially open buildings whose entrances, due to the use, remain open during long periods of time: warehouses, open wide while loading and unloading, or cars and trucks repair workshops with continuous vehicular traffic are two examples of industrial buildings of the type. Refurbishment in this case needs to be faced with different criteria than those which are customarily used for the remainder of buildings.

Firstly, the poor indoor environmental conditions achieved, very close to outdoors climate due to the high permeability of the building, represent a significant economic cost to the owners, as the activity of workers, and therefore the length of the working day, is linked to acceptable indoor higrothermal conditions.

INTECH
open science | open minds

© 2012 The Author(s). Licensee InTech. This chapter is distributed under the terms of the Creative Commons Attribution License http://creativecommons.org/licenses/by/3.0), which permits unrestricted use, distribution, and reproduction in any medium, provided the original work is properly cited.

Article 7 and Annex III of R.D. 486/1997 (1997) set out the environmental requirements that are mandatory in work places for health and safety reasons: temperature must be kept between 17°C and 27°C if sedentary work is performed, and within a range of 14°C up to 25°C in the case of light work. Air velocities must be less than 0.25 m/s when working in cool environments, below 0.5 m/s in warm environments for sedentary jobs and lower than 0.75 m/s if the work done is light with a warm environment. In all cases the relative humidity has to be maintained between 30% and 70%.

Even if the extreme values of the range are considered (temperatures in the range of 14ºC and 27°C with air velocities below 0.75 m/s) the possibility of achieving these higrothermal indoor conditions without mechanical equipment in an open building exposed to a severe climate, as it happens in most of European countries, is very low.

Nevertheless, in this type of buildings, HVAC systems that control temperature and humidity inside the premises are not feasible, for the high rate of infiltration would lead to extreme cooling and heating loads, as it will later be demonstrated by some examples.

If, despite what has been exposed, it is found necessary to project a technical system that improves thermal conditions for labourers while increasing the working hours, the first problem to be solved is to establish an effective climate separation of indoor climate that allows to obtain thermal comfort at a reasonable cost. In this sense, the provision of air curtains in fixed openings is an essential strategy prior to air conditioning the premises.

It should be noted that the climate separation, though reducing building demand of energy by nearly 90 %, causes yet another problem: the indoor air quality worsens, as natural ventilation due to air infiltration through the openings tends to be neglected. At this point, when the need for mechanical ventilation becomes clear, the inevitably decision of air conditioning is derived, and so is the necessity to establish the basis for choosing the best technical system that ensure the higrothermal comfort of the occupants and the air quality of the premises as well as minimize both the installed thermal power and the energy consumption associated with conditioning.

Therefore, this chapter seeks to propose a methodology for facing the refurbishment of wide-open industrial buildings. Firstly, by establishing a climatic separation via air curtains and, finally, by choosing a high efficiency air conditioning system. The expected benefits of the proposed system will be:

- An increase in hours of work, taking into account the existing labour legislation;
- An improvement of comfort, compared with a conventional system; and
- A reduction of energy consumption.

2. Strategies for carbon emissions reduction in the refurbishment of wide-open industrial buildings

As it was mentioned in the previous section, in the rehabilitation of wide-open industrial buildings, at least three problems, related one to another, must be faced. Namely:

- Energy consumption;
- Higrothermal comfort; and
- Indoor air quality.

A holistic approach to building design suggests a long list of possible strategies to improve its energy efficiency. However, to guide the development of a more efficient HVAC system, the concept of adaptive comfort criteria was used (Clark & Edholm, 1985; Nicol, 1993).

What this means in practice is that less fossil fuel is used to maintain comfortable temperatures if the building can be kept to a relatively constant level through an interactive control system that adapts the internal environment conditions in response to carbon dioxide levels and air velocities. In order to reduce the carbon emissions of the heating and ventilation system the following steps need to be taken:

2.1. Thermally isolate the building

This includes solar shading as well as an improvement of the insulation materials, as it derives from the following reasons.

With respect to buildings heating and cooling energy demand, Spanish legislation (R.D. 314/2006, 2006a) states:

> *" Buildings shall have an enclosure that adequately limit the energy demand required to achieve thermal comfort, depending on the local climate, the use of the building during summer and winter, as well as on the characteristics of isolation and inertia of the materials, the air permeability and the exposure to solar radiation, and adequately considering thermal bridges, in order to properly limit heat gains or losses and to avoid the higrothermal problems related to them"*

Thermal characterization of the opaque elements of the building enclosure (walls, roofs and floors) is made by the thermal transmittance *U* ($W/m^2 \cdot K$), which is defined:

$$\frac{1}{U} = R_T = \frac{1}{h_i} + \sum_{i=1}^{n} \frac{e_i}{\lambda_i} + \frac{1}{h_e}. \tag{1}$$

with h_i, surface heat transfer coefficient for the inside air layer ($W/m^2 \cdot K$); h_e, surface heat transfer coefficient for the outside air layer ($W/m^2 \cdot K$); *e*, thickness of the layers that forms the enclosure (m); and λ, thermal conductivity of the material of each layer ($W/m \cdot K$).

With respect to the openings (windows, doors and skylights), thermal transmittance is also used. In this case it is obtained from the respective transmittances of glass, U_v, and window frame, U_m, according to the expression.

$$U_H = (1\text{-}FM)\, U_v + FM\, U_m \tag{2}$$

being FM (%) the fraction of the opening taken up by the frame.

Due to the high contribution of the openings to the heat gains, an additional coefficient is used in order to characterize its response to the solar radiation. It is the modified solar factor, F (-), defined as:

$$F = F_s \,.\, [(1 - FM) \,.\, g + FM \,.\, 0{,}04 \,.\, U_m \,.\, \alpha] \tag{3}$$

where F_s (-) is called shadow factor, which is defined as the percentage of the solar radiation incident on a vertical plane that eventually reaches the opening. Its value is affected by remote obstacles, the self-shadowing of the building, facade obstacles like setbacks, overhangs or projections, and the sun control devices, fixed or movable exterior shades included.

The whole expression represents an average solar factor of the opening, taking into account the aforementioned effect of shadowing, and the weighted contribution of glass and frame in the response to solar radiation. The contribution of the glass is expressed by its total solar thermal transmittance, g (-), determined, for a quasi-parallel radiation and for a quasi normal inclination, with the definition formula (BS EN 410:1998, 1998).

$$g = \frac{\alpha\, h_i}{h_e + h_i} + \tau \tag{4}$$

where α is the absortivity and τ the transmissivity, both dimensionless. This factor is obviously much higher than the frame one, which is representative only when having small openings or thick window frames.

Finally, with respect to the thermal inertia of the envelope, in buildings such as those discussed in this chapter, with light walls, its influence can be neglected.

2.2. Produce hot water more efficiently

In addition to the energy demand, which has been analyzed in the previous section, the average performance of the HVAC systems is the determining factor in the final energy consumption of buildings. From all the subsystems that takes part in the air conditioning (heat emission, distribution and production), the latter has the higher incidence in the energy efficiency of the building.

When producing heat by means of combustion, two aspects of boiler design have to be considered: the heat losses and its efficiency. What is desired is a highly efficient boiler system which minimizes the heat losses, especially those associated to combustion gases, so that less fuel is needed to heat the water. On the other hand, the boiler needs to be as effective as possible at transferring heat from the energy source to water.

The method used to evaluate the final energy consumption of the heat production system is:

- Annual energy needed by the boiler to meet the demand = Annual energy demand x (1+Boiler losses)
- Total annual energy consumption for the boiler and its fuel = Annual energy needed by the boiler to meet the demand / (Calorific potential x Seasonal efficiency)

Boiler losses depend on the type of heat generator and on the range of power output. According to the European legislation (BS EN 15603:2008, 2008), boilers must comply with the heat conversion efficiency requirements, always referred to the fuel net calorific value, that are set out in the following Table 1:

Efficiency at rated output					
Type of boiler	Range of power output (kW)				
	50	100	200	300	400
Standard	87.4	88.0	88.6	89.0	89.2
Low temperature	90.0	90.5	91.0	91.2	91.4
Gas condensing boilers	92.7	93.0	93.3	93.5	93.6
Efficiency at partload (30%)					
Type of boiler	Range of power output (kW)				
	50	100	200	300	400
Standard	85.1	86.0	86.9	87.4	87.8
Low temperature	90.0	90.5	91.0	91.2	91.4
Gas condensing boilers	98.7	99.0	99.3	99.5	99.6

Table 1. Normative heat conversion efficiency requirements for different types of boilers

On the other part, seasonal efficiency is difficult to determine, as it is related to the size of the boiler, the burner type and the method of operation all over the heating session. A simple way to approach the problem can be consulted in (R.D. 275/1995, 1995). To obtain more accurate results, the use of energy simulation programs is strongly recommended.

2.3. Reduce the amount of carbon from the energy source used by the system

Primary energy is an energy that has not been subjected to any conversion or transformation process. Thus, to determine the primary energy required to provide the final energy demanded by a technical system, it is taken into account the energy content associated with the extraction, processing, storage, transport, generation, transformation, transmission, distribution and any other operation necessary to supply energy to the area where it is used. Primary energy is the essential energy indicator to determine the net heat balance of a building (when heat production involves different energy vectors), or to compare the energy performance of different technical systems. As such is considered by European official methodology of calculation (Moss, 1997), and has been accordingly incorporated into rating procedures for energy building performance of different member countries, including Spain (R.D. 47/2007 (2007).

It is calculated by multiplying the energy supplied to the system by a factor greater than unity. If, however, the carbon emission is preferred as indicator, a second conversion factor has to be applied.

According to it, the method used to evaluate the carbon emission due to the energy source for the heat production is:

- Annual energy consumption of the primary fuel used in the boiler = Annual energy needed by the boiler x Coefficient for the primary energy used.
- Amount of carbon emitted = Annual energy consumption of primary fuel x Carbon conversion factor.

The following Table 2 shows the conversion coefficients used by the official program for energy rating process in Spain, CALENER (IDAE, 2009). Each different energy source has two energy conversion factors, for carbon emissions and primary energy.

Type of energy	**Final energy (kWh)**	**Primary energy (kWh)**	**Emissions (kg. CO_2)**
Electricity	1	2,603	0.649
Natural Gas	1	1,011	0.204
Coal	1	1	0.347
Liquefied petroleum gas	1	1,081	0.244
Diesel oil	1	1,081	0.287
Fuel oil	1	1,081	0.280
Biofuel	1	1	0
Renewable energy	¿1?	¿1?	0

Table 2. Carbon emission and primary energy conversion factors used by CALENER

Therefore, a significant reduction of carbon emissions can be carried out in two ways. The first is to choose a primary source of energy with the lowest carbon footprint, eg. natural gas. The second is to use a source of renewable energy.

2.4. Separate outdoor and indoor environments by means of a thermal barrier if no physical limit exists between them

Through the openings of building walls and partitions there are inlets and outlets of air, whose direction and value depend, for each point of the envelope, on the pressure differences that exists at each of its sides. In the absence of mechanical systems, the pressure difference is due to the temperature difference between indoor and outdoor air (thermal draft) or to the wind pressure.

The airflow through an opening due to thermal buoyancy has been studied by different authors (Allard & Utsumi, 1992). The air flow rate Q in m³/s which enters through an opening h meters high and with a surface of S m² when there is a temperature difference T_i - T_e between the outside air and the inside air can be obtained from the following expression:

$$S = \frac{Q}{g\,\rho_0\,T_0\left(\frac{1}{T_e}-\frac{1}{T_i}\right)z^{1/n}} \tag{5}$$

where $z = h/2$ (considering that the neutral pressure level passes through the centerline of the side opening), and n is the flow exponent that indicates the degree of turbulence. An n value of 0.5 represents fully turbulent flow and 1.0 represents fully laminar flow.

From the volume flow rate obtained, the sensible and latent losses (or gains) are calculated using the well known expressions:

$$\phi_S = Q \,.\, (T_e - T_i) \,.\, \varrho_a \,.\, c_a \tag{6}$$

$$\phi_{Lv} = Q \,.\, (w_e - w_i) \,.\, \varrho_a \,.\, L \tag{7}$$

Heating (or cooling) loads laws distribute at both sides of the neutral pressure level according to the Figure 1:

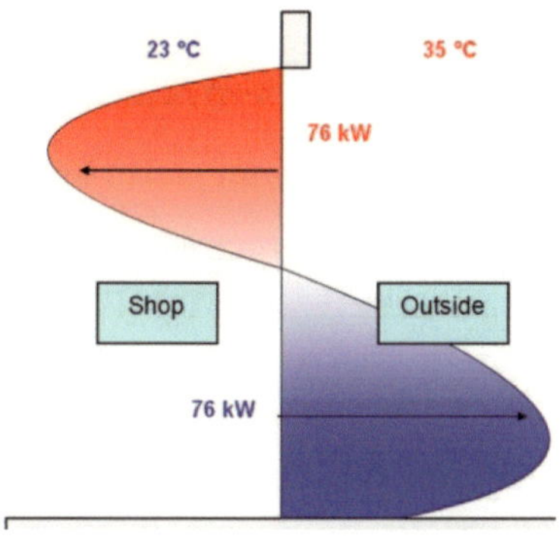

Figure 1. Heating loads due to thermal buoyancy through openings

According to the technical services prevailing code (R.D. 1027/2007, 2007), the design of a partially open building, with 8 side openings of 4x4 m^2 each, which keeps the indoor environment between 19°C and 28°C throughout the year, would require a heating power close to 2.5 MW if the outdoor temperature were 5°C. In summer, sensible cooling power reaches 1.5 MW for an outside temperature of 35°C.

At this point it should be emphasized the importance of the latent loads in warm and humid climates. In these cases, the use of direct expansion cooling machines leads to high electrical consumption for hardly maintaining the desired internal temperature conditions. This fact alone would justify the use of climate separators in these climates, for all type of buildings, provided that a high frequency of doors opening and closing is expected.

It should also be noted that the infiltration air flow rate, and consequently the thermal load, is usually increased due to wind pressures acting on opposite faces of the building. In this case, even more so air curtains reveal as the essential strategy to reduce the thermal load of this type of buildings, for they create a barrier between the two environments, the inside working area of the workshop and the outside air. The air curtains also are highly energy efficient terminal units, and their coils permit a choice of the type of energy they run on, hot water included.

2.5. Reduce the causes of discomfort

The degree of user acceptance of the environmental conditions of the premises is a function of their air quality, which includes higrothermal comfort and adequate pollution levels.

There is enough information about the influence of higrothermal conditions on the predicted percentage of dissatisfied (PPD) in a room (Fanger, 1993a). The expressions that relate the PPD with different causes of higrothermal discomfort are the following:

a. PPD due to vertical temperature gradient

The *PPD* as a function of temperature gradient is given by the equation (ASHRAE, 2009a):

$$PPD = \frac{100}{1 + e^{(5.76 - 0.856\,TG)}} \tag{8}$$

where *TG* is the vertical thermal gradient between head and feet.

b. Synergistic effect on PPD of air velocity, temperature and turbulence

The changes in the *PPD* by the synergistic effects of air velocity, temperature and turbulence, are defined by the expression (BS EN ISO 7730:2005, 2006a, 2006b):

$$PPD = (34 - T_a)\cdot(v - 0.05)^{0.62}\cdot (0.37\cdot v\cdot T_u + 3.14) \tag{9}$$

where: T_a is the air temperature (ºC), v is the air velocity (m/s) and T_u is the turbulence intensity (%), considering that the air turbulence intensity in a point is defined by the equation:

$$T_u = \frac{\sigma_v}{\bar{v}} \tag{10}$$

where σ_v is the standard deviation and $\bar{v}$ is the mean air velocity of a random sample of velocities.

c. PPD due to inadequate floor temperature

The percentage of dissatisfied by warm or cold floor can be deduced through the following expression (BS EN ISO 7730:2005, 2006c), that relates the *PPD* to the floor temperature t_f.

$$PPD = 100 - 94\, e^{\left(-1.387 + 0.118\cdot t_f - 0.0025\cdot t_f^{\,2}\right)} \tag{11}$$

d. PPD due to asymmetric thermal radiation

Asymmetric radiation from warm or cold surfaces, created by high lighting levels, due to large glazed surfaces or direct sunlight can reduce thermal acceptability of the spaces (BS EN ISO 7730:2005, 2006d). Radiant temperature asymmetry is analysed to 1.1 m off the ground for standing and 0.6 m for seating conditions, and must be kept under the following limits (Table 3).

Element	Cold wall	Cold roof	Warm wall	Warm roof
Maximum temperature differences	11	16	35	6

Table 3. Values of asymmetric radiation for a PPD ≤10%

With respect to air quality, the effect on the PDP of the contaminants perceived by the sense of smell has also been extensively discussed (Fanger, 1993b), but it does not when it comes to pollutants that are not detected by humans, for it is unknown its influence on the PPD and how affects the workers productivity.

e. PPD due to human detected pollutants

Fanger established how the percentage of people dissatisfied is influenced by the presence of pollutants, provided that they can be perceivable. For low concentrations (usual when working indoors) the relationship is linear. For low concentrations (usual working indoors) the relationship is linear.

Of pollutants detected by humans, one of the most common is carbon dioxide (CO_2). The relationship between the *PPD* and the concentration of CO_2 in the air can be determined through the expression (Fanger, 1993c):

$$PPD = 395\, e^{\left[-15.15\cdot(C-350)^{-0.25}\right]} \tag{12}$$

where C is the carbon dioxide CO_2 concentration in ppm.

From the previous expression it can be deduced:

- For *PPD* = 100%, CO_2 concentration is 10.000 ppm (350 ppm supplied by the outside air).
- For *PPD* = 30%, *PPD*/CO_2 relationship is almost linear.
- For *PPD* = 15%, concentration is 850 ppm (500 ppm produced in the inside the premises, and 350 ppm supplied by the outside air).

f. PPD due to any other pollutant

Under the hypothesis that contaminants not perceived could be studied under the aforementioned theoretical frame, so that their effect on humans could be brought together and their contribution to the PPD could be analyzed, Gomez (2009) has developed a method for quickly and easily determine the PPD for any room with any type of pollution (detected or not by humans). It is based upon the concentrations of pollutants prescribed by international health agencies, for which there is adopted the type of curve derived by Fanger for CO_2 as an average value.

ACGIH (1999) and the existing standards on health and hygiene at work suggest the maximum allowable concentration of pollutants for different periods of time TLV (Threshold Limit Values). According to them, being exposed to a specific concentration of pollutants for 8 hours a day leads to a *PPD* of 100%. This concentration is known as the TLWA value.

By analogy with the case of CO_2, whatever the type of pollutant studied, it can be constructed a function that relates its concentration with the caused *PPD*, since three points of it are known. Firstly, the curve passes through the origin. Secondly, it is also known the concentration which produces a 100% *PPD*. And finally, according to the experience for CO_2, a *PPD* of 15% is reached with a concentration 11.76 times lower than the one which causes a *PPD* of 100%.

The error made with this assumption depends on the shape of the curve that relates *PPD* to concentration of each pollutant. In the case of the substances for which ACGIH provides the permissible limit for TLV-STEL (concentration at which users may be exposed continuously for 15 minutes without chronic or irreversible damage) and TLV- TLWA, they are related within a range from 1.5 up to 6.

NIOSH REL provides a value of 6 for the relation between permissible concentrations of CO_2 for 8 hours and 15 minutes (ASHRAE, 2009b), while German list of MAK (Maximale Arbeitsplatzkonzentration) values (Deutsche Forschungsgemeinschaft, 2007) gives a relation value of 2.

It then follows that it can be adopted for the rest of pollutants the type of curve derived by Fanger for CO_2 without making significant errors. It is expected that further investigations in this field could provide new data. For permanent flow, as the *PPD* of the interior of the premises takes low values, it is only used the straight part of the curve, in which for a *PPD* of 15%, concentration is 1/11.76 of that corresponding to 100%.

Through "in situ" measurements, or prediction by means of CFD models, of the concentration of different contaminants for specific points of the room, it can be deduced the PPD achieved under any circumstance. This method can be used whatever the air diffusion system considered, though in the next section it will be discussed the advantages and drawbacks of the usual systems.

When it comes to know the indoor air quality of a room it is essential to determine the influence of ventilation air flow rate in the PPD. Considering that its effect on higrothermal conditions, and eventually on the PPD achieved, is opposite to pollutants concentration (Figure 2), the authors Castejon et al. (2011) and Galvez-Huerta et al. (2012) have recently studied the problem of settling the ventilation rate that minimizes the predicted percentage of dissatisfied in a room. The final aim is that this result in energy savings, for it minimizes the ventilation air flow rate.

In wide-open buildings, draughts and unwanted air currents can be a constant feature when the door opens automatically, affecting higrothermal comfort. At the same time, discomfort if there is a significant difference in temperature of their feet and their head can be fairly controlled through an air diffusion system.

With respect to the existing sources of contaminants, the rate of air change will depend on the carbon dioxide and monoxide emissions.

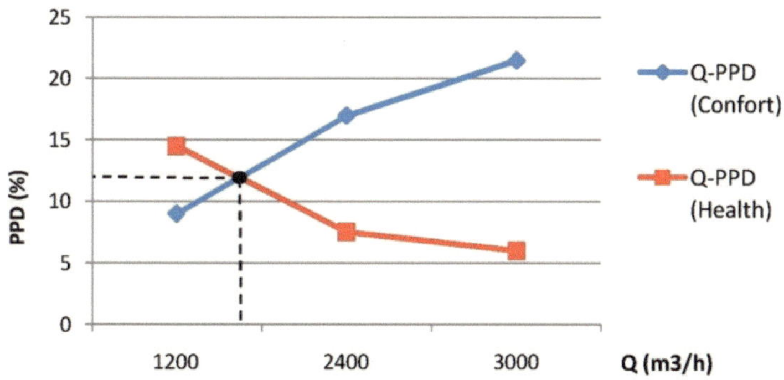

Figure 2. PPD due to combined action of higrothermal and health factors

2.6. Election of an adequate diffusion air system

The effectiveness of air renewal provided by the ventilation system is the essential parameter for the design of an air diffusion system. Effectiveness, r_v (-), of a ventilation system is defined with the expression:

$$r_v = \frac{C_{extr} - C_{imp}}{C_{zona} - C_{imp}} \quad (13)$$

with C_{imp}, contaminant concentration in supply air; C_{ext}, contaminant concentration in exhaust air; y C_{zona}, contaminant concentration in the occupied zone.

As it is shown in Table 4, among the systems used in air diffusion, , used in its early stages only in industrial premises (Baturin, 1972), shows as the air diffusion system that eliminate sources of contamination most efficiently, especially when air is driven at a lower temperature than the indoor air. This always occurs during cooling period, and also with ventilation air in winter.

Mixed mode		Displacement mode	
ΔT	r_v	ΔT	r_v
< 0	0.9 … 1	< 0	1.2 … 1.4
0 … 2	0.9	0 … 2	0.7 … 0.9
2 … 5	0.8	> 2	0.2 … 0.7
> 5	0.4 … 0.7		

Table 4. Ventilation effectiveness dependence on air diffusion mode and difference of temperatures

With this systems air is supplied at very low velocity near the floor, within the occupied zone (Nielsen, 1993). This is, then, the coolest zone of the room, and forces a profile of temperature and pollutant concentration which vertically increases up to the roof (Figure 3), where the return takes place.

When the emission of pollutants is uniformly distributed, as in places where human presence is dominant, the system design is based on the control of the temperature gradient (Mundt, 1995) to maintain comfort conditions in the occupied area: Namely, air temperature at 1.80 m high, maximum air velocity in the occupied zone, radiant asymmetry from the ceiling, walls and floor, and maximum temperature difference between the head and feet.

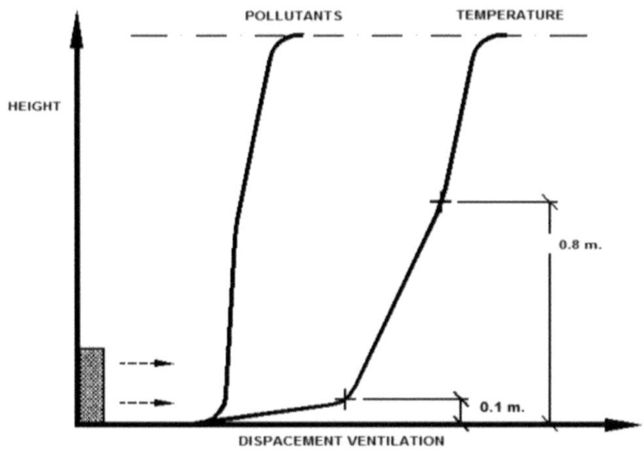

Figure 3. Temperature and concentration of contaminants gradient

When, on the contrary, the emission of pollutants is concentrated, as in areas that have car engines running, in this case the cycle of air change will be greater than in those areas where less strenuous work is carried out. This allows to reduce the ventilation air to the minimum required, thus saving energy.

2.7. Use of Dedicated Outdoor Systems (DOAS)

As some authors (Mumma, 2001) have expressed, a new paradigm in the design of HVAC systems is in its early stages. Amidst its requirements, the more remarkable are: separating outdoor air from the air conditioning system to ensure adequate ventilation and the use of energy recovery strategies.

Despite the fact that the new systems can also handle the space latent load and part of its sensible load, the transition from mixed ventilation systems to DOAS always compels to use a second air conditioning system (be it passive or active beams, water or direct expansion fan coils, ceiling cooling panels, radiant floors and thermoactive surfaces). This drawback of having duplicated systems is usually compensated with the use of high efficiency terminal units, which run on low temperature water, closer to the indoor air conditions.

However, it is usually forgotten that the present way towards more efficient buildings, with the commissioning Nearly Zero Energy Building by 2020 (Directive 2010/31/EU, 2010), has produced substantial improvements in the envelope and lightning systems, which have consequently reduced their contribution to cooling loads. But these improvements, while reducing consumption, also involve a decrease of the thermal load met by the coils.

Consider that, with an average occupancy of the building of 5 m^2/person, the Spanish mandatory ventilation air flow rate, supplied at 10ºC less than indoor air, can met a cooling load of 30 W/m^2. This may seem certainly low, but is close to the expected for the coming years.

In the case of partially open workshops, the primary air handling unit can guarantee the quality of air in the building with a reasonable low energy consumption (for they supply air with indoor conditions) only if climate separation in openings has been previously solved. When extreme winter conditions have to be faced, primary air conditioning can be supplemented with water terminal units in the weakest points of the perimeter, the doors. In this sense, the placement of an air curtain in each opening can accomplish both tasks, reducing air movement within the premises to the necessary to remove contaminants.

2.8. Maximize the cost-effectiveness of heat recovery equipment

The Spanish legislation (R.D. 1027/2007, 2007) makes it mandatory for air conditioning systems of a certain size the following specific strategies for energy recovery, framed in a more general requirement of energy efficiency:

- Free cooling by outside air, which applies to constant volume or VAV air conditioning systems with rated cooling output greater than 70 kW.
- Recovery of heat from ventilation air with adiabatic cooling of the extract air. This requirement is applicable to buildings in which the exhaust air flow is greater than 0.5 m^3/s. It also establishes minimum efficiency of recovery of sensible heat depending on the air flow rate and the number of hours the system is operating throughout the year (Table 5).

Annual operating hours	**Outdoor air flow rate (m3/s)**				
	> 0.5...1.5	> 1.5...3.0	> 3.0...6.0	> 6.0...12	>12
	%	%	%	%	%
≤2,000	40	44	47	55	60
> 2,000...4,000	44	47	52	58	64
> 4,000...6,000	47	50	55	64	70
> 6,000	50	55	60	70	75

Table 5. Minimum efficiency of sensible heat recovery system, according to I.T. 1.2.4.5.2

In the previous section Dedicated Outdoor Air Systems (DOAS) have been identified as the most appropriate systems to ensure adequate air quality in rooms with high concentrations of pollutants.

For this type of systems, energy recovery strategies should focus on the recovery of sensible heat, where the following relationship is satisfied:

$$T_{salida} = T_{ext} - \eta\,(T_{ext} - T_{local}) \tag{14}$$

with η (-), efficiency of sensible heat recovery, according to Table 4.

The temperature of the exhaust air, T_{exp}, can be reduced by means of an adiabatic cooling process, in which the air supplied to the coils is cooled down to the temperature T_{adiab}, that can be calculated with the expression:

$$T_{adiab} = T_{local} - \varepsilon\,(T_{local} - T_h) \tag{15}$$

with ε (-), efficiency the evaporative cooling. As it is a function of the pad geometry and the air flow rate, it is difficult to establish a reliable average value.

The capacity for heat recovery in the primary air handling unit significantly increases when using a displacement air diffusion system. Because of extracting air near the roof, where maximum temperatures are reached, the system has a great potential for heat recovery in winter, but also in summer, when adiabatic cooling processes can be used, the more effective the higher the extract air temperature.

3. An outlook on high efficient air conditioning systems for wide-open workshops

3.1. Climate separation via air curtains

Air curtains have the function of neutralizing outside air infiltration through the doors, reducing up to 90% heating and cooling thermal demand. In winter period, their use as hot water terminal unit makes also possible to meet the thermal load due to the infiltration rate not eliminated by the curtain. Reduction of latent loads demand is crucial in warm humid climates. Furthermore, they allow to control indoor environment regardless of external conditions.

To effectively carry out its function of climate separation, the curtains should maintain a proper discharge length whatever the external conditions of wind are. If the air curtain jet is too weak and the throw distance is short it does not prevent infiltrations. By contrast, excessive throw distance due to a strong jet can reduce efficiency by almost 50%. In this case, high velocity and turbulent flow make the air curtain partially mix with outside air.

The parameter that best characterizes the operation of a curtain is the momentum of jet, I_0, that indicates the strength of the curtain. It is defined by the expression:

$$I_0 = \rho_0 \cdot d_0 \cdot U_0^2 \tag{16}$$

being d_0 the outlet width and U_0 the outlet velocity.

Modern air curtains vary the air flow driven maintaining a constant flow rate by adapting the geometry of the discharge outlet (Figure 4).

By choosing a relatively large outlet width, it can be achieved an optimum momentum, as needed to reach the floor, but with low velocity that keeps the flow in laminar regime.

Air curtain strength and heating can be controlled independently according to the needs (Figure 5).

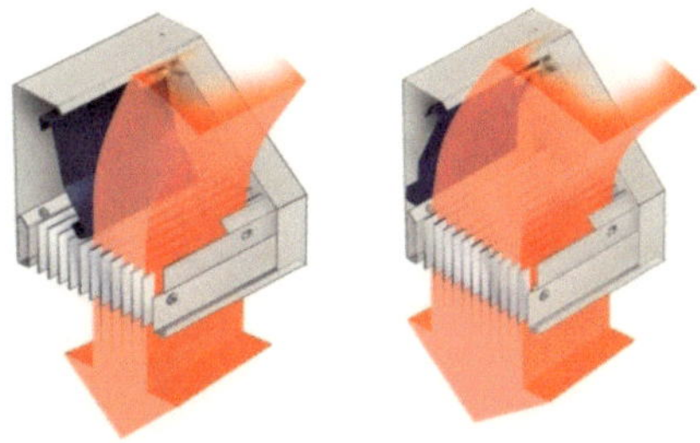

Figure 4. Variable outlet width positions in an air curtain

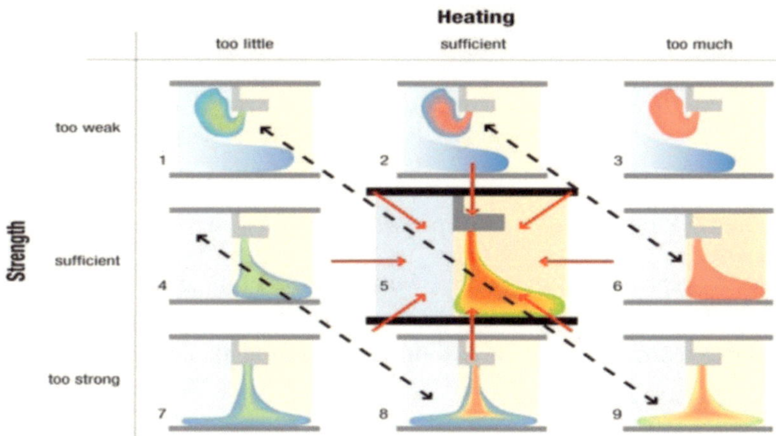

Figure 5. Air curtain strength and heating control possibilities

As an additional advantage, when used as heating terminal units they run on low temperature water, making them particularly suitable for being used with condensing boilers and solar thermal production. These options are discussed in the next two sections.

3.2. Heat production

Air curtains and air handling unit heating coils run on low temperature hot water. Under such circumstances, thermal production can be provided by a solar thermal system. For the auxiliary energy supply a condensing boiler modular gas is projected. As with any solar installation, the energy collected is transferred to storage tanks, also connected to the boiler that takes charge of heating water when solar coverage falls. The control of the group of curtains and the primary air handling unit is done by varying the water flow rate with a three-way valve (Figure 6).

a. Conventional energy contribution

Condensation gas boilers use the heat content of vapor from the combustion, which is transferred to the heating system. As the heat conversion efficiency of the boiler is referred to the fuel net calorific value, performance values are reached greater than unity. As condensation inside the boiler begins when flue gases drop to about 54ºC, the boilers are particularly suitable when the facility is operating at part load or when using terminal units operating at low temperature.

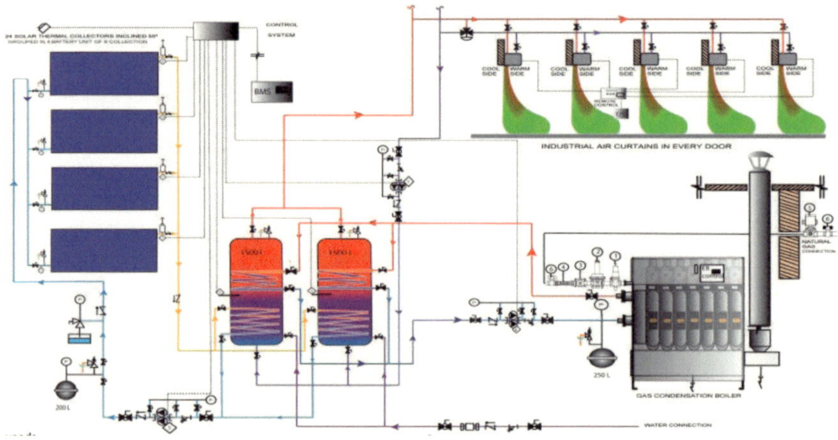

Figure 6. Proposed heating system for wide-open workshops (diagram)

The effect of this high efficiency boilers, with seasonal efficiency of 0.97, on the overall efficiency of the heating system as opposed to normal gas or electric boilers (Table 6) has been assessed in a recent work by the authors (Gil-Lopez et al., 2011).

b. Renewable energy contribution

The savings achieved with the use of solar energy in thermal plants is sufficiently well known. However, when considering the impact on the carbon emissions avoided by the provision of supplementary carbon free energy by solar heating, an interesting result is obtained: the largest amount of carbon emissions avoided occurs when the solar installation provides supplementary energy to less efficient energy sources (Table 7).

This problem, similar to that experienced when using economic and investment indicators that favour consumption and not savings has been fully studied by the authors (Gil-Lopez et al., 2011). The question that arises is how can be assessed the impact of the solar installation in a way that reflects the value of the savings being made and not the energy being consumed. This can be easily seen by the impact the solar installation has on the energy certification (Table 8a and b).

As it was expected, the electric emersion heater boiler has the lowest certification value of G, the diesel and condenser option a C level, and the modular gas condensation boiler a certification rating of B. When the same calculations are conducted but with the solar installation providing supplementary power during part of the year, the certification ratings for the air curtains powered by hot water both receive a value rating of A, whereas that for the electric emersion heater boiler, although it has an improved indicator value, retains its G rating certification. Therefore, to obtain an A rating energy certification the air curtains need to be powered by hot water supplied through a combination of either diesel with condenser unit or a modular gas condensation boiler, with a solar installation providing supplementary source of energy heating the water demanded by the air curtains.

	Boiler type			Reference	
	Diesel boiler and condenser	Electric boiler	Modular gas condensation	Normal gas boiler	Direct electricity
Name	TRISTAR	General	MODULEX	General	General
Boiler option number	1	2	3		
Fuel source	Diesel -C	Electrical emersion heater	Natural gas	Natural gas	Mains electricity supply
Annual energy demand (kWh)			35,459		
Conversion factor for the carbon emitted from energy consumed (kgCO_2/kWh)	0.280	0.649	0.204	*0.204*	*0.649*
Annual energy consumption by the boiler (kWh)	39,005	42,551	39,005	*223,394*	*56,735*
Annual energy consumption from the fuel (kWh)	42,165	110,761	39,434	*62,737*	*147,681*
Total energy consumed by the system	4 m^3	113,021 kWh	3,902 m^3	*7,526 m^3*	*150,695 kWh*
Total carbon emitted (kg)	11,806	71,884	**8,045**	*12,798*	*95,845*

Table 6. Carbon emission levels for an ITV workshop in Madrid when energy demand is met only by the boiler

Impact on Energy and Carbon emissions for energy supplied by the solar installation, rather than the boiler			
Boiler option number	1	2	3
Annual energy savings provided by the solar installation (kWh)		21,054	
Annual energy demand not provided by boiler (kWh)	23,160	25,265	23,160
Annual amount of fuel not consumed (kWh)	25,036	65,766	23,415
Energy that was not consumed	2.5 m^3	67,108 kWh	2,317 m^3
Amount of carbon emissions avoided (kg)	7,010	**42,682**	4,777

Table 7. Impact of solar collectors for an ITV workshop in Madrid

Certification value without the solar installation		**Main energy supply options**		
		Diesel boiler and condenser	Electric boiler	Modular gas condensation
Name		TRISTAR	General	MODULEX
Boiler option number		1	2	3
Type of fuel		Diesel-C	Electricity	Natural gas
Quantity of carbon emitted without the solar installation (kg)		11,806	71,884	8,045
CO_2 emissions per m^2 of surface area of the building (kgCO_2/m^2)	1,620 m^2	7.29	44.37	4.97
Emission reference boiler: for natural gas	7.90 kgCO_2/m^2			
Emission reference boiler: for electrical boiler	59.16 kgCO_2/m^2			
Indicator value (kgCO_2/m^2) Natural Gas		0.923	5.600	0.629
Energy certification value		**C**	**G**	**B**

(a)

Certification value with the solar installation		**Main energy supply options**		
		Diesel boiler and condenser	Electric boiler	Modular gas condensation
Name		TRISTAR	General	MODULEX
Boiler option number		1	2	3
Type of fuel		Diesel-C	Electricity	Natural gas
Quantity of carbon emitted with the solar installation (kg)		4,796	29,202	3,268
CO_2 emissions per m^2 of surface area of the building (kgCO_2/m^2)	1,620 m^2	2.96	18.03	2.02
Indicator value (kgCO_2/m^2)		0.37	2.28	0.26
Energy certification value		A	G	A

(b)

Table 8. a. Rating without solar energy for an ITV workshop in Madrid, b. Rating with solar energy for an ITV workshop in Madrid

3.3. Displacement ventilation system

It helps significantly reduce the level of pollution in the occupied zone and the required air flow compared with ventilation with mixing with turbulent flow. It also allows the integration of direct and indirect adiabatic cooling systems. Furthermore, it optimizes the air conditioning system efficiency, as air can be driven to higher temperatures achieving the same degree of thermal comfort for the workers through a well calculated design for terminal air velocities and temperature.

According to Skistad (1994), the design procedure is the following:

a. The following variables are known:
- Specific cooling load, Φ_r, (W/m²).
- Supply air temperature, T_s. Due to the fact that air is thrown directly to the occupied zone, air is only cooled to a temperature 3 ºC lower than indoor air conditions.
- Difference of temperatures between extract and supply air, ΔT. It is related to the supply air flow rate, Q_s by means of the thermal load. Whether this parameter is restricted to 3ºC or is increased to meet greater thermal loads, the resulting flow can lead to excessive diffuser sizes, problems of noise and inappropriate air velocities.

b. Resolution of air temperature value at 0.1 m height.

$$T_{0.1} = T_s + \theta_{0.1} (T_R - T_S) \qquad (17)$$

where $\theta_{0.1}$, dimensionless temperature, is obtained with Mundt law:

$$\theta_{0.1} = \frac{1}{Q_s c\rho \left(\frac{1}{\alpha_{c,f}} + \frac{1}{\alpha_{r,c}} \right) + 1} \qquad (18)$$

in which $\alpha_{c,f}$ is the convective heat exchange between air and floor, that usually adopts a value of 4.5 W/m². With respect to radiant heat exchange between air and ceiling, $\alpha_{r,c}$, it is obtained, from the initially unknown ceiling temperature T_c, by means of an iterative process. Although usually is considered equal to $\alpha_{c,f}$, its value is far from being constant.

c. Calculation of floor temperature, T_f

$$T_f = T_{0.1} - \alpha_{c,f} / h_i \qquad (19)$$

with h_i, surface heat transfer coefficient between floor and air. For a horizontal heat flow with air at low velocity, it can be considered $1/h_i = 0.13$ m²K/W

d. Assuming that the temperature gradient in the room is constant, air temperature at 1.8 m height, $T_{1.8}$, is obtained:

$$T_{1.8} = T_{0,1} + 1.7 \cdot \frac{T_R - T_{0.1}}{H_R - 0.1} \qquad (20)$$

as well as the corresponding temperature gradient:

$$grad\, T = \frac{T_{1.8} - T_{0.1}}{1.7} \qquad (21)$$

With the obtained values, accomplishment of normative comfort conditions is tested, paying special attention to the temperature difference between head and feet (ASHRAE, 1992).

3.4. Primary air conditioning unit

Ventilation is provided by a primary air handling unit with indirect adiabatic cooling heat exchanger section (Figure 7). The unit has two water coils. The heat coil is supplied by the heat production system described in section 3.2. The cooling load is met by water from a compression chiller. An alternative system with a solar absorption chiller for cold production is also suggested, but not discussed in this chapter.

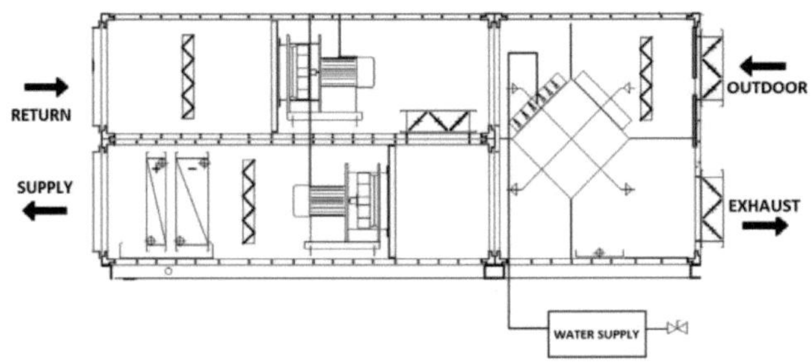

Figure 7. EQUAM adiabatic air handling unit

Air transformations in the air handling unit components are shown in the psycrometric chart (Figure 8), where O stands for outdoor air, S for air supplied to the coil, R for room conditions and EX for exhaust air.

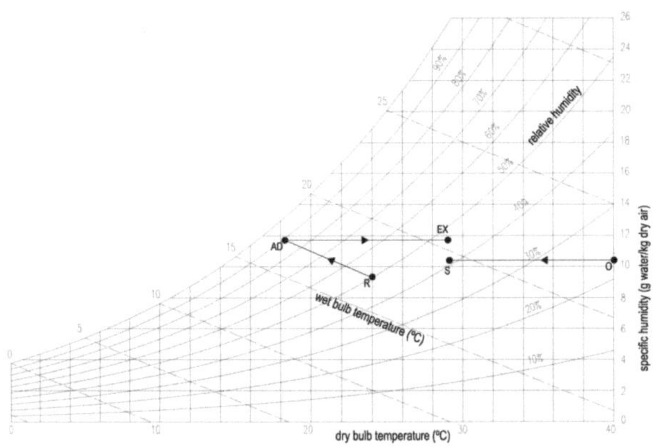

Figure 8. Psycrometric processes of indirect adiabatic cooling

4. Analysis of the proposed options for an ITV workshop

This section compares the energy performance of various options for air conditioning a wide-open workshop. Starting from an initial situation (option 0) of a non conditioned building, its energy consumption once refurbished is studied and compared for each of the three following options: an air curtain system functioning as a climate separator (option 1); the previous system with a conventional air handling unit (option 2); and finally, a comprehensive air conditioning system of high efficiency that includes all aspects covered by section 2 and 3 of this chapter.

The design of this system and the selection of components for the case study are due to Eng. Paul Gerard O'Donohoe, who has used technology developed by TAYRA S.L.

To analyze the performance of the technical systems for each of the proposed situations, a representative industrial building has been selected. Such is the case of a wide-open workshop, where technical assessments of roadworthiness are carried out on cars and trucks (ITV workshop). This type of buildings is of rectangular shape, with a surface area of 1620 m^2 and a height of 7.5 m. It has no windows, but five entrances and exits, three of them at 5 m x 3 m and two at 5 m x 4.5 m. The exit doors are controlled to limit the flow of vehicles leaving the premises. These buildings, that outnumber 1,000 all around Spain, are usually located in remote areas just off main motorways. For the case study, an ITV workshop located in Carmona (Sevilla) was chosen. See attached plan in Figure 9.

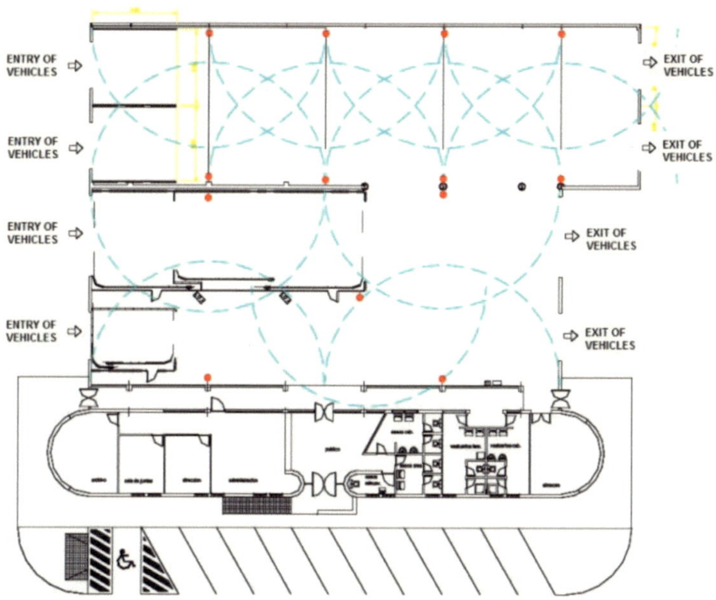

Figure 9. ITV workshop in Carmona (Seville). Plan

Using a computer simulation of the building with the e-Quest program (Figure 10), a prediction of its performance for the aforementioned options has been obtained. E-Quest uses DOE2 engine to perform an hourly simulation of the building for a whole one-year time period. For each hour, heating and cooling loads are calculated. The performance of pumps, fans, boilers, chillers and every energy consuming equipment within the building is also simulated. Finally, the energy use of every end use, including lightning, is tabulated. Authors such as Crawley et al. (2008) provide a sound comparative study of the potential offered by the most common simulation tools, DOE2 included.

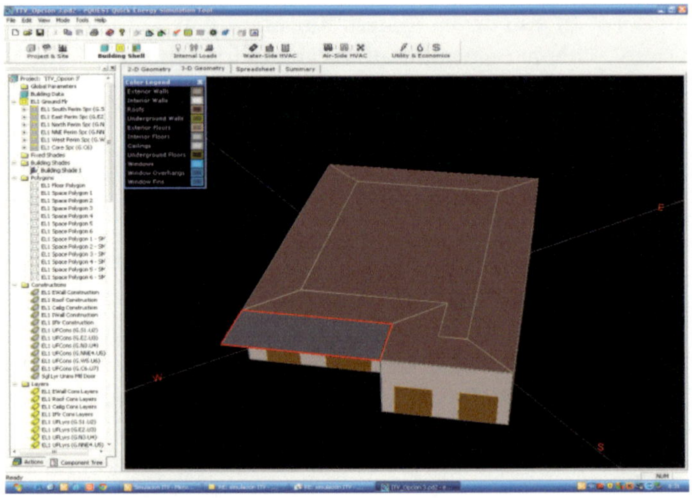

Figure 10. ITV workshop in Carmona (Seville). E-Quest simulation

In the simulations, synthetic meteorological data have been used. 8.762 records, including hourly data, have been generated by the program CLIMED 1.3 (IDAE, 2009) from the normal data of AEMET for the meteorological station of Seville.

The aforementioned cases are described next:

4.1. Case 0: Original building

ITV workshops, for the reasons stated in the introduction, do not have any higrothermal conditioning system. Not even a mechanical ventilation system, as room air is changed by means of natural ventilation through permanently open doors. The energetic simulation in this conditions, only seeks to analyze the evolution of indoor temperatures throughout the year, in order to identify the times when it exceeds the limits allowed by labor legislation (16ºC in winter and 28ºC in summer).

The simulation of the building in its original condition reveals that the 56% of annual hours, indoor temperature remains outside the limits marked by labour legislation (Table 9).

TEMPERATURE (° C)	HOURS																								
	AM												PM												
	1	2	3	4	5	6	7	8	9	10	11	12	1	2	3	4	5	6	7	8	9	10	11	12	TOTAL
ABOVE 29,4	0	0	0	0	0	0	0	0	0	0	0	0	5	14	17	23	34	38	31	19	12	0	0	0	193
26,7 - 29,4	0	0	0	0	0	0	0	0	0	0	2	14	32	44	44	50	41	37	43	44	29	0	0	0	380
23,9 - 26,7	0	0	0	0	0	0	2	9	12	18	26	38	53	50	30	25	26	28	24	28	36	0	0	0	405
21,1 - 23,9	0	0	0	0	0	0	24	32	36	44	60	53	37	31	37	35	38	35	38	28	29	0	0	0	557
18,3 - 21,1	0	0	0	0	0	0	43	66	59	53	40	36	42	43	40	41	40	41	42	50	40	6	0	0	682
15,6 - 18,3	0	0	0	0	0	0	42	40	42	41	46	45	44	58	54	41	38	40	38	42	50	30	0	0	691
BELOW 15,6	0	0	0	0	0	0	38	113	131	124	106	94	67	40	23	10	8	6	9	14	29	40	0	0	852

Table 9. Annual distribution of hourly temperatures

4.2. Case 1: Air curtains as climate separators and mechanical extraction ventilation system

The climate separation is achieved by means of the following equipment:

- Two air curtains Indac S150 (Biddle) placed in series for each of the six existing doors for small vehicles.
- Two air curtains Indac S200 (Biddle) placed in series for the two doors for large vehicles.

Air curtains characteristics have been selected (Figures 11a and 11b) by using a commercial simulation program (Biddle Innovative Klimatechnik, 2012).

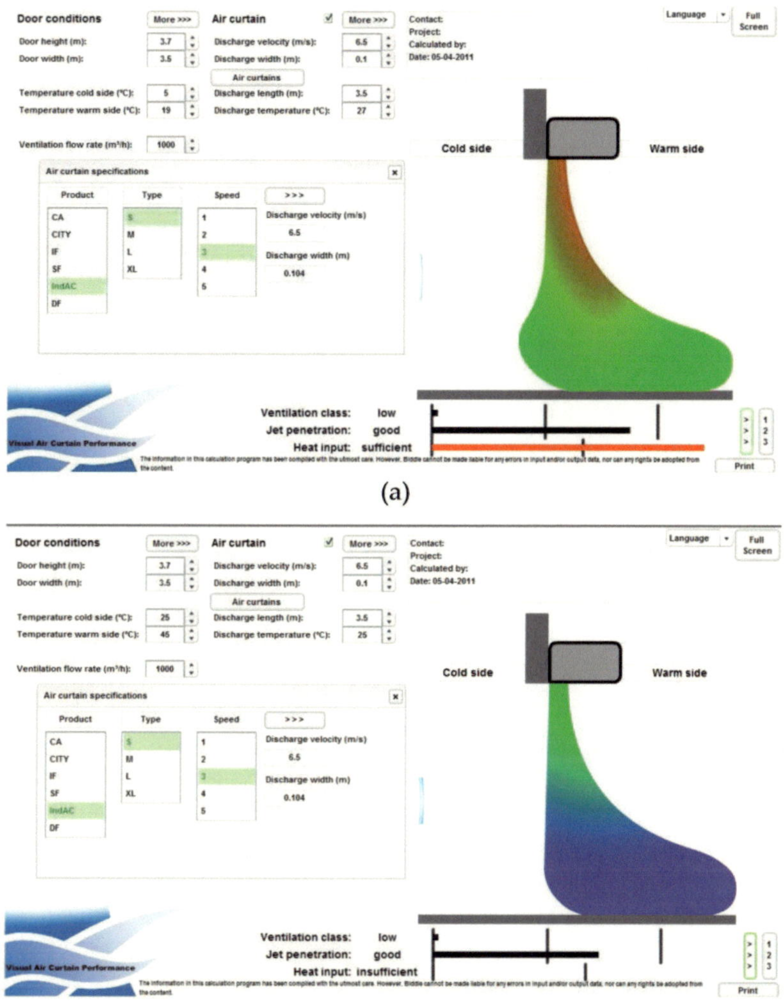

Figure 11. a. Winter conditions air curtain selection, b. Summer conditions air curtain selection

They recirculate air at the room temperature throughout the year. The supplied air flow rate is fixed (5,000 m³/h), and the equipment adapts its velocity of discharge speed by varying the geometry of the outlet. Thus, an identical download length is achieved whatever the external conditions of wind are. The estimated electric power is 0.44 kW per curtain, giving a total power of 7.04 kW.

- A mechanical extraction system with an air flow rate of 70,000 m³/h, which is equivalent to 7 ACH, according to the most restrictive rate of ventilation indicated by current Spanish legislation (R.D. 314/2006, 2006b, 2006c; Ayuntamiento de Madrid, 1985).

The simulation of the case 1 yields the monthly evolution of electric consumptions that are shown in Figure 12:

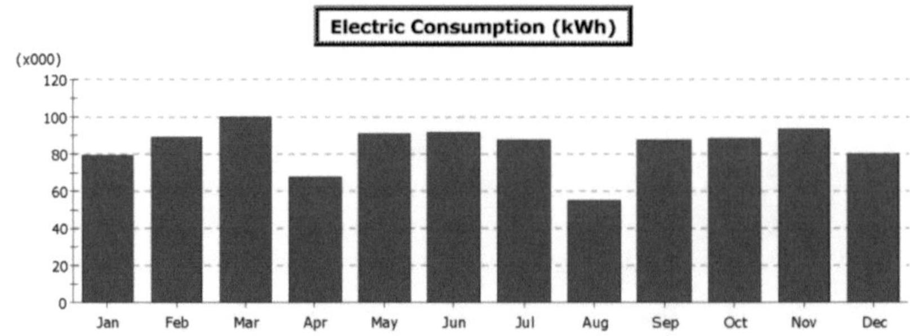

	Jan	Feb	Mar	Apr	May	Jun	Jul	Aug	Sep	Oct	Nov	Dec	Total
Run 1.	79,0	89,3	100,2	67,8	90,5	91,8	87,8	55,4	87,3	88,6	93,0	79,8	1.010,6
Run 2.													
Run 3.													
Run 4.													
Run 5.													

Figure 12. Case 1: Monthly electric consumption

4.3. Case 2: Air curtains and a conventional HVAC system

The climate separators described in the previous option also function as terminal heating units that supply air at 27°C in winter conditions. They are completed with a conventional air conditioning system comprising:

- A primary air handling unit that supplies a volume air flow rate of 7 ACH which. Considering that the whole volume of air in the building is renewed, the air flow rate comes to 70,000 m³/h. Winter indoor temperature is 18ºC, while, during the summer, temperature is kept to 26ºC. They are obtained with supply air from the unit at 30ºC and 14ºC, respectively.
- A gas boiler with a power output of 700 kW (estimated seasonal efficiency: 86%), meet the heating loads demanded by the air curtains (47,7 kW each) and the primary air handling unit heating coil (183 kW).
- A VRV air condensed water chiller with a power output of 218.1 kW, with an estimated seasonal EER of 3.5.

The energy simulation of the building and its technical system leads to results of monthly electric and fuel consumption that will be compared in the next section with the results for the more efficient option 3.

The simulation of the case 2 yields the monthly evolution of electrical and fuel consumption that is shown in Figures 13a and b:

4.4. Case 3: Air curtains and a high efficiency HVAC system

The air curtains also function as terminal heating units that supply air at 27°C in winter conditions. Air conditioning is achieved in this case by a high efficiency system that includes all the strategies that have been described in section 2 and 3 of the chapter:

- A modular condensation gas boiler Tayra/Unical Eco-pacQ-Mx-550-CS2-EM with a power output of 228 kW (heat conversion efficiency: 108 %) meets the thermal loads demanded by air curtains (25 kW each) and the air handling unit coil (28 kW).
- A solar thermal installation is design to provide 70% of the energy requirements for heating. It has been chosen a 24 m^2 surface of collectors ASTERSA AT020 (optical efficiency: 0.748 and heat loss factors: 3.718 and 0.014 W/m^{2}K) oriented South and inclined at 35º, that transfer the collected thermal energy to a 3,000 liters storage tank.
- A VRV air condensed water chiller with a power output of 57.1 kW, and estimated seasonal EER is 3.5.
- A EQUAM primary air handling unit (100% outside air) with the following characteristics: Supply and return fans air flow rate is 30,000 m^3/h (7 ACH for a 3 m height, that corresponds to the occupied zone); A 28 kW heating coil, running on hot water (50/35ºC), supplies air at 19ºC, maintaining the outdoor specific humidity conditions; A 46.7 kW cooling coil, running on chilled water (7/12ºC) supplies air at 24ºC and 78% relative humidity; A cross flow heat recovery unit (70% average efficiency), with a heat recovery capacity of 151 kW for nominal flow in winter conditions (0.6 ºC y 90% relative humidity), provides a supply air at 16.2 ºC from the extract conditions (19ºC and 26% RH); in summer conditions heat recovery capacity is 161.6 kW, for it includes an adiabatic cooling by water sprays of extract air at 24ºC and 78% relative humidity. By means of this indirect adiabatic cooling, outside air at 40ºC y 24% de HR, is brought to 28.7ºC and 59% relative humidity before entering the water coil.
- A displacement air diffusion system, with 13 diffusers VA-ZDA DN 355, of Kranz Componenten, suspended at 3 m height in the position indicated in Figure 9. The throw distance is 9 m, when nominal flow is supplied, with a sound power level of 63 dBA. They incorporate a device for regulating the discharge, which can change its direction upwards or downwards depending on the regime conditions (cooling or heating). The return is by ceiling, but some grids at floor level help to remove heavier contaminants.

With this system, air is supplied at 26ºC in winter to keep indoor conditions close to 28ºC. This indoor project temperature, combined with residual air velocities higher than usual but acceptable due to the physical activity of the workers, allows to keep comfort conditions in the occupied zone. Furthermore, high temperatures that are obtained in the ceiling may be used with high efficiency in the heat recovery process.

The simulation of the case 3 yields the monthly evolution of electrical and fuel consumptions that are shown in Figures 14a and b:

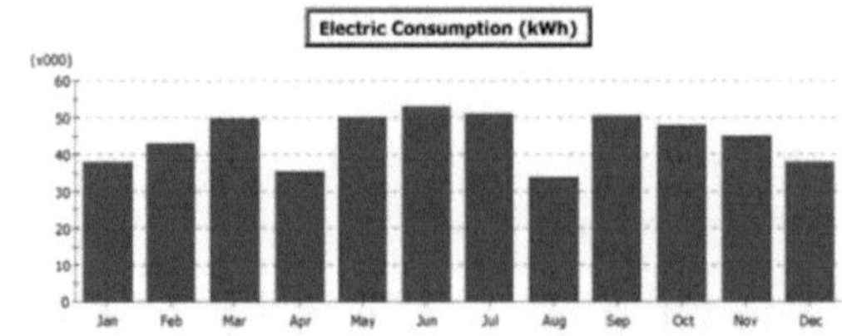

	Jan	Feb	Mar	Apr	May	Jun	Jul	Aug	Sep	Oct	Nov	Dec	Total
Run 1.	37,95	42,97	49,68	35,43	50,14	52,94	50,76	33,84	50,22	47,76	45,19	37,94	534,81
Run 2.													
Run 3.													
Run 4.													
Run 5.													

(a)

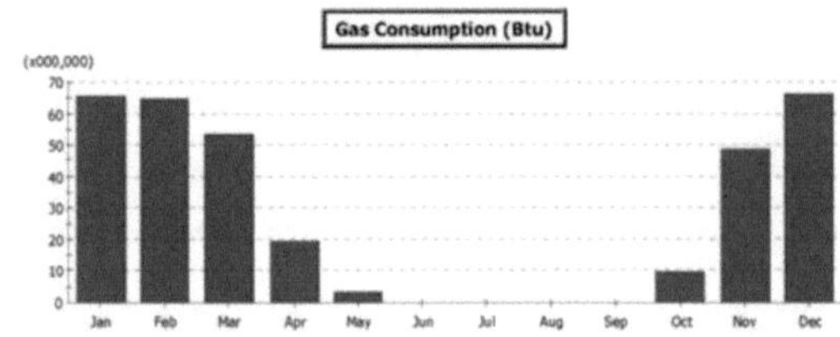

	Jan	Feb	Mar	Apr	May	Jun	Jul	Aug	Sep	Oct	Nov	Dec	Total
Run 1.	65,54	64,84	53,67	19,56	3,57	-	-	-	-	9,91	48,76	65,93	331,79
Run 2.													
Run 3.													
Run 4.													
Run 5.													

(b)

Figure 13. a. Case 2: Monthly electric consumption, b. Case 2: Monthly gas consumption

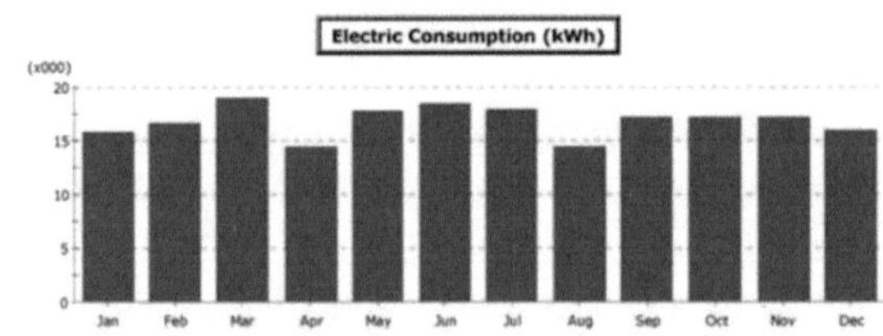

	Jan	Feb	Mar	Apr	May	Jun	Jul	Aug	Sep	Oct	Nov	Dec	Total
Run 1.	15,85	16,64	19,03	14,50	17,84	18,46	17,88	14,40	17,23	17,20	17,17	16,02	202,21
Run 2.													
Run 3.													
Run 4.													
Run 5.													

(a)

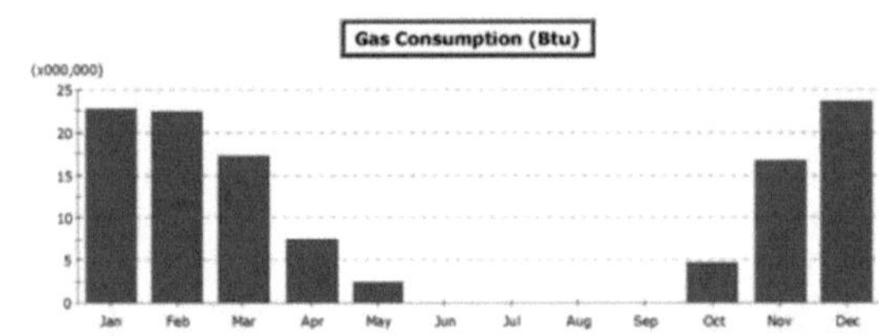

	Jan	Feb	Mar	Apr	May	Jun	Jul	Aug	Sep	Oct	Nov	Dec	Total
Run 1.	22,76	22,34	17,16	7,52	2,38	0,08	-	-	-	4,60	16,65	23,54	117,04
Run 2.													
Run 3.													
Run 4.													
Run 5.													

(b)

Figure 14. a. Case 3: Monthly electric consumption, b. Case 3: Monthly gas consumption

5. Conclusion

In the absence of air conditioning system, the predicted indoor temperature evolution throughout the year suggests that there is significant potential for intervention to improve the performance of the activities in ITV workshops.

For what respects to the more efficient system to air conditioning a wide open workshop, the analysis of the energy consumption during a year simulation allows to conclude that high efficient system like the proposed, composed by a solar heating installation with a condensing gas boiler auxiliary power option, air curtains for environmental thermal control and displacement ventilation supplied by a dedicated outdoor air handling unit with evaporative indirect cooling, reduces electric consumption to 37% and gas consumption to 35%, compared to a conventional HVAC system.

From the point of view of its efficiency, it is highlighted the importance of the general conception of the system, instead of placing design efforts in improving the performance of its components.

The comparison between systems should be done in terms of primary energy and carbon emissions, rather than based on economic criteria. European legislation states that a system of energy performance certification is a mandatory requirement for rented, sold or constructed buildings. In response to this, Spain's recent building regulations have established the process of certification of new buildings. A series of tools are provided to calculate whether the new buildings are highly efficient (class A) or least energetically efficient (class G). The official program used for this process of certification is called CALENER and provides an average carbon value of the overall building energy performance, and classifies it by means of a comparison of the performance of the building systems and construction materials with their corresponding reference values.

Author details

Tomas Gil-Lopez, Miguel A. Galvez-Huerta, Juan Castejon-Navas and Paul O'Donohoe
Madrid Polytechnic University ,Tayra S.L., Spain

Acknowledgement

We wish to thank Virginia Gomez-Garcia and TAYRA S.L. for the invaluable contribution to this chapter (the whole conception of the HVAC proposed system and the selection of design parameters during the simulation). In this respect we specially thank David García (Plenum Ingenieros S.L.P.), who has meticulously carried out the simulation process.

6. References

ACGIH (1999). Threshold Limit Values and biological exposure indices, *American Conference of Governmental Industrial Hygienists,* Spanish version: Valores límite para sustancias químicas y agentes físicos en el ambiente del trabajo, Editorial Consellería d'economia, i ocupació. Generalitat Valenciana, p. 19, Valencia, Spain.

Allard, F. & Utsumi, Y. (1992). Airflow through large openings. *Energy and Buildings,* Vol.18, No.2, pp. 133-145, ISSN 0378-7788

ASHRAE (1992). *Standard 55-1992,* Atlanta, USA.

ASHRAE (2009a). Fundamentals, In: *2009 ASHRAE Handbook,* Ch.9, Fig.12, p. 9.15, ISBN 978-193-3742-54-0, USA.

ASHRAE (2009b). Fundamentals, In: *2009 ASHRAE Handbook,* Ch.10, p. 10.10, ISBN 978-193-3742-54-0, USA.

Ayuntamiento de Madrid (1985). Ordenanza General de Protección del Medio Ambiente Urbano. B.O. Ayuntamiento de Madrid, No.4636, (December 1985), Madrid, Spain.

Baturin, V. V. (1972). *Fundamentals of industrial ventilation,* Pergamon Press, ISBN 978-008-0158-28-0, New York.

Biddle Innovative Klimatechnik (2012). *VACP (visual air curtain performance) – Luftschleier - Simulation,* Gesellschaft mit beschränkter Haftung, Germany.

BS EN 410:1998 (1998). Glass in building. Determination of luminous and solar characteristics of glazing, *British Standards Institution,* (November 1998), ISBN 0-580-30154-0.

BS EN ISO 7730:2005 (2006a). Ergonomics of the thermal environment. Analytical determination and interpretation of thermal comfort using calculation of the PMV and PPD indices and local thermal comfort criteria, *British Standards Institution,* Art. 6.2, (July 2006), ISBN 0-580-47207-8.

BS EN ISO 7730:2005 (2006b). Ergonomics of the thermal environment. Analytical determination and interpretation of thermal comfort using calculation of the PMV and PPD indices and local thermal comfort criteria, *British Standards Institution,* Art. 6.3, (July 2006), ISBN 0-580-47207-8.

BS EN ISO 7730:2005 (2006c). Ergonomics of the thermal environment. Analytical determination and interpretation of thermal comfort using calculation of the PMV and PPD indices and local thermal comfort criteria, *British Standards Institution,* Art. 6.4, (July 2006), ISBN 0-580-47207-8.

BS EN ISO 7730:2005 (2006d). Ergonomics of the thermal environment. Analytical determination and interpretation of thermal comfort using calculation of the PMV and PPD indices and local thermal comfort criteria, *British Standards Institution,* Art. 6.5, (July 2006), ISBN 0-580-47207-8.

BS EN 15603:2008 (2008). Energy performance of buildings. Overall energy use and definition of energy ratings, *British Standards Institution,* (September 2008), ISBN 978-0-580-59028-3.

Castejon, J.; Gil, T.; Galvez, M.A. & Gomez, V. (2011). Environmental protection and energy savings in HVAC systems from the higrothermal and air quality standpoints. *Proceedings of International Conference on Management and Service Science (MASS 2011),* ISBN 978-1-4244-6579-8,Wuhan, China, August 2011.

Clark, R. P. & Edholm, O. G. (1985). *Man and his thermal environment,* Edward Arnold, ISBN 978-071-3144-45-1, London, UK.

Crawley, D.B.; Hand, J.W.; Kummert, M. & Griffith, B.T. (2008). Contrasting the capabilities of building energy performance simulation programs. *Building and Environment,* Vol.43, No.4, pp. 661-673, ISSN 0360-1323.

Deutsche Forschungsgemeinschaft (Ed.). (2007). *List of MAK and BAT Values 2007,* Wiley-VCH, ISBN 978-3-527-31955-8, Weinheim, Germany.

Directive 2002/91/EC (2003). European Parliament and of the Council of 16 December 2002 on the energy performance of buildings. *Official Journal of the European Communities,* 4.1.2003, (December 2002), pp. L1/65-L1/71, ISSN 1561-1280.

Directive 2010/31/EU (2010). European Parliament and of the Council of 19 May 2010 on the energy performance of buildings (recast). *Official Journal of the European Union,* 18.6.2012, (May 2010), pp. L153/13-L153/35, ISSN 1725-2555.

Fanger, P. O. (1993a). Calidad del aire y fuentes de contaminación en los edificios, In: *Calidad del aire interior,* Cristalería Española, pp. 13-30, ISBN 978-84-604-4724-5, Madrid, Spain.

Fanger, P. O. (1993b). Calidad del aire y fuentes de contaminación en los edificios, In: *Calidad del aire interior,* Cristalería Española, pp. 16-19, ISBN 978-84-604-4724-5, Madrid, Spain.

Fanger, P. O. (1993c). Calidad del aire y fuentes de contaminación en los edificios, In: *Calidad del aire interior,* Cristalería Española, Fig.2, pp. 13-30, ISBN 978-84-604-4724-5, Madrid, Spain.

Galvez-Huerta, M.A.; Castejon-Navas, J.; Gil-Lopez, T. & Gomez-García, V. (2012). Influence of the number and location of the luminaries in the "Predicted Percentage Dissatisfied" of an operating room. *Proceedings of International Conference on Environmental Pollution and Public Health (EPPH 2012),* Shanghai, China, May 2012.

Gil-Lopez, T.; Galvez-Huerta, M.A.; Castejon-Navas, J. & Gomez-García, V. (2011). Analysis of carbon impact using mix energy sources for industrial heating applications. An assessment of options to refit services in an existing building. *Fuel Processing Technology,* DOI 10.1016/j.fuproc.2011.11.002, ISSN 0378-3820.

Gomez, V. (2009). *La Edad del Aire: Un nuevo parámetro para optimizar la relación ventilación-calidad del aire en el interior de los locales,* Doctoral Thesis, Madrid Polytechnic University, Madrid, Spain.

IDAE (2009). Condiciones de aceptación de Procedimientos alternativos a LIDER y CALENER, In: *Calificación de Eficiencia Energética de Edificios,* Ministerio de la Vivienda, Spanish Government, Spain.

Instituto Nacional de Seguridad e Higiene en el Trabajo (1997). *Guía Técnica para la evaluación y prevención de los riesgos relativos a la utilización de los lugares de trabajo,* Spanish Government, Spain.

Ley 31/1995 (1995). Prevención de Riesgos Laborales. *B.O.E,* No.269, (November 1995), pp. 32590-32611, Spanish Government, Spain.

Moss, K. J. (1997). *Energy management and operating costs in buildings,* E& FN Spon, ISBN 978-041-9217-70-1, London, UK.

Mumma, S. A. (2001). Designing Dedicated Outdoor Air Systems. *ASHRAE Journal,* May 2001, pp. 28-31, ISSN 0364-9962.

Mundt, E. (1995). Displacement ventilation systems. Convection flows and temperature gradients. *Building and Environment,* Vol.30, No.1, pp. 129-133, ISSN 0360-1323.

Nicol, J. F. (1993). *Thermal Comfort: a handbook for field studies toward an adaptive model,* University of East London, ISBN 978-187-4210-20-7, London, UK.

Nielsen, P. V. (1993). Displacement ventilation: Theory and design, *Aalborg University Department of Building Technology and Structural Engineering,* Vol.18, ISSN 1395-7953.

R.D. 275/1995 (1995). Requisitos de rendimiento para las calderas nuevas de agua caliente alimentadas con combustibles líquidos o gaseosos. *B.O.E,* No.73, (February 1995), pp. 9414-9421, Ministerio de Industria y Energía, Spanish Government, Spain.

R.D. 486/1997 (1997). Disposiciones mínimas de seguridad y salud en los lugares de trabajo. *B.O.E,* No.97, (April 1997), pp. 41429-41430, Spanish Government, Spain.

R.D. 314/2006 (2006a). Código Técnico de la Edificación. DB HE1. Ahorro de energía. *B.O.E,* No.74, (March 2006), pp. 11816-11831, Ministerio de Vivienda, Spanish Government, Spain.

R.D. 314/2006 (2006b). Código Técnico de la Edificación. DB HE3. Salubridad. Calidad del aire interior. *B.O.E,* No.74, (March 2006), pp. 11816-11831, Ministerio de Vivienda, Spanish Government, Spain.

R.D. 314/2006 (2006c). Código Técnico de la Edificación. DB SI4. Seguridad en caso de incendio. Detección, control y extinción del incendio. *B.O.E,* No.74, (March 2006), pp. 11816-11831, Ministerio de Vivienda, Spanish Government, Spain.

R.D. 47/2007 (2007). Procedimiento básico de certificación de eficiencia energética de edificios de nueva construcción. *B.O.E,* No.27, (January 1995), pp. 4499-4507, Ministerio de la Presidencia, Spanish Government, Spain.

R.D. 1027/2007 (2007). Reglamento de Instalaciones Térmicas de Edificios. *B.O.E,* No.207, (August 2007), pp. 35931-35984, Ministerio de la Presidencia, Spanish Government, Spain.

Skistad, H. (1994). *Displacement ventilation,* Research Studies Press, ISBN 978-086-3801-47-1, Taunton, Somerset, England.

Chapter 6

Energy Efficient Control of Fans in Ventilation Systems

Bjørn R. Sørensen

Additional information is available at the end of the chapter

http://dx.doi.org/10.5772/48597

1. Introduction

Ventilation fans are energy-demanding equipment that stands for a significant share of a building's total energy consumption. Improving energy efficiency of ventilation fans is thus important. Fans used for demand controlled ventilation (DCV) are intended to deliver a specific flow rate derived from the actual load in the building. This chapter is about how to deliver the required air flow rate to all rooms, without wasting energy on throttling. There are several ways to control the flow rate in a ventilation system. The most common ways are (1) pure throttling, (2) constant differential pressure and (3) constant duct static pressure. In some situations, (4) direct fan control are used as a control strategy (DDCV). The latter is known to be efficient, as it will provide only the required air flow at all times, without the need for any throttling. It needs however more extensive instrumentation, and flow rates to and from all zones in a building have to be continually recorded and fed back to the central flow controller. A simulation study shows that by introducing a relatively simple control procedure, the fan can be controlled more efficient and energy consumption can be reduced to a significant degree, also for alternative (2) and (3), without the need for added control equipment.

Pressure energy used for throttling of air in the ventilation system is transformed to heat by friction. Thus the ventilation air is heated by the 'wasted' pressure energy. For this reason it can be argued that throttling is just as good as pure speed regulation. However, this makes sense only if a demand for heating exists. During periods with cooling demands, throttling is a true waste of energy since the heat produced from it cannot be exploited. If there is local cooling of air within the zones, throttling is even worse since energy must be used to remove the 'throttling' heat. Throttling also produces noise. Hence, the degree of throttling in ventilation systems should be as low as possible.

INTECH
open science | open minds

© 2012 The Author(s). Licensee InTech. This chapter is distributed under the terms of the Creative Commons Attribution License http://creativecommons.org/licenses/by/3.0), which permits unrestricted use, distribution, and reproduction in any medium, provided the original work is properly cited.

In this study a different approach of controlling the static pressure difference of a fan is suggested. The study is based on a detailed dynamic simulation system that was presented in (Sørensen, 2006, 2010). Controlling static pressure difference of the fan to a fixed set point is a compromise between pure throttling and pure speed regulation. Seen from an energy consumption perspective, control is not optimal. To make control more optimized energy-wise, while still retaining proper pressure control, the static pressure set point should be altered automatically, along a predefined path.

The developed simulation systems used as basis for this study cannot at this stage account for heating of air due to friction and throttling in the system. The error introduced by this is assessed not to be very significant due to other more dominant loads.

The following sections will explain the models and simulation systems used to demonstrate the possible benefits of the reset algorithm in practice. Furthermore, a case study will be carried out and discussed.

2. Procedure to reset of static pressure difference set point

The study presented in this section is based on (Sørensen 2011). Fig. 1 shows three different paths from one flow rate to another in the fan diagram. In this diagram, we consider pressure difference along the y-axis, and flow rate along the x-direction. Path 1-2 is the normal path to follow during constant fan pressure difference control. Pressure variations due to throttling in the duct system are then compensated for by adjusting the fan speed. Path 1-4 is obtained while there is no throttling going on in the duct system and thus no static pressure control is needed. This is the optimal path from an energy consumption perspective. Path 1-4 is used in a variable air volume ventilation system addressing a DDCV control strategy. Path 1-3 represents a compromise between path 1-2 and 1-4. This means more speed regulation and less throttling than 1-2. The path 1-3 was used in this study to reset the set point as a function flow rate.

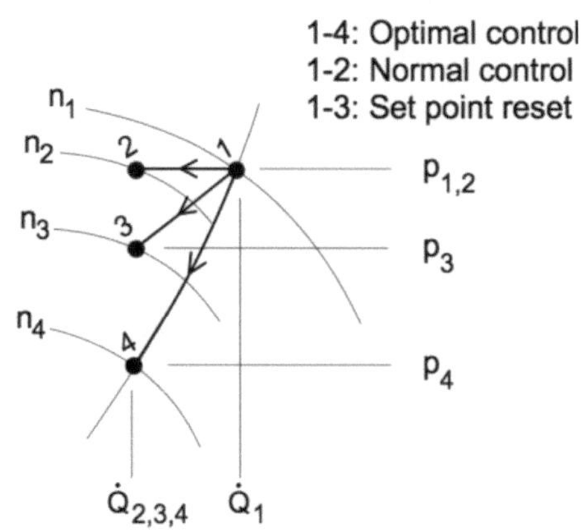

Figure 1. Three different paths of getting from one flow rate to another in the fan diagram (n=fan speed, Q=flow rate, p=pressure)

The direction of path 1-3 must be very carefully assessed. Choosing a too slack slew rate is not critical, but then there will be room for more energy savings still. Choosing a too steep curve is however critical. A too steep curve will affect the available maximum air flow rate for the local zones in the building. The risk of getting too little air to some zones is then present.

The algorithm for resetting the set point along path 1-3 is presented below. The following parameters have to be found or known:

- The set point to which the fan pressure difference normally would have been controlled - $\Delta p_{1,2}$.
- Design airflow rate (all local dampers open) - Q_1.
- Lowest airflow rate at $\Delta p_{1,2}$ (all local dampers closed) - $Q_{2,3,4}$. This will be the minimum flow rate from the fan during operation.
- Pressure difference at minimum fan speed (all local dampers open) - Δp_4.

The parameters can be found for instance during the commissioning process, by measurements or from the fan characteristics. In addition to these parameters, a gain (K_{path}) must be carefully selected. K_{path} determines the steepness of the path 1-3 (refer to Fig. 1) and is in the range between 0 and 1. A value of zero means no adjustment of the set point, and a value of one means nearly pure fan speed control. The reset algorithm was obtained through a linear curve fit of the above parameters, and requires continuous measurements of the main air flow rate:

$$\Delta p_{SP} = \Delta p_{1,2} + K_{path} \cdot \frac{\Delta p_{1,2} - \Delta p_4}{\dot{Q}_1 - \dot{Q}_{2,3,4}} \left(\dot{Q}(t) - \dot{Q}_1 \right) \quad (1)$$

To avoid that the set point moves to a very low or high value, it should be saturated at a maximum value. Stopping the set point from moving to a too high or too low value can be achieved by implementing the following algorithm in the controller:

if $\Delta p_{SP} > \Delta p_{1,2}$

then $\Delta p_{SP} = \Delta p_{1,2}$

if $\Delta p_{SP} < \Delta p_4 + (1-K_{path}) \cdot (\Delta p_{1,2} - \Delta p_4)$

then $\Delta p_{SP} = \Delta p_4 + (1-K_{path}) \cdot (\Delta p_{1,2} - \Delta p_4)$

3. Simulation system description

3.1. Fan model

The pressure and flow characteristics produced by the fan model, are crucial for the simulation results. Thus quite much effort has been put into the development of a proper fan model. The model is based on a backward curved, inclined radial fan, and takes into account fan efficiency and mechanical losses.

The model presented below refers to a fan which has no mechanical losses, which applies to incompressible flow and on which the mathematical description is purely based on the fan laws. Losses, in terms of fan efficiency, is added to the model later, thus enabling computation of the required fan power and energy usage and also the fraction of fan power which goes into heating of ventilated air.

For simulation of flow distribution throughout the ventilation system, a function which expresses the relationship between pressure difference, air flow rate and fan speed is required. This function can be derived, using a second order parabolic polynomial, as follows:

$$\frac{\Delta p_2}{\Delta p_1} = \left(\frac{n_2}{n_1}\right)^2 = \left(\frac{\dot{Q}_2}{\dot{Q}_1}\right)^2 = K_A \cdot \left(\frac{n_2}{n_1}\right)^2 + K_B \cdot \left(\frac{\dot{Q}_2}{\dot{Q}_1}\right)^2 + K_C \cdot \left(\frac{\dot{Q}_2}{\dot{Q}_1}\right) \cdot \left(\frac{n_2}{n_1}\right) \tag{2}$$

This relation is developed from the general polynomial model presented by (Lorenzetti, 1993). The factors K_A, K_B and K_C can be found by specifying certain boundary conditions of the fan characteristics. If assuming that the maximum pressure difference Δp_{max} is present at an air flow rate of Q_{pmax}, and that the maximum air flow rate Q_{max} is found at the minimum pressure difference (that is a pressure difference of zero), the boundary conditions of the characteristic are:

$\Delta p_2 = \Delta p_{max}$ for $Q_2 = Q_{pmax}$ and $n_2 = n_{max}$

$\Delta p_2 = 0$ for $Q_2 = Q_{max}$ and $n_2 = n_{max}$

In addition, the following constraint must be fulfilled:

$$\left.\frac{\partial(\Delta p_2)}{\partial \dot{Q}_2}\right|_{n_2=const} = 0 \quad \text{for} \quad Q_2 = Q_{pmax} \quad \text{and} \quad n_2 = n_{max}$$

Solving eq. NN for the factors K_A, K_B and K_C gives:

$$K_A = \frac{\Delta p_{max}}{\Delta p_1} \cdot \left(\frac{n_1}{n_{max}}\right)^2 \cdot \left(1 - \frac{\dot{Q}_{pmax}^{\,2}}{\left(\dot{Q}_{max} - \dot{Q}_{pmax}\right)^2}\right) \tag{3}$$

$$K_B = -\frac{\Delta p_{max}}{\Delta p_1} \cdot \left(\frac{\dot{Q}_1^{\,2}}{\left(\dot{Q}_{max} - \dot{Q}_{pmax}\right)^2}\right) \tag{4}$$

$$K_C = \frac{\Delta p_{max}}{\Delta p_1} \cdot \frac{n_1}{n_{max}} \cdot \frac{2 \cdot \dot{Q}_1 \cdot \dot{Q}_{pmax}}{\left(\dot{Q}_{max} - \dot{Q}_{pmax}\right)^2} \tag{5}$$

where for instance Δp_1, Q_1 and n_1 refer to the point at which Δp_{max}, Q_{pmax} and n_{max} occur. Hence, the governing (ideal) fan model can be expressed by:

$$\frac{\Delta p(t)}{\Delta p_{max}} = \left(\frac{n(t)}{n_{max}}\right)^2 - \left(\frac{\dot{Q}(t) - \dot{Q}_{p\,max} \cdot \frac{n(t)}{n_{max}}}{\dot{Q}_{max} - \dot{Q}_{p\,max}{}^2}\right)^2 \tag{6}$$

Q_{max}, Q_{pmax} and n_{max} can be found from fan product catalogues.

The total efficiency of the fan must be known to compute the total fan power and energy usage. In this section, a simplified model of fan efficiency is presented.

It is based on the following assumption:

- Fan efficiency scale to the system pressure (or working) characteristic.

This means that if the flow obstacles in the ventilation system are kept unchanged, the fan efficiency is constant, even if the fan speed is altered. Note that since the mechanical losses are directly proportional to the fan speed, the total efficiency of the fan is not constant.

Below, the terms fan efficiency and total fan efficiency are addressed. For a VAV system in which the air flow is controlled by a fan only (a single zone system or a DDCV system), the fan efficiency is approximately constant. In more complex systems, control dampers and VAV boxes continuously change the flow resistance and hence the system pressure (or working) characteristics. Then the fan efficiency is a variable which depends on both the flow rate and the fan speed.

Fan efficiency can be modelled in a similar manner as the pressure drop shown earlier, using a second order polynomial to fit two arbitrary chosen data points from the fan characteristics; $\eta_1 = f(n_1, Q_1)$ and $\eta_2 = f(n_1, Q_2)$. The resulting expression for the fan efficiency then becomes:

$$\eta_f(t) = \frac{\eta_1 \cdot \dot{Q}_2^{\,2} - \eta_2 \cdot \dot{Q}_1^{\,2}}{\dot{Q}_1 \cdot \dot{Q}_2 \cdot (\dot{Q}_1 - \dot{Q}_2)} \cdot n_1 \cdot \frac{\dot{Q}(t)}{n(t)} - \frac{\eta_1 \cdot \dot{Q}_2 - \eta_2 \cdot \dot{Q}_1}{\dot{Q}_1 \cdot \dot{Q}_2 \cdot (\dot{Q}_1 - \dot{Q}_2)} \cdot n_1^{\,2} \cdot \left(\frac{\dot{Q}(t)}{n(t)}\right)^2 \tag{7}$$

The total efficiency η_t at any speed n can be related to a reference mechanical efficiency $\eta_{mech,0}$ at a reference speed n_0 as follows (Eck, 1973):

$$\eta_t(t) = \frac{\eta_f(t) \cdot \eta_{mech,0}}{\eta_{mech,0} + (1 - \eta_{mech,0}) \cdot \left(\frac{n_0}{n(t)}\right)^2} \tag{8}$$

Since fans are installed and connected to the air handling unit or ducts in different manners, entrance and discharge losses are normally not included in the fan efficiency.

Furthermore, to simulate fan acceleration, a rate limiter has been added to the relative speed input (i.e. input 2). The fan drive speed input is in itself equal to the controller output (or the frequency inverter output), and must be limited. Typical acceleration times for common HVAC fans are in the range of 5 - 20 sec (Daly, 1988).

3.2. Fan model validation

Most of the component models used in the study have been validated through comparison with either measurements or product data, and show good agreement with real life equipment. See for instance (Sørensen, 2006, 2008, 2010) for a complete overview.

The non dynamical qualities of the pressure/flow fan model were examined by comparing the model output with data from fan product catalogues (fan characteristics) Such comparisons was made for two non ducted real fans. Both were backward inclined radial fans, but of different size and capacity. Fig. 2 and Fig. 3 show comparisons of the output from the model and data from the real fan characteristic.

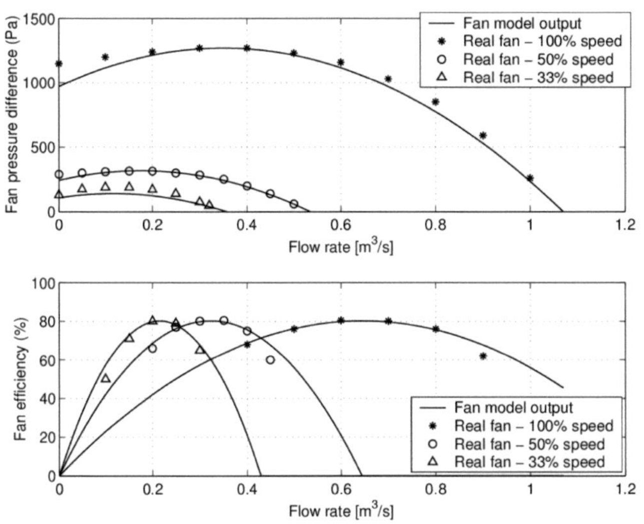

Figure 2. Comparison of pressure and flow rate produced by model to real fan data (small fan).

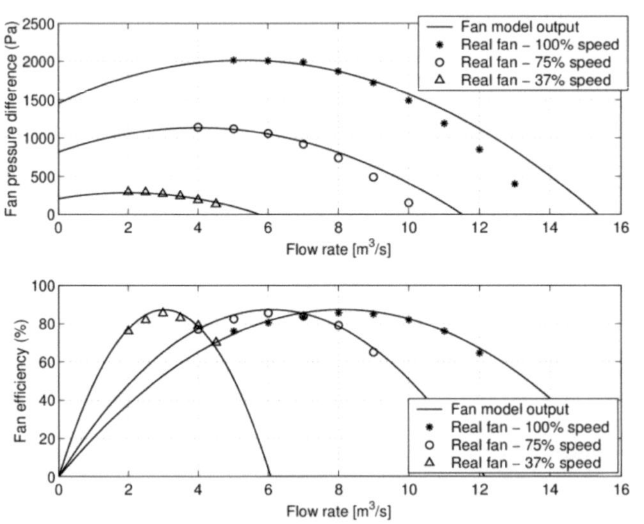

Figure 3. Comparison of pressure and flow rate produced by model to real fan data (large fan).

3.3. Simulation system

The study addressed a CO_2 controlled VAV system serving four similar classrooms in a school (Fig. 4).

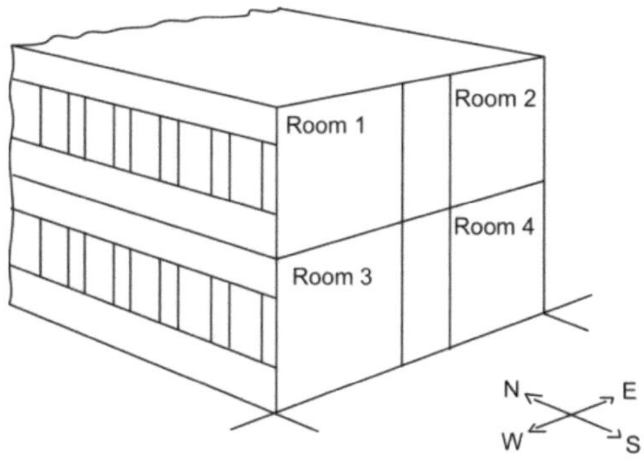

Figure 4. The investigated scenario; four classrooms in a school. Two rooms are facing west and two are facing east

The tool used for modeling and simulation was Matlab Simulink, developed by Mathworks. Matlab Simulink enables a visual programming technique, and is well suited for dynamic (time dependent) modeling. The program facilitates a modular approach, giving the user possibility to re-use models in other simulations.

The dynamic simulation system (Fig. 5) was built of four main subsystems:

- a steady state flow/pressure system
- a dynamical thermal building system
- a dynamical thermal air handling unit (AHU) system
- a dynamical contaminant building system.

The flow/pressure supply air subsystem is shown by Fig. 6. A steady state approach was chosen for the flow/pressure subsystem because changes to the parameters will have almost immediate effect on other connected subsystems. While for instance the thermal systems use considerable time to stabilize its outputs after a change on the inputs, the flow/pressure system outputs will change without any delay caused by inertia. The only dynamic elements of the flow/pressure subsystems are pure time delays caused by transportation of air in the ducts, time dependent ramp functions to model opening/closing time of dampers and ramp functions modulating fan speed.

Each of the blocks of Fig. 6 represents a component model describing the air flow and pressure loss through that component. To calculate pressure loss of ducts, duct fittings, dampers and so on, flow (velocity) dependent functions of friction and single loss factors have been implemented in the various component models. Flow rates are fed forward through the systems (from main duct to terminals), while pressures are summarized backwards. The total pressure loss is compared to the set point of the fan controller, and based on this, fan speed is either increased or decreased. This creates a simulation system which has to be solved by iteration for each time step. The computed airflow rates from the pressure/flow system were fed into the other systems, and the room CO_2 and temperature responses were computed.

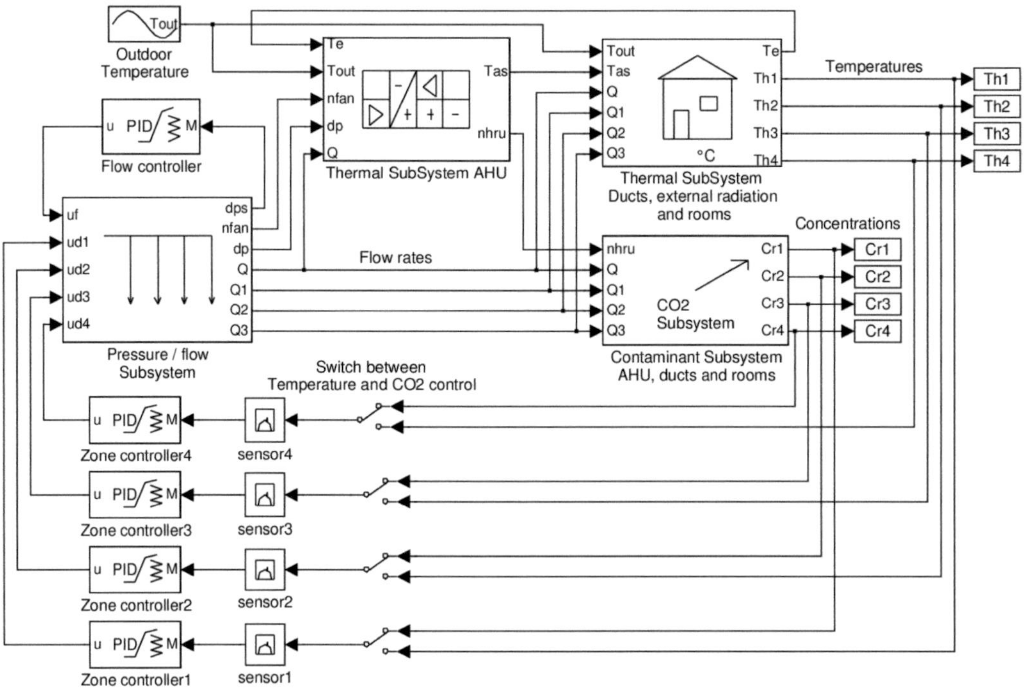

Figure 5. Main simulation system.

The thermal subsystem (AHU – air handling unit, ducts and rooms) was implemented to be able to calculate energy consumption from the dynamical system states. The AHU simulation subsystem, shown in Fig. 7, contains fan, water based heating coil, cooling coil and rotary heat recovery unit (HRU). It also contains thermal models for pipes, ducts, shunt, actuator and controls. Heating coil, het recovery unit and cooling coil are controlled by a sequential PID controller, which ensures that operation of these units is not overlapping. The sequence PID has three sets of controller parameters to ensure proper control of all the units.

The contaminant subsystem considers CO_2 room responses from human presence. The model is a simple dynamic mass balance of CO_2 in a room that accounts for infiltration/exfiltration, ventilation efficiency, and through a contaminant model of a rotary heat exchanger, also for leakage of exhaust air to the supply. Based on the supply air CO_2 concentration and the chosen set point of the rooms, flow rates will by varied. For instance, if room concentration is too high in a room, i.e. above set point, flow rate to the room will be increased until the concentration falls below the set point again. If concentration continues to fall, flow rate will be decreased accordingly. At zero occupancy, flow rate will decrease to a minimum level. For more details, refer to Sørensen 2006, 2008, 2010.

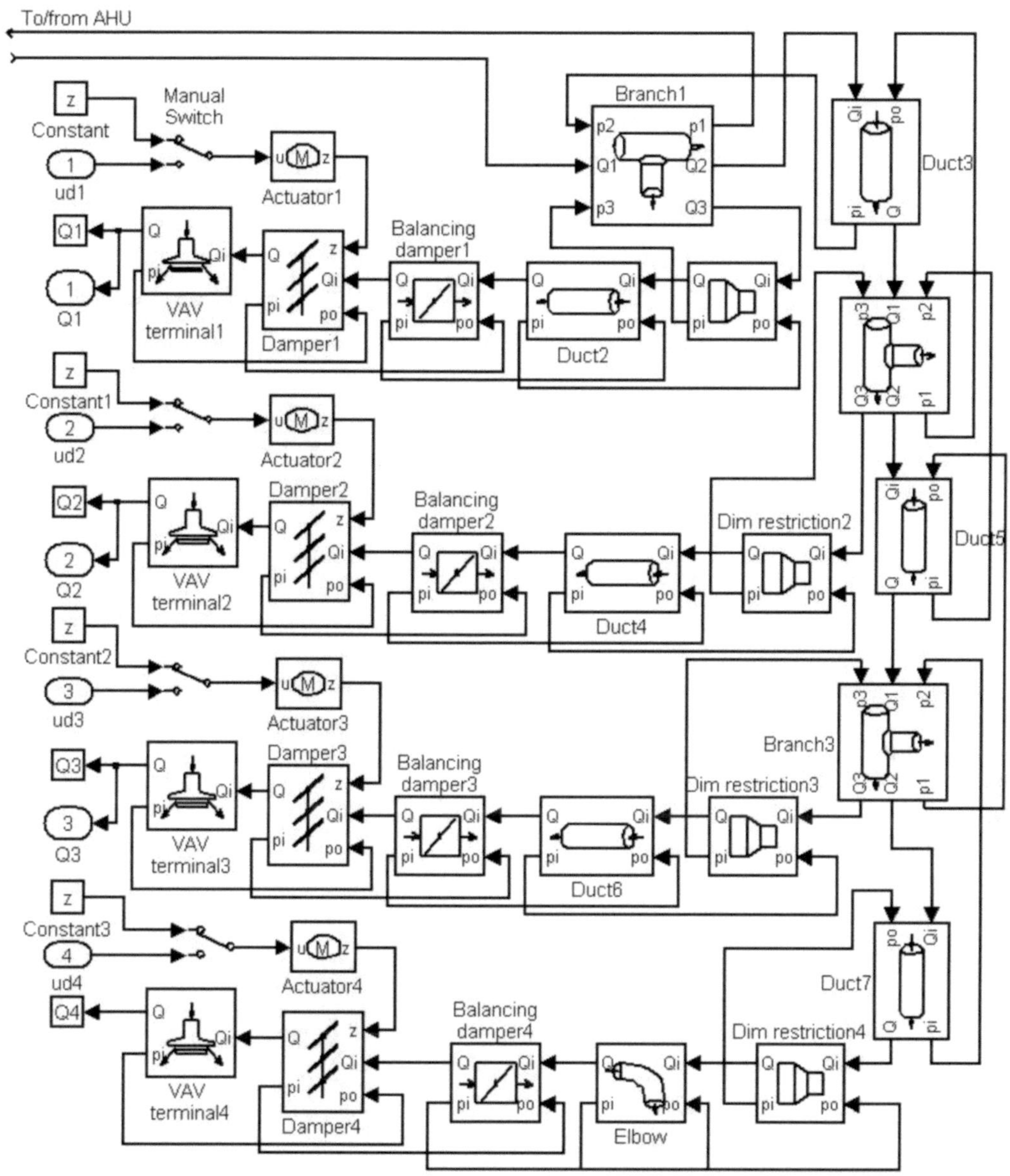

Figure 6. The pressure/flow supply system model (steady state). Each block represents a component model.

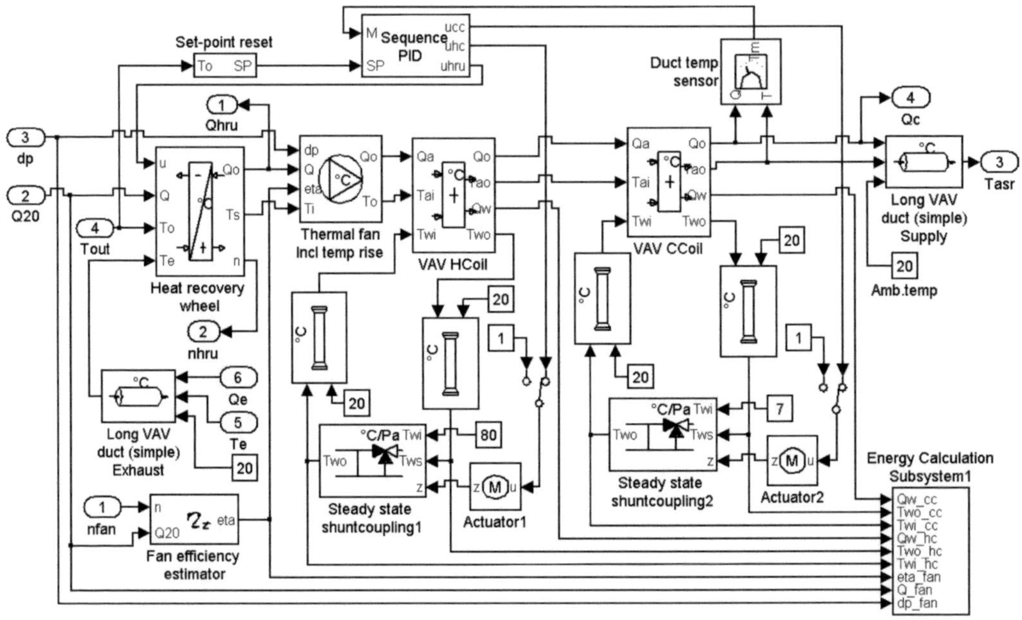

Figure 7. Simulation model (block diagram) of the air handling unit model.

3.4. Case study

The ventilation system served four classrooms in a school. The ventilation system was designed for a maximum CO_2 level of 800 ppm, which approximately gives the same flow rate as mandated by the Norwegian building regulations. Minimum flow rates were determined from 2 liters/s per m^2 floor area. Design values rates for each room thus as follows:

- Maximum occupant load: 28
- Ventilation flow rate (m^3/h): 1150
- Minimum flow rate (m^3/h): 430
- Room volume (m^3): 180

Ventilation was turned on at 07.00, and shut down at 18.00. The main ducts and AHU were designed to handle 4600 m^3/h of air. Infiltration ratios were 0.1 air changes per hour. The maximum flow rate from the fan characteristic was 2.2 m^3/s and the pressure peak was 2100 Pa at 0.6 m^3/s. The HRU was sequentially controlled with water-to-air heating and cooling coils, using PID controllers. At design conditions, the heating coil was able to increase air temperature by 15°C, with supply and return water temperatures of 80 and 60° C. The cooling coil gave a maximum air temperature drop of -15°C at 7/13°C water temperatures.

Temperature efficiency of the HRU at design conditions was 60%. The HRU had a purge sector and the leakage factor was specified to 0.05 (Sørensen, Riise, 2010). It should be noted that the leakage was considered to be constant over the HRU, even though this may vary for a VAV system (Sørensen, 2008).

The outdoor CO_2 level was considered to be constant at 400 ppm. In all cases the occupant load varied as shown in Tab. 1. It was assumed that the time used by occupants to enter or leave the rooms was 5 seconds per occupant.

The building was considered to be multi storey, wherein the simulated rooms had only one exterior surface each. Moreover, terrain was assumed to be completely flat on the west side of the building. On the east side there was an obstruction of angle 20° in the two middle 45° view sectors (the 180° horizontal view from the windows was divided into four 45° sectors) (Sørensen, 2006). As shown in Fig. 2, classroom 1 and 2 were facing west and classroom 3 and 4 were facing east. There was no external shading. Windows were internally shaded (light curtains, 50% reflection) between hour 09.00 and 12.00 in classroom 2 and 4, and from 14.00 to 17.00 in classroom 1 and 3. Window U-value and solar factor were respectively 2.0 W/m^2K and 0.7. Total window area of each room was 9 m^2. The external wall had a U-value of 0.3 W/m^2K and a time constant of 15 hours. Internal walls, floor and ceiling had U-values of 0.7 W/m^2K and a time constant of 10 hours (light inner structure). Heat capacity of the interior was set to 10000 J/m^2K. The shares of radiation onto the walls from persons, internal equipment and external radiation (both atmospheric and solar) were respectively 0.5, 0.5 and 0.8. Heat from lights was specified to 10 W/m^2 and the lights were on between 07.00 and 18.00 only.

Hour of day	Room 1	Room 2	Room 3	Room 4
07.00	0	0	0	0
08.00	25	0	25	15
10.00	0	0	0	0
10.30	25	25	30	30
12.00	0	0	0	5
13.00	15	25	25	25
15.00	8	3	5	25
17.00	0	0	0	0
18.00	0	0	0	0

Table 1. Occupant load of selected four classrooms

4. Control strategy

Flow regulation was achieved through static pressure difference control of the fan and CO_2 control of the room airflow via local dampers. To account for steady state offset and to avoid too aggressive control, a PI controller was used on the fan. Pure P control was used locally. Maximum set point for the fan pressure difference control was 450 Pa. At this point the fan provided the design flow rate (at which the system was balanced and all local dampers were fully opened). To reduce throttling, and to ensure sufficient minimum flow rates to all zones, minimum damper positions were set to 30% of maximum. The correction gain of the pressure set point (K_{path}) was calculated from a minimum flow rate of 0.8 m³/s at design pressure difference, and from a minimum pressure difference of 200 Pa. These values were chosen from the fan characteristics. Hence, the steady state paths of control were as shown in Fig. 8.

Room CO_2 set points were 700 ppm and supply air temperature was controlled between 15 and 18°C, dependent on outdoor temperature.

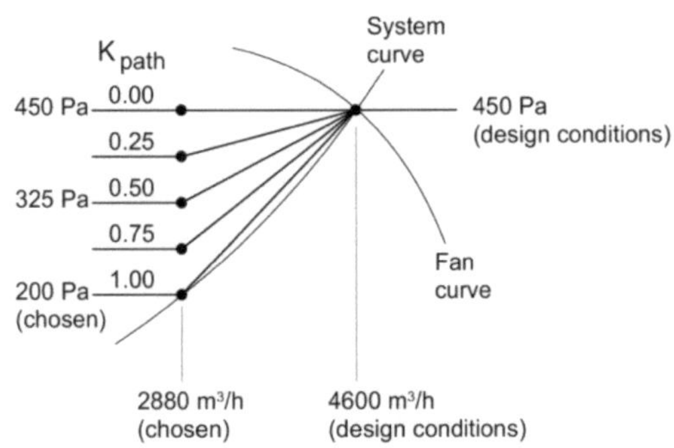

Figure 8. Paths of the pressure difference set point of the fan as a function of the correction gain K_{path}.

5. Results and discussion

Considering the limited data sets that were simulated, making any generalizations based on the simulation results in this case study would seem ambitious. The results are nevertheless important and can provide the basis for a more general representation of using variable set points on static pressure control.

Simulations were performed for different values of the gain K_{path} (0 to 1 with a step of 0.25). Three mean outdoor temperature levels were considered (-20, 0 and +20°C). Outdoor temperature varied as a ±4°C sinusoidal swing during a 24 hour period. Simulations included heat gain from solar radiation, and the simulated dates were in January, April, July and October. The energy consumption of 0°C outdoor conditions was determined as a mean between the usage of April and October.

The two upper diagrams of Fig. 9 show the pressure difference set points (left) and ventilation flow rates (right) during a day for different values of K_{path}. The remaining diagrams show how CO_2 responses in each of the rooms became for different values of K_{path}.

- Minimum flow rates to the zones decreased with an increasing K_{path}. However, due to a relatively wide minimum position on the dampers, all zones got a sufficient minimum flow rate (to maintain IAQ at no occupant load).
- For K_{path} approximately equal to 0.75, the flow rates to room 3 and 4 were marginally insufficient. The CO_2 level then rose slightly above the set point level (above the offset). It was at this point the maximum energy savings for this system were obtained.
- A K_{path} value equal to or below approximately 0.70 provided proper CO_2 control of all rooms.

- Although minimum flow rates were sufficient for a K_{path} of 1.0, the system was unable to provide enough air during occupancy. This shows in the CO_2 responses. It can be explained by first looking at a condition where all dampers are closed and secondly a condition where dampers are not closed. In the first case the fan delivers the minimum flow rate at 200 Pa pressure difference. The dampers are in their minimum positions (30% open). Then some dampers start to open, and as a result, the fan pressure difference drops. To compensate for the pressure drop, the fan increases speed to maintain the 200 Pa set point. The new flow rate is not sufficiently large to change the set point to a larger value (2880 m³/h). Thus, no changes in set point, even while the amount of air varied.

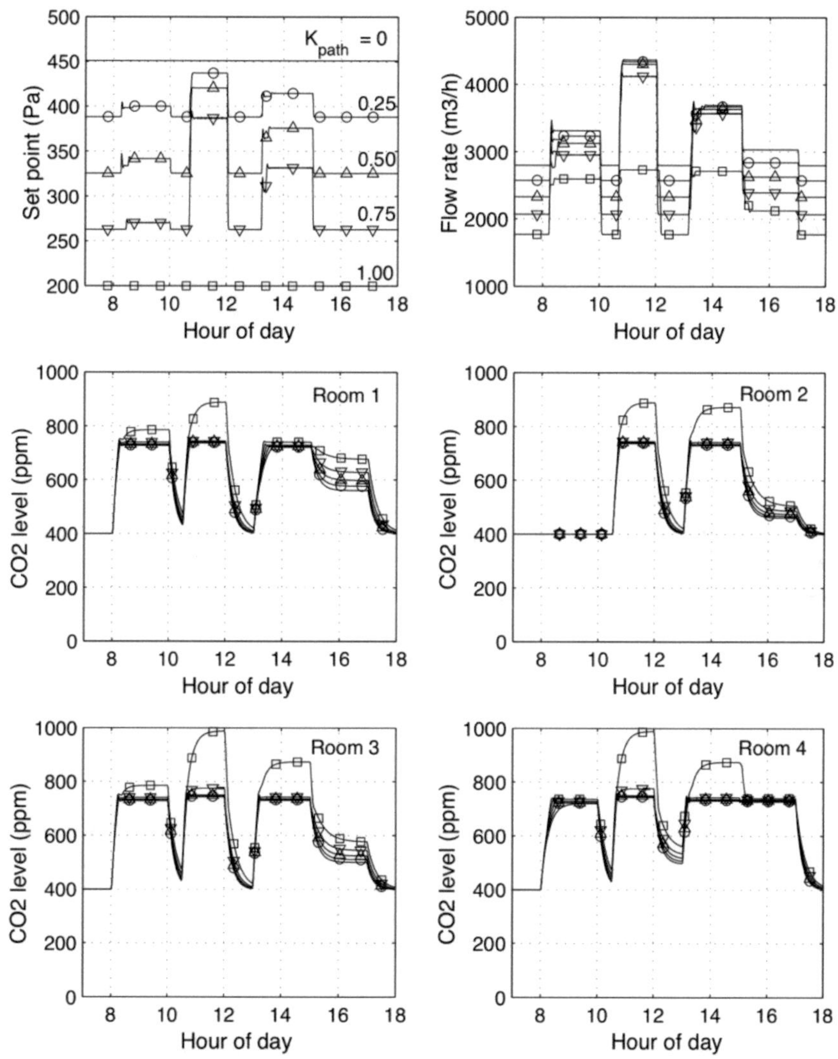

Figure 9. Simulation results of the system with pressure difference reset. The two upper diagrams show the SP and flow rate during a day. The remaining diagrams show CO_2 concentrations of the rooms.

In figure Fig. 10, the corresponding room temperature responses are shown for the different values of K_{path}. The effects from solar radiation on rooms with windows orientated in different directions can be seen here. Rooms with windows facing east naturally got high solar radiation early in the day, while windows facing west got high radiation later in the day. As flow rates were decreased, temperatures of course increased. Consequently, while deciding on a value of K_{path}, also thermal conditions should be assessed.

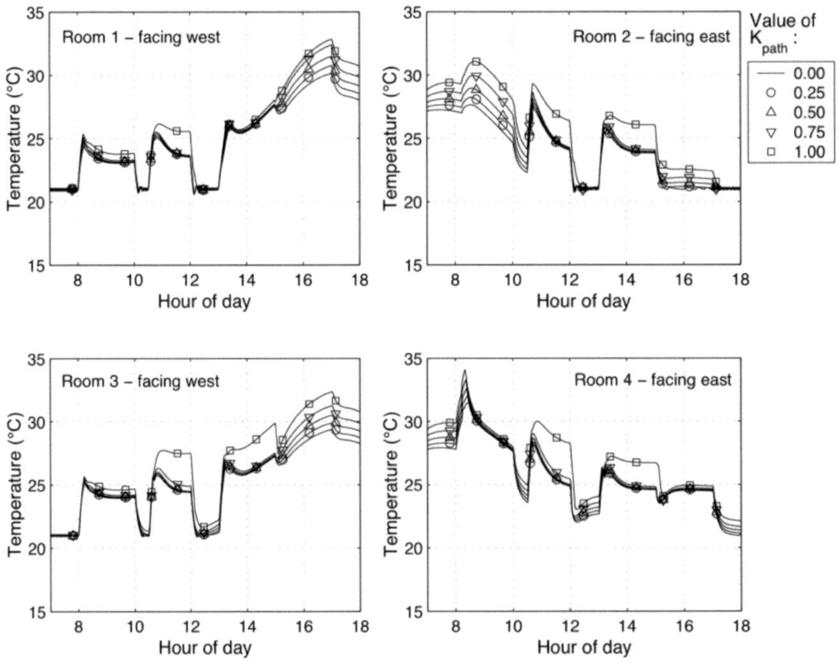

Figure 10. Room air temperatures during a day in July (+20°C outdoor temperature) in Narvik, Norway

Energy usage was compared to an identical VAV system with K_{path} gain equal to zero. In Fig. 11, percentage energy usage is shown as a function of K_{path}, for different outdoor temperatures. Energy usage was of course also affected by the degree of variation to the set point, i.e. the variation of K_{path}. The results generally indicate that energy usage for the AHU (and for the complete building) can be reduced considerably by utilizing variable set point control of the fan. The lack of integral function of the local zone controllers showed as a positive offset (up to 40 ppm). To account for the offset and to provide the correct flow rate to the rooms (specified in terms of a set point), the integral term should be utilized also for local control. In the simulations, the offset during full loads caused the flow rate never to reach design conditions.

Another important factor is damper authority. Low authority means that the dampers must produce a relatively high pressure drop before noticeably affecting flow rate. The dampers must then alter positions quite a bit to obtain the required flow regulation. In some cases this can lead to poor control and risk of instability, and will also affect flow rates to other rooms. Pressure independent VAV boxes represent a more sophisticated alternative. Other rooms are not affected to the same degree as for pure damper control. However, even these will suffer if damper authority is low.

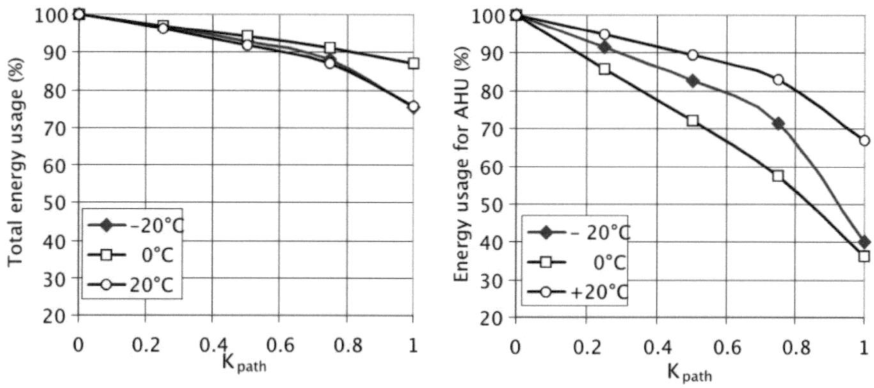

Figure 11. Energy usage (%) as a function of K_{path} and outdoor temperature (relative to a VAV system with K_{path} =0). The left diagram shows total energy usage (local heaters and lights included), and the right diagram shows energy usage for the AHU only. Location was Narvik, Northern Norway.

6. Conclusions

Reducing the energy consumption of the AHU using variable set point for control of fan pressure difference is possible. This study has suggested reductions up to 40% for the air conditioning process, as a result of implementing a more efficient control procedure for the fans. The main criteria to achieve this, is to choose the optimum value of K_{path}. Too low values will generate lower energy reductions. Too high values may lead to instability of the control loop, or no control at all. Influence on the results from other factors may also be important. The controller used for fan control is particularly important. Not so much which controller function that is addressed (P, PI or PID), but more how the controller parameters are determined. In this study, a PI controller was used on the fan, and parameters were determined from a modified Ziegler Nichols step response method. It was assumed that this gave fairly optimized control. It is likely that for larger ventilation systems using CO_2 DCV (or other types of DCV), the maximum value of the correction factor K_{path} (=0.7) will be lower than suggested in this case study, and other controller functions and parameters must be used to achieve the best possible results.

It is thus difficult to produce generalized numbers for K_{path}. The allowable level of correction of the set point must be found and assessed for each individual case.

Since the factor K_{path} obviously depends on occupancy, its maximum value will vary with the various load patterns that occur in the rooms, improved result can be obtained by adjusting it automatically. This can be done as follows (also refer to Fig. 12):

- The measured CO_2 concentrations (or any other control variables) in the rooms can be used as measures of inadequate control. A too large value of K_{path} will lead to augmented CO_2 levels above set point in the rooms. If levels are augmented, K_{path} is simply decreased by a controller.
- If concentrations can be maintained at the set points, K_{path} will be increased until concentrations starts to rise above set point.

Most commercial controllers allow for programming and are thus able to incorporate the algorithm presented above without major implications.

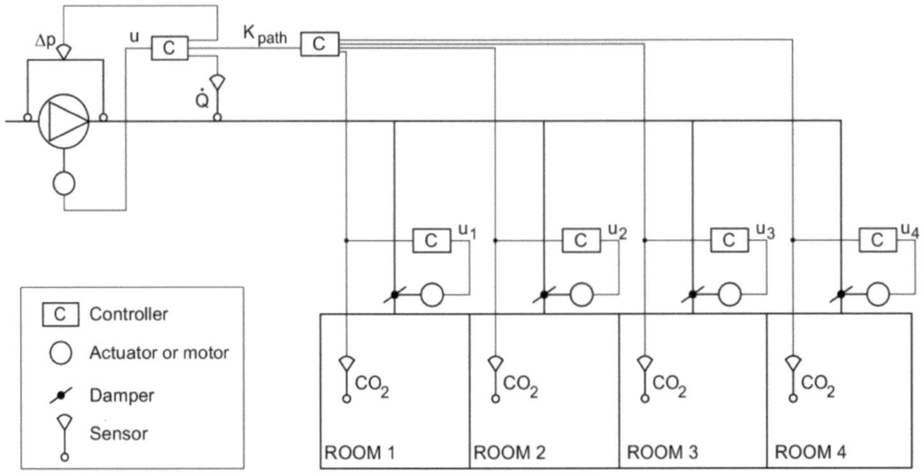

Figure 12. Fan static pressure difference control system (supply side only). The CO_2 concentrations of the zones (or exhaust) are used to determine the degree of set point variation (represented by K_{path}).

Author details

Bjørn R. Sørensen
Narvik University College, Norway

7. References

Daly, B. B. (1988). *Woods Practical Guide to Fan Engineering.* Woods of Colchester Limited.

Eck, B. (1973). *Fans: Design and Operation of Centrifugal, Axial-flow and Cross-flow Fans.* 1st English Edition, Pergamon Press, ISBN 0 08 015872 2.

Lorenzetti D. (1993). Modelling Adjustable Speed Drive Fans to Predict Energy savings in VAV Systems. *Proceedings of the 14th AIVC Conference*, Copenhagen, Denmark.

Sørensen, B. R. (2011). An Energy Efficient Control Strategy for Fan Static Pressure Difference, *Proceedings of the International Conference on Information Engineering and Mechanical Engineering (IEME)*, Xianning, China.

Sørensen, B. R. (2010). A Model Library for VAV Systems, *Proceedings of the 10th REHVA World Congress – Clima 2010*, Antalya, Turkey. ISBN 978-975-6907-14-6.

Sørensen, B. R., Riise, R. (2010). Rotary heat exchanger model for control and energy calculations. *Proceedings of AIVC 2010 conference*, Seoul, Korea.

Sørensen, B. R. (2008). Rotary heat exchangers and VAV systems. *Proceedings of the 11th international conference on indoor air quality - Indoor Air 2008*, Copenhagen, Denmark.

Sørensen, B. R. (2006) Demand Controlled Ventilation - A detailed study of energy usage by simulation. *Proceedings of the 27th annual AIVC conference*, Lyon, France.

Sørensen, B. R. (2006) Modelling and simulation of solar and atmospheric irradiation of Buildings. *Proceedings of the Cold Climate HVAC 2006*, Moscow, Russia.

Chapter 7

Increasing the Energy Efficiency in Compressed Air Systems

Dragan Šešlija, Ivana Ignjatović and Slobodan Dudić

Additional information is available at the end of the chapter

http://dx.doi.org/10.5772/47873

1. Introduction

The increase of energy efficiency is a general trend in a worldwide relation. According to the Kyoto Protocol from 1997, the EU has to reduce greenhouse gas emission by 8% below the level from 1990 by the 2008 - 2012 period. To achieve these reductions, substantial efforts have to be undertaken in all branches of human enterprise.

One of the important industry utilities that has to be encompassed by this energy policy are compressed air systems (CASs). The application of compressed air has had a growing trend due to its easy and safe generation, manipulation, and usage. In previous years, the research efforts in this domain were concentrated on the CASs development and application aimed at boosting the productivity regardless of the energy consumption. With increased awareness of the energy costs as well as the effects of greenhouse gas emission, the attention has been recently placed on the energy efficient use of compressed air.

The experience gained in numerous CAS optimisation projects, as well as the opinions of the experts in the field, indicated that many industrial systems are missing the chance to improve energy savings with the relatively low costs of projects for increasing energy efficiency. Energy saving measures in CASs that have been identified in the course of energy audits in the small and medium industrial enterprises may yield an average energy saving of nearly 15%, with a payback of two years, the energy saving potential in some of them amounting from 30% up to even 60% (USDOE, 2001). The basis for all decisions concerning energy efficiency of the existing CASs is the understanding of the way of their functioning and existence of appropriate data. In that sense, it would be necessary to make measurements of consumed electricity of compressors, airflow, system leakage and pressure drop in the system.

Besides energy savings, increasing energy efficiency of CASs may ensure other significant benefits for the enterprise. Energy saving measures imply a high monitoring level of CASs and

INTECH
open science | open minds

© 2012 The Author(s). Licensee InTech. This chapter is distributed under the terms of the Creative Commons Attribution License http://creativecommons.org/licenses/by/3.0), which permits unrestricted use, distribution, and reproduction in any medium, provided the original work is properly cited.

appropriate maintenance. That leads to decreased breakdowns of production equipment, avoiding the loss of raw materials or other inputs, longer life cycle of pneumatic devices and higher reliability of CASs. Reduction in energy consumption will also lower the emissions of dangerous and polluting substances, which will lessen the influence on the environment. Often, these benefits are more valuable than the energy savings.

This paper is concerned with the identification of the current state of energy efficiency in the production and usage of compressed air and possibilities for improvements that would yield the corresponding energy saving.

2. Current state of energy efficiency of compressed air systems

It is often present a misconception that the costs of compressed air are so low that they do not justify the expenditure of expensive managerial time for optimising all the parameters included in this problem. However, the air is free of charge only before compression. But, after the compression, it has significant price so it is justified to invest efforts in increasing energy efficiency of CAS. Firstly, it is necessary to identify the current state of CAS from the aspect of energy efficiency. In that sense, a brief outline of the state of CASs in the USA, European Union, China, and Serbia were given. The USA was taken because of its leading position in the industry; European Union because of its overall significance; China as an example of the fast-growing industry, and Serbia as a typical example of a small, developing country.

2.1. State of energy efficiency in USA

A key finding of a survey carried out within the "Motor Challenge Program" (MCP), launched in 1993, was that 20% of all US electricity was used to operate industrial motor-driven systems, a large portion of this being associated with pumps, fans and blowers, and air compressor systems. The reported potential savings were over 1 TWh of electricity or USD 3 billion per year, with the existing and new technology by 2010 (or 10% of the total energy cost of industrial motor-driven systems). System improvement opportunities were recognized in motor sizing and proper matching to load, improvements in the system layout, updating and well-maintaining controls, improving operation and maintenance, and use of adjustable speed drives (McKane et al., 1997).

The MCP was followed by the project "Compressed Air Challenge" of the US Department of Energy (USDOE, 2001). One of the most significant findings of this project was that the CASs energy consumption in a typical manufacturing facility could be reduced by 17% through appropriate measures, with a payback of 3 years or less. Apart from these energy savings, improvements to the energy efficiency of CASs could also yield some other important benefits to the end users, such as reliable production, less rejects, higher quality, etc.

2.2. State of energy efficiency in European Union

The European Commission launched the "Motor Challenge Program" with the aim to overcome energy efficiency barriers. Of the total electricity consumption in the EU-15 in

2000, of the overall 2,574 billion kWh, 951 billion kWh were used in the industry. Of this, 614 billion kWh, or 65%, was consumed by motor-driven systems. It was estimated that the potential saving could be 181 billion kWh, (29%), or seven percent of the overall electricity consumption (De Keulenaer et al., 2004).

According to the study "Compressed Air Systems in European Union" (Radgen and Blaustein, 2001), the EU-15 was spending 10% of the total electricity consumed in the industry for the production of compressed air. The most important potential energy savings are related to the system installation and renewal (the overall system design, improvement of drives, use of sophisticated control systems, recovering heat waste, improved cooling, drying and filtering, reducing frictional pressure losses, etc.) and system operation and maintenance (reducing air leaks, more frequent filter replacement, etc.). The percentages of potential saving varied from country to country. For instance, Germany spent seven percent, United Kingdom 10%, Italy, France and the rest of EU 11% (Radgen and Blaustein, 2001). Details on potential energy savings can be found in the corresponding references: for Germany in (Radgen, 2003; Radgen, 2004), for Switzerland in (Gloor, 2000), for Sweden in (Henning, 2005), and for Austria in (Kulterer and Weberstorfer, 2007).

2.3. State of energy efficiency in China

The electricity consumption of CASs in Chinese enterprises goes from 10% up to 40% (Li et al., 2008) of the total industrial electricity consumed. According to (Li et al., 2008), the most widely used compressors in China are reciprocating compressors, often several decades old. To meet increased compressed air demands, many enterprises have undertaken retrofits of their CASs, yielding increased compressor capacity, improved system piping, etc. The most frequently implemented energy saving measures are: purchasing rotary screw compressors, application of variable speed drives and changes to the piping system to allow centralized production of compressed air, etc.

2.4. State of energy efficiency in Serbia

Based on data of the Electric Power Industry of Serbia (EPS, 2009a), the amount of electricity consumed in Serbia in 2008 was 27,639 GWh. Industrial CASs installed in Serbia consume about 8% of the electricity used by industry (Šešlija et al., 2011). Although this percentage is low compared to the values reported for some other countries (Radgen and Blaustein, 2001; Radgen, 2003; Radgen, 2004), this does not mean that the CASs in Serbia are more efficient. This low consumption percentage is a consequence of the inefficient electricity utilization in the industry, the value of energy intensity being three times higher than in the developed European countries (USEIA, 2006). Besides, the price of electricity in Serbia is relatively low, and it does not exist appropriate attention to its economic utilization.

There is a high potential for increasing energy efficiency of CASs in Serbia. One of possible ways of increasing the energy efficiency of CASs is the replacement of the reciprocating single-acting compressors with rotary screw compressors, which would reduce the CAS energy consumption by about 2.8%. On the other hand, the introduction of frequency regulation would result in the saving of 10% (Wissink, 2007). If this is combined with the potential saving that could be achieved by eliminating air leak in CASs, which is in average 30%, and if this mode of saving is applicable in about 80% of companies (Radgen and Blaustein, 2001), the additional reduction would be 24%. This would result in the potential saving of 36.8% of the total energy consumed by CASs. With the current price, which is regulated by the government, of approximately 0.04 €/kWh (EPS, 2009b), Serbia could save at least € 8.07 million every year.

Energy saving measures should be applied in all developing countries, also as in developed countries, because the process for increasing energy efficiency is continuous and never ending.

3. The possibilities for the energy efficiency increase in the CAS

The establishment of measures for efficient production, preparation and distribution and rational consumption of compressed air is important in order to increase the energy efficiency. By applying the procedures for pneumatic system optimisation, rational consumption, compressed air preparation and appropriate equipment selection, with skilled management and software support, and proper maintenance, it is possible to significantly improve the energy efficiency of CAS. Production and distribution of compressed air is one of the most expensive and least-understood processes in a manufacturing facility. The costs of compressed air are often unknown or hidden within other operation costs.

In the majority of plants only a portion of total produced compressed air is used in an efficient manner. The system's operation depends on characteristics of each element but even more on the design of the entire system. Identification of possibilities for increasing energy efficiency in compressed air systems are very important step in overall optimisation procedure (Šešlija et al., 2009). The following technical measures can improve the functioning of the entire process of a compressed air system with the return of investment of less than 3 years:

- Power drive improvement: usage of high efficiency drives and integration of variable speed drives,
- Optimal choice of compressor type, as a function of specific needs of end users,
- Improvement in compressor technology, particularly in the segment of multistage compressors,
- Application of sophisticated control systems, for compressed air production,
- Regeneration of the dissipated heat and using it in other functions,
- Improvement of compressed air preparation: reduction of pressure and energy lost in processes of cooling, drying and filtering; optimisation of filtering and drying as a function of consumer needs and temperature conditions,
- Overall system design, including the systems with multiple pressure levels,
- Reduction of pressure losses due to friction in the pipeline,
- Air leakage elimination,
- Reduction of operation pressure,

- Optimisation of devices that consume compressed air: application of more efficient, better adjusted devices or, in some cases, replacement of compressed air with an electrical drive,
- Optimising the control systems at the point of use,
- Measuring and recording of the system performance.

3.1. Power drive improvement

Usage of high efficiency drives increases the energy efficiency. Integration of variable speed drives (VSD) into compressors can lead to energy efficiency improvements with respect to characteristics of the load. Application of high efficiency drives renders the largest savings to new systems, because the chances of users installing high efficiency drives into existing compressor systems, without changing the compressor itself, are rather small. Integration of speed controllers (frequency inverters) into compressed air systems is a very cost effective measure, under the conditions of variable demands, and it is estimated that such systems participate in the industry with 25%. In compressor rooms where several compressors are installed, variable speed drives are integrated into only one machine and are usually coupled with more sophisticated control system for the whole compressor station that powers on and off individual compressors with a constant speed and also varies the speed of one compressor in order to adjust the production of compressed air to instantaneous requirements of consumers.

3.2. The optimal selection of compressor type

The segment of the market covering power range from 10 to 300 kW is now dominated by rotary screw compressors with oil injection – it is estimated that around 75% of compressors sold in EU belong to this category (Radgen and Blaustein, 2001). Besides, there are other compressor types available that have other advantages within certain exploitation characteristics. In order to make an optimal selection of compressor it is necessary to consider the users' demands. The choice of compressor can greatly influence the energy efficiency of the system, with respect to compressor performance, but also regarding multiple interactions with other elements in the system. The advantages of multiple compressor systems are especially emphasized in production systems with the high workload that operates almost continuously.

3.3. Improvements in compressor technology

A whole array of efforts is directed towards improving the existing compressor lines but also the development of new types, which are usually customized to different segments of industry. Another aspect of research is concerned with improving production methods such as applying narrower tolerances in order to reduce the leakage within the compressor.

It must be taken into consideration that the laws of thermodynamics limit the further improvements of compressor so that only minor improvements can be made in the area of energy efficiency, while the greatest potentials lie in adequate design of the entire system and procedures for system control and maintenance.

3.4. Application of sophisticated control systems for compressed air production

Sophisticated control systems are applied in order to adjust the compressor outlet flow to the requirements of the consumers. They save the energy by optimising the transition between non-loaded working state, loaded working state, and non-operating state of compressor. Sequencers optimise the operation of multiple compressor system and can be combined with applications of variable speed drives. Predictive control uses fuzzy logic and other algorithms to predict the future behaviour of consumers, assuming history of system behaviour. Since the price of control technologies is decreasing and industrial familiarity with its usage is simultaneously increasing, their usage is rapidly expanding and their applications on compressors are rising in the occurrence. This kind of control can be purchased along with new machines but can also be applied onto existing systems.

3.5. Regeneration of dissipated heat

Compressors intrinsically generate heat, which might be used for other functions. The recommendations for its usage depend on the presence of those consumers of thermal energy whose characteristics comply with the amounts of generated heat, whose usage is enabled by adequate equipment (heat exchangers, pipelines, regulators etc.). The price of that equipment should be favourable in comparison with alternative solutions. The design of the heat regeneration system must provide appropriate compressor cooling.

The heat dissipated by the compressor is in most cases too low in temperature, or too limited by its quality to adequately respond to the needs of industry regarding their main processes or heating. The climate and seasonal changes also influence the ratio between investments and yields. Typical application is heating the space close to the location of compressor, when needed. Possibilities for using the compressor recycled energy are:

- Use in buildings (water heating and building heating),
- Compressed air preparation (integral dryers with compressor, standard dryer regeneration),
- In processes (heating, drying),
- Boiler preheating (drinking water, boilers).

The cost efficiency of heat regeneration depends on available alternative energy sources. It could be very cost effective, only if it is alternative to electrical energy. However, if natural gas is available, or the process residual heat, the cost efficiency of regenerated heat is much smaller. However, the attention should be placed on renewable energy sources, instead of using fossil fuels.

3.6. Improvements in preparation of compressed air

Well prepared compressed air has the following purposes:

- *Prevents damaging of the production equipment.* The impurities contained in compressed air can cause malfunction of production equipment that uses it. The appropriate quality of compressed air increases the reliability of equipment that uses it.

- *Increases product quality*. In some production systems, compressed air enters the end product directly or comes into contact with the end product (for example in food and pharmaceutical industries or electronics). In these cases, poor compressed air quality leads to decrease in the end product quality.

The equipment for drying and filtering causes the pressure to drop while dryers often consume electrical energy or partially use compressed air for their operation and regeneration. Because of that, the optimisation of compressed air preparation as a function of the user needs is one of the main sources of energy savings. The possible measures are:

- The dynamical setting of the degree of drying in accordance with external temperature conditions. This is applicable only when the purpose of drying is to keep the air temperature above the dew point in order to prevent the condensation. This measure can be inappropriate if drying is required to fulfil the precisely defined needs of a process with respect to compressed air quality.
- To optimise the degree of particle filtering as well as oil and oil vapour filtering, so it can precisely match the needs of the system. Excessive filtering leads to unnecessary waste of energy. This problem is explained in detail in chapter 3.6.1.
- Increase the filter capacity. The increase of the number of filters in parallel operation decreases the speed of air and thus reduces pressure loss. This investment can be very cost effective for new as well as for existing systems.

3.6.1. Optimisation of filtering process

In order to optimise the filtering process it is necessary to (Golubović et al., 2007; Mitrović et al., 2006; Šešlija, 2002a; Šešlija, 2002b; Šešlija et al., 2008):

- Identify the possible filter locations,
- Define the flows, pressures, temperatures, allowed pressure drops, compatibility and needs for validation in critical places,
- Define the types and concentrations of contaminants (particles, water, oil, oil vapours, etc.),
- Determine the needed filtration stages for each characteristic location,
- Choose the adequate filter elements for each location,
- Choose the housings for each characteristic location.

Every aspect of the previous algorithm deserves special attention. When the selection of filter elements is in question, they are generally expected to have a high throughput, large filtration surface, high mechanical resilience, high thermal resistance, high contamination capacity, long operation period between the services, low price and low exploitation costs as well as appropriate certification, the possibility to fulfil quotas, standards and legislature requirements in their area of application. Proper dimensioning of filters is a precondition for energy efficient functioning of pneumatic systems (Golubović et al., 2007). Improperly dimensioned filter can cause either disabling the system from fulfilling its task, or partial usage with higher investment costs. Each filter should be dimensioned in accordance with exploitation conditions and final conditions of filtration.

It is possible to present general guidelines for the selection of the right filter. However, it is advisable to comply with the filter manufacturer's requirements. If the filter manufacturer gave no recommendations, it is recommended to follow the general guidelines listed in table 1.

Filter type	Removal	Max. ΔP at operating pressure of 7 bar	Special demands
Regular filters	Particles	0,14 - 0,5	No
Coalescent	Particles and fluids	0,17 - 0,7	Regular pre-filter
Absorption	Fumes and odours	0,0017 - 0,13	Regular and coalescent pre-filters
Microbiological	Biological load	3,0 - 5,3	Regular, coalescent and absorption pre-filters

Table 1. General guidelines for the selection of the filter type concerning energy efficiency

The maintenance of all components of the pneumatic system is a precondition for its energy efficient functioning. The malfunctioning of one component in a pneumatic system can generate new pollutants that emerge from component wear and tear (valves, distributor pistons, sealing etc.). In such system filters are subjected to additional load, which is not accounted for in filtration design and their life cycle would be shortened.

If filtering elements are not changed within a predetermined period an increased pressure drop may occur, which directly influences the increase of energy consumption. The basis of proper maintenance of filtering components is to track their operation. For this purpose, it is necessary to increase or optimise the frequency of filter replacements. The maintenance procedures should involve regular filter inspections and, when needed, their replacements. It is advisable to install the filters containing a visible indicator of a condition of filtering element, and numerous systems have been developed for automatic registering and alarming that indicates that the pressure drop has exceeded the allowed value. An especially interesting possibility is the application of wireless technology where a filter is equipped with wireless communication, which receives the pressure drop data from the differential pressure sensors which measures filter contamination and transmits a warning in case excessive contamination has occurred. Equipment maintenance personnel need not, in this case, to check for every individual filter but to carry a wireless receiver which receives the information about contaminated filter.

Wireless filter monitoring system (WFMS) for compressed air filters based on a very low power consumption microcontroller was developed on the Faculty of Technical Sciences in Novi Sad. This system is intended for decreasing of the energy loss due to pressure drop caused by filter clogging. The proposed system consists of two separate units (sensor and base), which constantly monitor the filters, see Fig. 1. WFMS is very easy to install in the existing manufacturing systems. It is simple, low cost, flexible, portable and efficient for production, installation in the existing plant, and use (Ignjatović et al., 2012).

Figure 1. Wireless filter monitoring system

3.7. Designing the overall system

The goal of a proper system design is to adjust the pressure, quantity and quality of compressed air to the needs of different users at their points of use. Although this can be a very simple task, complications are possible in cases when different end users have different or varying consumption needs. Arising problems in system design are:

- One or multiple pressure levels within a system. Typical systems are designed to deliver the air according to the highest pressure and quality required by an end user. This approach can cause unnecessary expenses of energy if air prepared in such a way is required only by a small portion of consumers. The alternative solutions may be:
    - To build a system that delivers lower pressure and to install a pressure amplifiers for those consumers which require higher pressure.
    - To provide and install separate compressor for devices that require higher pressure.
- Limitation of the pressure variations. Inadequate control systems may produce large pressure oscillations which in turn consume an excessive amount of energy. When certain consumers have stochastic demand, the solution could be found in installation of an additional reservoir close to those consumers in order to reduce pressure variations.

3.8. Reduction of pressure loses due to pipeline friction

Pressure loses in compressed air distribution network mostly depend on several factors: topology (ring or network, etc.), geometry (pipeline diameter, curvature radius), materials used, etc. The proper designing and realisation of distribution network can optimise the friction loses. Regardless of the importance of a network, a majority of the existing compressed air systems has poor distribution networks due to various reasons:

- During the period of factory construction, the compressed air distribution network is often designed and installed by the companies that perform all other fluid related installation works. These companies are often poorly qualified for designing and installation of compressed air distribution network.
- Under-dimensioned pipelines are occurring very often. Even the systems that were initially well designed, become the "energy devourers" if the compressed air consumption is constantly increased and exceeds the level for which the system was initially designed.

- The lack of valves for interrupting the compressed air supply to the parts of the system being no longer in use or for machines that are not operated in second or third shift.

Since it is difficult and expensive to improve the existing network, proper designing and installation, which encompasses predictions for future system expansions, represent a significant factor for building of a good system.

3.9. Reduction of air leakage

Reduction of air leakage is probably the most important measure for obtaining energy savings that are applicable to most of the systems. The awareness regarding importance of introducing regular leakage detection programs is on a very low level, partially because these spots are difficult to visualize and partially because they do not cause direct damage. Leakages can lead to requirements for additional increase of compressor capacity and to increased compressor operating time. If pressure within a system drops below minimum level, the devices utilizing compressed air can be less efficient and equipment life cycle can be shortened, and in some cases breakdown of production lines may occur. In typically well maintained plants, leakages range between 2 and 10% of total capacity, but can amount up to 40% in the plants that are not maintained properly. It is considered that leakage can be tolerated while being less than 10% of total production. An active approach that involves permanent leakage detection and appropriate maintenance work can reduce the leakage to this level. The causes of air leakages are: employees' negligence, poor system design and poor system maintenance.

Table 2 can serve as a guideline in evaluating the scope of loses that arise due to leakage. In this example, it is assumed that the price of electrical energy is 0.1 €/kWh (costs of industrial electrical energy in EU average to 0.09 – 0.12 €/kWh) and that system is operated at 8,000 hours/year, while the price of compressed air preparation is 0.02 €/m³.

Orifice diameter Actual size		Air loss with 600 kPa (6 bar) l/min	m^3/h	kW (approx.)	Production costs €/year	Costs of production, preparation and distribution €/year
	1	80	4.8	0.4	320	768
	3	670	40	4	3,200	6,432
	5	1,857	111	10	8,000	17,827
	10	7,850	471	43	34,400	75,360

Table 2. Costs of compressed air leakages (Lau, 2006)

Proper design and installation of network can eliminate leakage spots to a great extent, for example, with application of contemporary devices for condensate removal without air loss or by specifying high quality fast decomposing junctions. Awareness must be kept towards the fact that leakage continuously increases after the reparation has been made. The leakage is increasing with the same rate, regardless of whether the reparations are executed or not.

Removing leakage sources is based on detecting and repairing locations of leakage and removing the root causes that generated leakage within the system. Proper maintenance is of essential importance when fighting leakages and a good program for leakage detection can prevent unexpected failures from happening and reduce downtimes and loses. In many cases leakage is easily detected because large leakages are audible. Small and very small leakages are hard to detect and are hardly audible. In those cases, the elements of the system should be checked by some of the methods for leakage detection. The methods for detection of compressed air leakage are:

- Leakage detection via sense of hearing,
- Leakage detection via bubble release,
- Ultrasonic detection and
- Infrared leakage detection (Dudić et al., 2012).

The most significant of all these methods is an ultrasonic method that utilizes a special detector that is shown on Figure 2. Figure 3 shows the examples of operating the ultrasonic leakage detector.

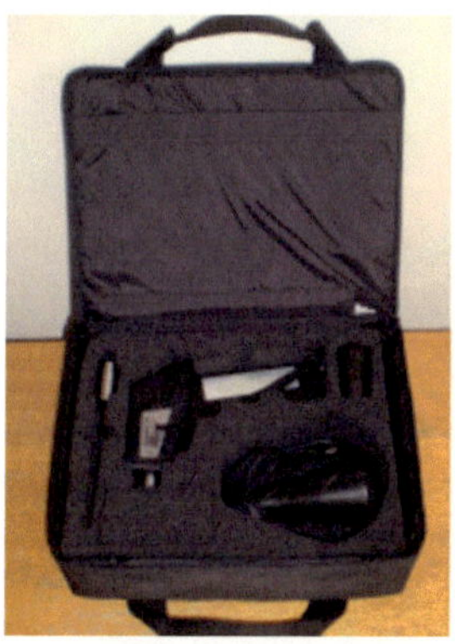

Technical characteristics:

- Indicator: 10 segment LED Bar Graph (red)
- Sensitivity selection: 8 position precision attenuation
- Operating temperature range: -30° to +75° C
- Frequency range: 300 tp 3,000 Hz
- Response time: 300 ms

Figure 2. The ultrasonic detector kit for Ultraprobe 100

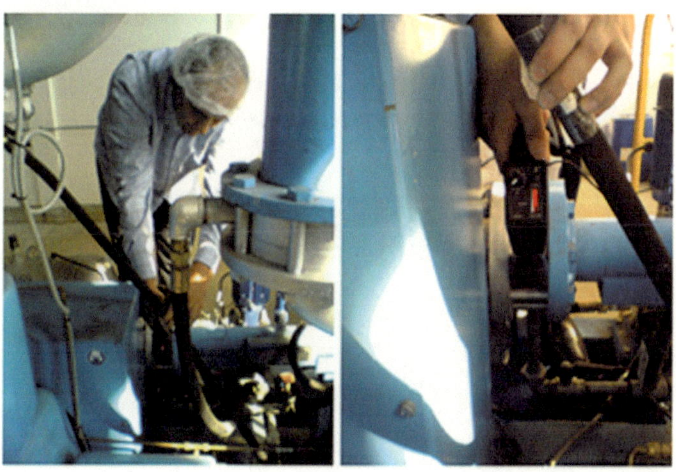

Figure 3. Examples of utilisation of ultrasonic detector

3.10. Reduction of operating pressure

Higher pressures increase leakage, and thereby the expenses. Usually, an increase of operating pressure is used to compensate for lack of capacity. The actual effect is quite opposite to the desired one. The higher pressure, higher is leakage, while the irregular consumers consume more air, and thus more energy. Each 1 bar of the pressure increase is followed by an increase in electrical energy consumption required to compress the air in a range between 5% and 8% (Šešlija et al., 2011).

If the consumers are allowed to independently determine the amount of their need for compressed air, this system will never operate in an efficient manner, because everybody will be misled by the fact that they can obtain the pressure of any desired amount in any desired quantity. Higher air flow and higher pressure impose higher costs. The characteristic situation in which it is possible to solve this problem is one in which there is one or a small number of consumers in a requirement for higher pressure. In this case, it is suggested to install a secondary, smaller, high pressure unit or an appropriate amplifier (pneumatic booster), instead of operating the compressed air system of the whole factory on the higher level of pressure.

3.11. Unsuitable applications of compressed air

Compressed air is extensively misused for applications in which it is not energy efficient, for which better solutions exist or, its implementation is incorrect in the places where its usage is justified. Compressed air is the most expensive form of energy in a plant but its good characteristics, such as simplicity in application, safety of operation and availability in the whole plant area, often lead people to apply it even where more cost effective solutions exist. The users should therefore, always primarily consider the cost effective energy sources before applying the energy of compressed air. Table 3 gives examples of unsuitable usage of compressed air and alternative solutions that should be applied instead.

Figure 4 shows an example of unsuitable application of compressed air. Nozzles are positioned above the line for bakery products in order to clean the product from the powder present and in order to cool it.

The nozzles themselves present an energetically unfavourable solution and an effort should be made to replace it with another solution. In this case, a fan could be used. Furthermore, if usage of nozzles is insisted upon, it's more energy efficient version should be used.

Figure 4. Cleaning of bakery products with compressed air (Norgren, 2011)

Unsuitable application of compressed air	Alternative solution
Control cabinet cooling	Ventilating, air conditioning
Cooling, aspiration, agitation, mixing, packaging blow-out	Blowers (customized compressors that produce compressed air in large quantities)
Vacuum production by Venturi pipe	Vacuum pump or application of Venturi method with appropriate, energy efficient control
Cleaning of parts and processing residuals removal	Brushes, blowers, vacuum system
Removal of parts from the moving production line by nozzles	Blower, electrical actuator or pneumatic cylinder
Blower guns	Blowers, cups or application of reduced pressure air (installing the pressure regulators on guns or constructing low-pressure network)
Pneumatic tools	Electrical tools are more energy efficient although they have lowered torque control possibilities, shorter life time and are not inherently safe
Air knifes	High pressure blowers that are automatically turned off when cutting object is absent
Powdered materials transport (pneumatic transport)	Electrical blowers
Vibrating the walls of powdered and granulated materials	Electrical vibrators

Table 3. Unsuitable applications of compressed air and the alternative solutions

Fig. 6 shows that nozzles are positioned too high so that higher flow of compressed air is needed to accomplish the task, which means that for product of different heights the nozzle carrier frame height should be adjusted. Finally, it is necessary to set the active control over nozzles that will allow the air to flow based on sensor that signals the presence of a bakery product, in contrary to situation from Fig. 6 where it can be seen that the air is flowing and that no bakery product is present beneath. Other unsuitable uses of compressed air involve unregulated consumers, supplying abandoned equipment or equipment that will not be used for a prolonged period of time, etc.

Unregulated consumers - This covers all places of usage in which compressed air can be directly released by opening a valve, all places where leakages are present, etc. For example, in applications with pneumatic tools, if the pressure regulator is not installed, the tool will use the full network pressure and this pressure might be significantly higher than the level required for its operation (for example, 8 bars instead of 5.5 bar). Furthermore, this kind of pressure increase leads to greater equipment wear, which leads to greater maintenance costs and reduction of the equipment life cycle.

Abandoned equipment - From time to time, reconstructions occur in factories that often lead to abandonment of some parts of compressed air equipment leaving the air supply pipeline intact. The airflow going through the pipeline to the abandoned piece of equipment should be interrupted, as close as possible to the air supply source, because it will inevitably generate some leakage and create unnecessary loses.

3.12. Optimisation of devices that consume compressed air

Many devices that consume compressed air can be used in a more energy efficient manner. The optimisation of devices that consume compressed air is one aspect of systemic approach (Šešlija, 2003) to designing a compressed air system. The optimisation can be achieved by: replacing the existing components with more energy efficient ones; installing the additional elements, and better use of existing components.

For example, in the case of applying a vacuum generator, the savings in the compressed air are realised by using more energy efficient components, which have an integrated vacuum switch with an air saving function - example is Air saving circuit (Festo, 2011). The vacuum range is set on the vacuum switch. The switch generates a pulsating signal which actuates the solenoid valves for vacuum when the vacuum pressure has fallen below the selected upper limit value (due to leakage etc.). At all other times, the vacuum is maintained with the non-return valve, even when the vacuum generator is not switched on. Fig. 5 presents the operational diagram for vacuum pump and vacuum generator with implemented Air saving function. Since the price of vacuum produced in this way is too high and vacuum suction elements represent significant consumers of compressed air, this option contributes to the increase of energy efficiency of the system. The savings are proportional to participation of time Δt, shown in Fig. 5, within a total time of holding the working object. This solution is especially suitable for application in which time of holding an object significantly participates in the total cycle of material handling.

Using contemporary engineering tools for supporting the design of vacuum applications, it can be analysed the change of system parameters or parameters of devices that are installed in the system. For example, by using the FESTO vacuum engineering module, for manipulation of an object with a cylindrical shape whose dimensions are: diameter of 150 mm, height of 40 mm and weight of 200 g, a total of 6 vacuum cups are needed. The change of operation pressure enables the usage of highest vacuum level. Air consumption, in the case of 6 bar pressure (see Fig. 6 left) is 27.60 l/min, while with the pressure of 5.4 bar (see Fig. 6 right) is 23.28 l/min.

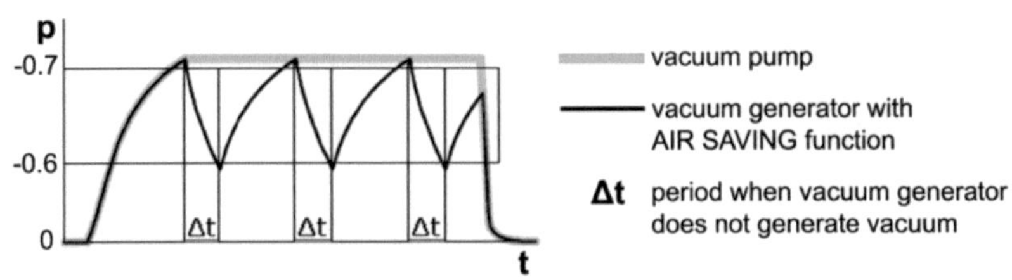

Figure 5. Operation of vacuum pump and vacuum generator with implemented Air saving function

In these way savings of 4.32 l/min, or 0.072 l/s of compressed air are accomplished. If vacuum cups operate for 30 s every minute, 8 h/day, 250 days/year, 259,200 l of compressed air savings could be achieved for a year. This represents the savings for engagement of one group of vacuum cups for manipulating one workpiece on one workplace. The production processes often use a larger number of vacuum cups. Therefore, the amount of compressed air that could be saved is significantly higher (Ignjatović et al., 2011).

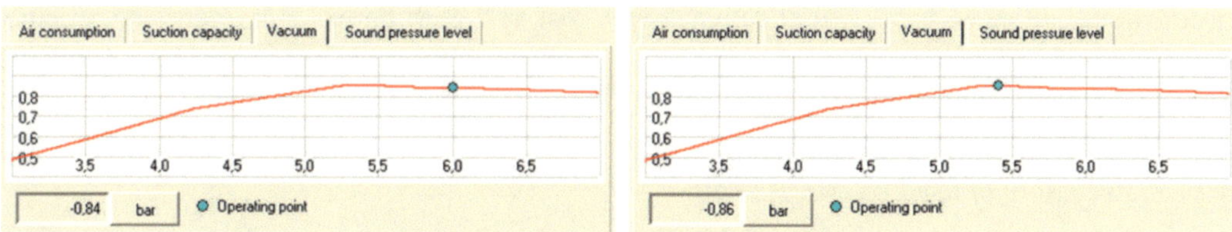

Figure 6. Vacuum level at operating pressure of 6 bar (left) and 5.4 bar (right)

3.13. Optimising the control system at the point of use

In traditional design of pneumatic control system there were no concerns about energy efficiency. Several approaches are developed for energy efficient control of pneumatic systems. Here we will stress only two: Optimising servo-pneumatic systems using PWM and Recycling of used compressed air.

3.13.1. Optimising servo-pneumatic systems using PWM

If servo control is required by means of pneumatic actuator, it is necessary to use proportional valve in order to control pressure in cylinder chambers. Regardless of the type, the proportional valve is the most expensive component of pneumatic servo system (Liu and Bobrow, 1988; Lai et al., 1990).

Instead of proportional valves and servo valves, on/off electromagnetic valves (2/2 or 3/2-way) are being investigated in order to develop cheap pneumatic servo systems. On/off electromagnetic valves take either entirely open or entirely closed position according to electric command. A pneumatic actuator with on/off electromagnetic valves can be controlled by Pulse Width Modulation (PWM) (Barth et al., 2003; Shen et all., 2004).

The control of pneumatic actuator by means of PWM enables servo control by on/off electromagnetic valves at significantly lower cost than the cost of the control done by proportional valves. If response rate and positioning accuracy are taken into account, the results obtained by PWM control are approximately the same as the results obtained by proportional control.

In the case of proportional valve based systems, the fluid flow is continuously varied. In the case of PWM-controlled systems, the valve is entirely open or entirely closed while the control is done by time of keeping the valves in final positions. Thus, the valve delivers discrete quantity of fluid mass whose size depends on control signal. If the frequency of valve opening and closing is much higher than boundary frequency of the system, the system responds to mean value of discrete flow which is the case of continuous flow, too. With on/off electromagnetic valves controlled by PWM it is possible to develop a multifunctional pneumatic system having acceptable price and performances (Šitum et al., 2007).

The results obtained in (Čajetinac et al., 2012) show that PWM principle of control gives a good quality of the signal tracking at the considerably lower costs and that PLC with standard program support can be used for its realization. It has been shown that it is possible to achieve control performances which are comparable to those achieved by proportional or servo valve but at rather lower cost and with better energy efficiency. Thus these methods of control may be applied as well as pneumatic servo systems since the described system is relatively new and expanding.

3.13.2. Recycling of used compressed air

Pneumatic actuators are widespread in many branches of industry and a lot of efforts are done in order to increase energy efficiency in their operation. It is a known fact that pressure inside a piston chamber can reach a final value that is equal to supply pressure, only after some time being at the end position of the cylinder piston has been reached. When the direction of movement of the cylinder piston is reversed, all the compressed air contained within a working volume is released into the atmosphere. This represents a significant loss of compressed air that possesses enough potential energy to perform some other kind of work. There is a possibility to reuse this kind of compressed air, see Fig. 7 (Blagojević et al., 2011).

The performance of the system with restoring energy can be improved in comparison with the traditional control system. Compressed air consumption saving of the conventional pneumatic system with restoring energy is from 33.3% to 44.3% or average 38.8%, for 200 to 600 kPa range of pressure supply. Compressed air consumption saving of the pneumatic servo system with restoring energy is from 20.6% to 33.3%, or average 28.6% for 200 to 600 kPa range of pressure supply. If the system is using pressure of nominal operating value that is 600 kPa than the average compressed air consumption saving is 41.9 % for conventional system and 30.6% for servo system. Return of investment periods of the proposed conventional pneumatic system with restoring energy are average 2.45 years. If the actuator of the conventional system is rodless or trough rod cylinders or semi-rotary drive, and if there is no load, which can be used for support of actuator movement, the saving is less than 5%.

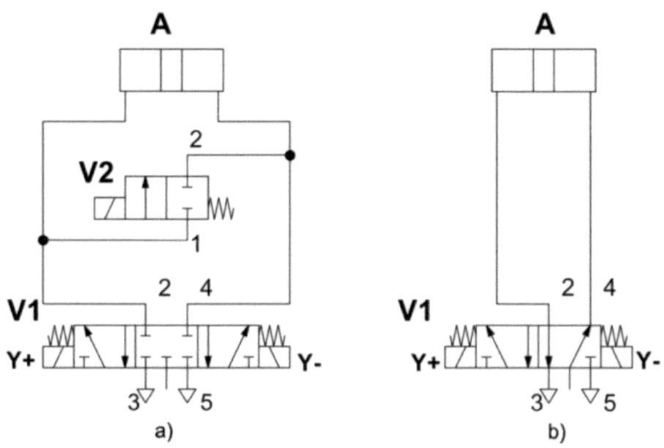

Figure 7. Pneumatic circle a) with restoring energy, b) without restoring energy (Blagojević et al., 2011)

3.14. Measuring and recording the system performance

Measuring and recording the system performance, by itself, does not increase the energy efficiency but usually presents a first step towards improvements in energy efficiency because of two primary reasons:

- Measuring the air consumption and energy used for its production is of essential importance for determination, whether the changes in maintenance practice or equipment investments have justified their costs. As long as the price of a unit of consumed compressed air remains unknown, it is difficult to initiate the managerial processes necessary for improving the system.
- Recording the system performance is a valuable tool for discovering the degradation in performance or changes in quantity or quality of used compressed air.

Three basic parameters – airflow, air pressure and consumption of electrical energy, must be measured and recorded in order to evaluate the system performance. Although this might look simple, in principle, the interpretation of these data can present a difficulty, especially under conditions of variable consumption. Measurement of flow of compressed air also involves certain technical problems and retrofitting reliable measurement instruments can be difficult if not impossible task, unless it was taken care of during the phases of system design and installation. For example, most frequently used type of flow meter must be installed into the pipeline sections that are free of turbulence and whose length must be 10 times greater than its diameter. In some systems, there is not an adequate place for installing of a flow meter. Because of that, it is suggested that large systems and medium size systems must be designed and installed in such a way to enable air flow measurements. If the flow data is not available, the cheaper alternative equipment used for pressure measurement can be very useful to, for example, measure the pressure drop over filters or pressure loss along the pipeline, or for the purposes of detecting larger variations of pressure within the system.

4. Increasing energy efficiency in complex robotic cell

Increase of energy efficiency in compressed air systems in often connected with optimisation of the whole automated system. That is particularly the case with the complex automated systems such as robotic cell and robotic line. In such applications are often integrated, besides robots, pneumatic and hydraulic systems. Minimisation of energy consumption is done in most cases up to now, separately for the each subsystem. There are numerous references dealing with trajectory optimisation of industrial robots, influence of the robot velocity on energy efficiency and design of the optimal components layout within the robotic cell (Šešlija, 1988; Tomović et al., 1990; Yoshimura, 2007; Guilbert et al., 2008; Kamrani et al., 2010; Wenger and Chedmail, 1991; Dissanayake and Gal, 1994; Aspragathos and Foussias, 2002).

As well for the optimisation of operation of hydraulic systems concerning efficiency increase it is of significans to monitor the system operating parameters and to increase the reliability of components (Jocanović et al., 2012).

Having in mind the fact that compressed air is widely used in robot applications, it could be naturally that there are many possibilities for savings and its efficient usage. In that manner we can describe an example of a robotic cell (Fig. 8) optimisation with installed electric and pneumatic devices (Mališa et al., 2011).

The key step is to define the parameters within a robotic cell that affect electricity and/or compressed air consumption, e.g. robot's velocity, device activity, movement trajectories, position of the robot relatively to working area; pressure of compressed air; suction capacity (in the case of vacuum); length of the supply tubes, etc. As various factors are present there are three different ways for optimisation of such complex robotic cell. First approach is the most common case when two independent experts (usually robotic expert and pneumatic expert) optimise the parameters in their domain in parallel. Second approach encompasses firstly the optimisation of the parameters that influence the compressed air consumption, and than parameters that influence the electricity consumption. The third method includes optimisation of the parameters influencing the electricity consumption, and after that, adjusting the parameters related to compressed air.

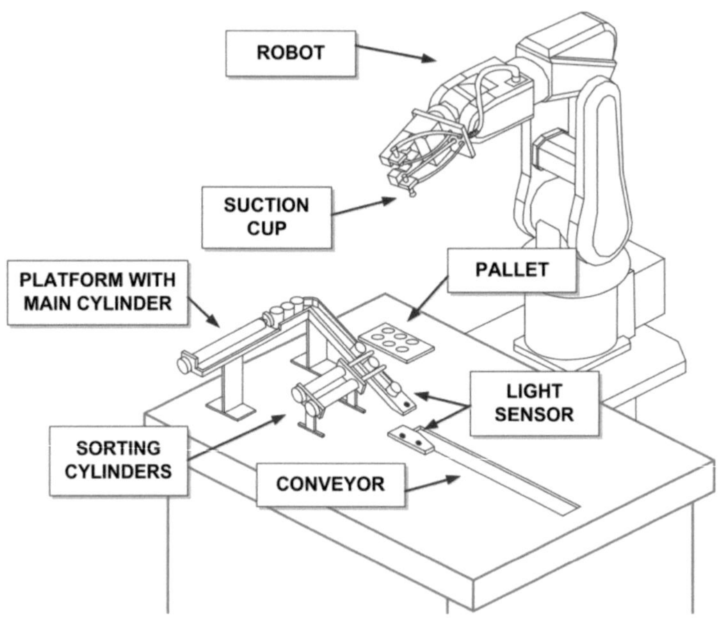

Figure 8. Complex robotic cell

Doing the experiments that would have confirmed one of the mentioned approaches, it was realised that, electric and compressed air parameters could not be observed separately, because most of them influence each other. For instance the adoption of the lowest robot velocity, as the lowest electricity consuming, would disregard the most important principle of a production system: productivity. On the other hand, the highest robot velocity would ensure the highest productivity of the robotic cell, but it would induce the problems with pneumatic devices which would not be able to serve the process, especially if pneumatic devices are set to be the lowest energy consuming. Theoretically, the minimum pressure gives the minimum energy consumption and this could be a pressure of 3 bar, because most of the nowadays devices are able to work on low pressures. The importance of problem could be described with example of vacuum application. For the high robot's velocities, working pressure of 3 bar could not provide a sufficient vacuum level for the amount of time determined by the process.

After the numerous experiments and analysis it is concluded that complex robotic cells that use electricity and compressed air for their operation should be optimised as follows. Firstly, parameters that influence the electricity consumption should be optimised. Secondly, parameters related to compressed air consumption should be adjusted according to constraints given by the robot and working regime. Applying the suggested optimisation method on the complex robotic cell, including the various parameters within it, is possible to significantly reduce the overall energy consumption. For example, in the considered complex robotic cell, with the reduction of the supply pressure from 6 to 4 bar, it is possible to decrease total air consumption by 15%.

5. The potential of applying the measures for energy efficiency increase

Measures for increasing energy efficiency of compressed air systems are related to different phases of the life cycle of a compressed air system:

- System design, gathering of offers or purchasing,
- System installation,
- Significant changes in components or system improvement,
- Preventive and corrective maintenance.

The greatest potential for achieving the savings exists in times of conceiving a new system because at that moment a great spectrum of energy saving measures, described in the table below, is available. This kind of situation is relatively rare, because new factories are not continuously built so even the best opportunity for systematic design becomes rarely available (first column in table 4). Table 4 gives approximate indications of phases in which each of described measures can be applied.

Much frequently encountered is the case of replacing the main components of the existing system. In this kind of situation, it is possible to implement many measures, some of which are faced with greater difficulty especially the ones that are related to system design: compressed air distribution network, systems with multiple pressure levels, selection of the type of end consumer, etc. It is estimated that the possibility for savings in the existing systems, in the time of replacement of main components, amounts to 2/3 of the efficiency increase that could be realized in a new system that would be designed with initially having energy efficiency in mind (Radgen and Blaustein, 2001).

	System design acquisition	Installatio n	Component replacement	Maintenanc e
Improvement of drives	++		++	+
Optimal choice of compressor	++		+	
Sophisticated control system	++		++	
Recuperating waste heat	++		++	
Improvement in air treatment	++		++	++
Overall system design	++		+	
Optimising end use devices	++		+	
Reducing frictional losses	++	+	+	
Reducing air leaks	+	+	+	++
Measuring system performance	++		+	++

Table 4. The applicability of energy saving measures in specified phases of a compressed air system life cycle (Radgen and Blaustein, 2001).

Some measures for energy savings can be implemented into the existing systems in any moment. These are, for example, implementation of sophisticated control systems or waste heat regeneration. The procedures related to maintenance and system operation, especially the frequency of filter replacement, represents one of the main sources for energy savings. These measures can also be implemented at any time during a life cycle of compressed air system components. Table 5 gives the estimate of applicability of these measures based on opinion of the larger number of experts (USDOE, 2001). Only the estimates of the average savings in relation to most significant measures for increasing energy efficiency are given, given that the return of investment time is less than 3 years.

Energy saving measures	Applicability %		Gain %	Potential contribution
	From – to	Average		
Reduction of overall system requirements	20 - 40	30	20	6.0
Match compressor and load	5 – 15	10	3	0.3
Improvements of compressor control	15 – 40	25	10	2.5
Improvement of compressor components	5 - 20	15	5	0.8
Operation and maintenance	50 – 85	75	10	7.5
Total savings				17.1

Table 5. The experts estimate of the average energy savings for compressed air systems in relation to most significant measures (USDOE, 2001).

6. Conclusion

Compressed air systems represent a significant segment of production and service systems. Therefore, it is necessary to pay attention to their energy efficient operation. The application of the measures for an energy efficiency increase in compressed air systems enables prolongation of the component's life cycle and the reduction of total operation costs that in turn increases the economic quality of working process. The procedure that was presented and explained in detail, containing cost effective activities illustrated with appropriate examples, can significantly increase the energy efficiency of compressed air systems.

Author details

Dragan Šešlija, Ivana Ignjatović and Slobodan Dudić
Faculty of Technical Sciences, University of Novi Sad, Republic of Serbia

7. References

Aspragathos, NA. & Foussias, S. (2002). Optimal Location of a Robot Path when Considering Velocity Performance. *Robotica*, Vol.20, No.2, pp.139-147, ISSN 0263-5747

Barth, J., Zhang, J. & Goldfarb, M. (2003). Control Design for Relative Stability in a PWM-Controlled Pneumatic System. *ASME Journal of Dynamic Systems, Measurement, and Control*, Vol.125, No.3, pp. 504-508, ISSN: 0022-0434

Blagojević, V., Šešlija, D. & Stojiljković, M. (2011). Cost Effectiveness of Restoring Energy in Execution Part of Pneumatic System, *Journal of Scientific & Industrial Research*, Vol.70, No.2, , pp. 170-176, ISSN (Online) 0975-1084

Čajetinac, S., Šešlija, D., Aleksandrov, S. & Todorović, M. (2012). PLC Controller used for PWM Control and for Identification of Frequency Characteristics of a Pneumatic Actuator, *Elektronika Ir Elektrotechnika*, (Article in press), ISSN 1392 – 1215

De Keulenaer, H., Belmans, R., Blaustein, E., Chapman, D., De Almeida, A., De Wachter, B. & Radgen, P. (2004). *Energy Efficient Motor Driven Systems*. European Copper Institute, Retrieved from <http://www.copperinfo.co.uk/motors/downloads/pub-176-energy-efficient-motor-driven-systems.pdf>

Dissanayake, M. & Gal, J. (1994). Workstation Planning for Redundant Manipulators. *International Journal of Production Research*, Vol.32, No.5, pp. 1105-1018, ISSN 0020-7543

Dudić, S., Ignjatovic, I., Šešlija, D., Blagojević, V. & Stojiljković, M. (2012). Leakage Quantification of Compressed Air Using Ultrasound and Infrared Thermography. *Measurement*, (Article in Press), ISSN 0263-2241

EPS-Electric Power Industry of Serbia (2009a). *Technical annual report*, Available from <http://www.eps.rs/GodisnjiIzvestaji/SRB%202009%20web1.pdf >

EPS-Electric Power Industry of Serbia (2009b). *Prices*

Festo. (2011). *VADM/VADMI product datasheet*, Festo AG&Co.KG, Retrieved from <http://xdki.festo.com/xdki/data/doc_ENUS/PDF/US/VADM_ENUS.PDF>

Lau, L. (2006). *Energy Saving, P.BE-FESS-01-EN*. Festo AG&Co.KG

Gloor, R. (2000). *Energy Savings in Compressed Air Systems in Switzerland,* Electricity Research Program (project number 33 564), Swiss Federal Office of Energy

Golubović, Z., Šešlija, D., Milovanović, B., Majstorović, B. & Vidovic, M. (2007). The Challenges in Sterile Pressurised Air Preparation, *Proceedings of PAMM – Conference*, pp. C152-153/2007, Balatonalmadi, Hungary, 31 May-3 June 2007

Guilbert, M., Joly, L. & Wieber, PB. (2008). Optimisation of Complex Robot Applications under Real Physical Limitations. *The International Journal of Robotics Research*, Vol.27, No.5, pp. 629-644. ISSN 0278-3649

Henning, D. (2005). *Electricity - What and How Much in Swedish Industry*. A Project of Energy Authority's Research Program, Swedish Energy Agency

Ignjatović, I., Šešlija, D., Tarjan, L. & Dudić, S. (2012). Wireless Sensor System for Monitoring of Compressed Air Filters, *Journal of Scientific & Industrial Research*, Vol. 71, No.5, pp. 334-340, ISSN (Online): 0975-1084

Ignjatović, I., Šešlija, D. & Dudić, S. (2011). Increasing energy efficiency of compressed air usage for sustainable production of food and beverage. *Acta Technica Corviniensis*, Faculty of Engineering Hunedoara, Vol.4, No.2, pp. 61-65, ISSN: 2067-3809

Jocanović, M., Šević, D., Karanović, V., Beker, I. & Dudić, S. (2012). Increased Efficiency of Hydraulic Systems through Reliability Theory and Monitoring of System Operating Parameters, *Strojniški vestnik - Journal of Mechanical Engineering*, (Article in press), ISSN 0039-2480

Kamrani, B., Berbyuk, V., Wäppling, D., Feng, X. & Andersson, H. (2010). Optimal Usage of Robot Manipulators. in *Robot Manipulators Trends and Development*, Jimenez A, Al Hadithi BM. pp. 1-26, INTECH, Retrieved from <http://www.intechopen.com/books/robot-manipulators-trends-and-development/optimal-usage-of-robot-manipulators>

Kulterer, K. & Weberstorfer, C. (2007). Motor Challenge Programme in Austria–Improving industrial energy efficiency, *Proceedings of eceee 2007. Summer study. Saving energy – just do it!*, pp. 1509-1517, ISBN 978-91-633-0899-4, La Colle sur Loup, France, June 4–9, 2007

Lai, JY., Menq, CH. & Singh, R. (1990). Accurate Position Control of a Pneumatic Actuator, *Journal of Dynamic Systems, Measurement, and Control*, Vol.112., No.4, pp. 734-739, ISSN 0022-0434

Li, L., Yuqi, L., McKane, A. & Taranto, T. (2008). *Energy efficiency improvements in Chinese compressed air systems*, Lawrence Berkeley National Library, University of California eScholarship Repository 2008; Paper LBNL-63415, Available from <http://escholarship.org/uc/item/0v72z2q0>

Liu, S. & Bobrow, JE. (1988). An Analysis of a Pneumatic Servo System and Its Application to a Computer-Controlled Robot, *ASME Journal of Dynamic Systems, Measurement, and Control*, Vol.110, No.3, pp. 228-235, ISSN 0022-0434

Mališa, V., Komenda, T. & Ignjatović, I. (2011). Energy Efficient Automation on an Example of a Demo Robot Cell, *Proceedings of International Scientific Conference on Industrial Systems*, pp. 97-100, ISBN 978-86-7892-341-8, Novi Sad, Serbia, September 14-16, 2011

McKane, A., Scheihing, P., Cockrill, C. & Tutterow, V. (1997). *US Department of Energy's Motor Challenge: Developed with Industry for Industry*, Lawrence Berkeley National Library,

University of California, Industrial energy analysis, Paper LBNL-43887, Available from < http://industrial-energy.lbl.gov/files/industrial-energy/active/0/LBNL-43887.pdf>

Mitrović, Č., Golubović, Z. & Šešlija, D. (2006). Implementation, Significance and Effects of Filtration in Industry, *Research and Design for Industry*, Vol.12, No.4, pp. 13-20, ISSN 1451-4117

Norgren. (2011). *Energy Saving, The Norgren guide to saving energy and cutting costs in compressed air systems*, Norgren Engineering Advantage, USA

Radgen, P. (2004). Compressed Air System Audits and Benchmarking. Results from the German Compressed Air Campaign "Druckluft effizient". *Proceedings of ECEMEI, Third European Congress on the Economics and Management of Energy in Industry*, Rio Tinto, Portugal, April 6-9, 2004

Radgen, P. (2003). The "Efficient Compressed Air" Campaign in Germany - Market Transformation To Activate Cost Reductions And Emissions Savings. *Proceedings ACEEE Summer Study on Energy Efficiency in Industry*, pp. 3-194-205, Rye Brook, New York, July 29 - August 1, 2003

Radgen, P. & Blaustein, E. (2001). *Compressed Air Systems in the European Union, Energy, Emissions, Saving Potentials and Policy Actions*. LOG-X Verlag GmbH, ISBN 3-932298-16-0, Stuttgart, Germany

Shen, X., Zhang, J., Barth, EJ. & Goldfarb, M. (2004). Nonlinear Averaging Applied to the Control of Pulse Width Modulated (PWM) Pneumatic Systems, *Proceeding of the 2004 American Control Conference*, pp. 4444 – 4448, Boston, Massachusetts, USA, June 30-July 2, 2010

Šešlija, D. (2003). The System Approach to the Pneumatic System Energy Efficiency, *Procesna Tehnika*, Vol.19, No.1, pp. 237-240, ISSN 0352-678X

Šešlija, D. (2002a). *Production, Preparation and Distribution of Compressed Air*, IKOS, Novi Sad

Šešlija, D. (2002b). Quality of Compressed Air in Advanced Automated Systems, *Proceedings of International Scientific Conference on Industrial Systems*, pp. 158-163, Vrnjačka Banja, Serbia, October 22-23, 2002

Šešlija, D. (1988). Industrial Robot Programming in Assembly Using GT Approach. *Strojarstvo*, Vol.30, No.5/6, pp. 243-247, ISSN 0562-1887

Šešlija, D., Ignjatović, I., Dudić, S. & Lagod, B. (2011). Potential Energy Savings in Compressed Air Systems in Serbia, *African Journal of Business Management*, Vol.5, No.14, pp. 5637-5645, ISSN 1993-8233

Šešlija, D., Stojiljković, M., Golubović, Z., Blagojević, V. & Dudić, S. (2009). Identification of the Possibilities for Increasing Energy Efficiency in the Compressed Air Systems. *Facta Universitatis, Series Mechanical Engineering*, Vol.7, No.1, 2009 pp. 37-60, ISSN 0354 – 2025

Šešlija, D., Stojiljković, M. & Golubović, Z. (2008). Increase of Energy Efficiency in HIPNEF Systems, *Proceedings of Congress with International Participation HIPNEF*, ISBN 978-86-80587-87-5, pp. 3-15, Vrnjačka Banja, Serbia, October 15-17, 2008

Šitum, Ž.; Žilić, T. & Essert, M. (2007). High Speed Solenoid Valves in Pneumatic Servo Applications, *Proceedings of the Mediterranean Conference on Control & Automation*, pp. T06-001, Athens, Greece, July 27-29, 2007

Tomović, R., Zelenović, D. & Šešlija, D. (1990). The Role of Intelligent Robots in Flexible Assembly. *Computers in Industry*, Vol.15, No.1–2, pp. 131–139, ISSN 0166-3615

USDOE-US Department of Energy (2001). *Assessment of the Market for Compressed Air Efficiency Services,* DOE/GO-102001-1197, Washington DC, USA, Available from <http://www1.eere.energy.gov/manufacturing/tech_deployment/pdfs/newmarket5.pdf>

USEIA-US Energy Information Administration (2006). *Independent Statistics and Analysis. International Energy Annual 2006. International and United States Total Primary Energy Consumption, Energy Intensity,* Washington DC, USA, Available from <http://www.eia.gov/emeu/international/energyconsumption.html>

Wenger, PH. & Chedmail, P. (1991). Ability of a Robot to Travel through Its Free Work Space in an Environment with Obstacles. *The International Journal of Robotics Research,* Vol.10, No.3, pp. 214-227, ISSN 0278-3649

Wissink, E. (2007). *Possibilities for Saving Energy in Part Load,* 's-Hertogenbosch, GEA Grasso B.V., The Netherlands, Available from <http://www.grasso.nl/en-us/News-and-Media/technical-articles-Grasso/Documents/Possibilities%20for%20saving%20energy%20in%20part%20load.pdf>

Yoshimura, M. (2007). System Design Optimisation for Product Manufacturing. *Concurrent Engineering: Research and Applications,* Vol.15, No.4, pp. 329-343, ISSN 1063-293X

Pumped-Storage and Hybrid Energy Solutions Towards the Improvement of Energy Efficiency in Water Systems

H.M. Ramos

Additional information is available at the end of the chapter

http://dx.doi.org/10.5772/50024

1. Introduction

The needs of water consumption, environmental targets and energy savings have become the main concerns of water managers over the last years, becoming more and more important goals for the sustainable development and energy efficiency in water systems (Ramos & Covas, 1999). The needs for water consumption, environmental targets and saving energy have become ones of the world's main concerns over the last years and they will grow to be more and more important in a near future (Refocus, 2006). The objective of these systems is to guarantee the delivery of enough water with good quality to populations. Although, in order to achieve that, energy for pumping is needed, representing the main cost for water companies who operate the systems. The evaluation of the energetic potential in water systems may become a common procedure to achieve energy improvements on these systems. This can be done by taking advantage of the possible environmental and economical benefits from the instalation of a water turbine as a clean energy converter.

The optimization of operations energy consumer or production systems has been investigated for some decades. The interest in this area is not only related to the complexity of the problem but mainly by the environmental, economical and social benefits by adopting this type of solution. The implementation of energy production components in water supply systems is a solution that intends to increase the energy efficiency by using local available renewable resources. With this kind of systems the external energy dependence and their costs can be reduced. The adaptation of water supply systems to produce energy is an advantageous solution because most of the system components already exist (e.g. reservoirs, pipe system, valves) and there is a guaranteed discharge continuous flow along each day. Pump hydro storage systems are used as energy and water storage on systems' networks. These systems consist of two reservoirs, where one is located at a low level and the other at

INTECH
open science | open minds

© 2012 The Author(s). Licensee InTech. This chapter is distributed under the terms of the Creative Commons Attribution License http://creativecommons.org/licenses/by/3.0), which permits unrestricted use, distribution, and reproduction in any medium, provided the original work is properly cited.

a higher elevation, with pump and hydropower stations for energy injection or conversion. During off-peak hours the water is pumped from the lower to the upper reservoir where it is stored. During peak hours the water is released back to the lower reservoir, passing through hydraulic turbines generating electrical power (Bose *et al.*, 2004).

When a wind park is combined with a pumped-hydro system, several advantages can be achieved:

- during low consumption hours, the wind energy that otherwise would be discarded, can be used to pump water to the upper reservoir and discharged whenever there is a need to produce energy;
- when the wind has high variability, these storage systems can be used to regulate the energy delivery;
- the energy stored in the pumped-hydro system can be utilized to generate electricity when wind power is not available;
- when there is a variable tariff applied, it is possible to achieve significant economical benefits by deciding optimal pumping / turbine schedules (Papathanassiou *et al.*, 2003).

The optimization of pump/turbine-operation with energy consumption/production has been investigated over the last decade (Allen, G., McKeogh, F.J., Gallachóir, B., 2006; Anagnostopoulos, J., Papantonis, D., 2006). An optimization problem is a mathematical model in which the main goal is to minimize or maximize a quantity through an objective function constrained by certain restrictions. The optimization models can use several methods which nowadays are becoming more efficient due to the computer technology evolution. In Firmino *et al.* (2006) an optimization model using linear programming was developed to improve the pumping stations' energy costs in Brazil. The study revealed that the energy costs can be reduced by 15%. In Gonçalves et al. (2011) a best economical hybrid solution is applied and the study showed the installation of a micro hydro in a real small water distribution system using water level controls and pump operation optimization by using genetic algorithms shows the improvement of the energy efficiency in 63%. In Castronuovo and Peças Lopes (2004) a model for the daily operation of a wind-hydro plant was developed using linear programming. They concluded that, for the test case presented, the predicted yearly average economic gain of including a pumped-hydro station in a wind farmer, is between 425.3 and 716.9k€.

In this paper an hourly discretized optimization model for the determination of operational planning in a wind pumped-hydro system is presented. Comparisons were made between cases with and without complementary wind energy. The economical profit for each case study is presented.

2. System description

A real case is analysed based on the "Multi-purposes Socorridos system" located in Madeira Island, Portugal. This system was designed to supply water to Funchal, Câmara de Lobos and Santa Quitéria, as well as to regularize the irrigation flows and produce electric energy.

In this system there is a pumping and a hydropower station located at Socorridos. It is a reversible type system which enables pumping (in one station) and power production (in a parallel station) of 40000 m^3 of water per day. Figure 1 depicts a scheme of Socorridos system. Figures 2 to 6 show the all elements of the system under real conditions.

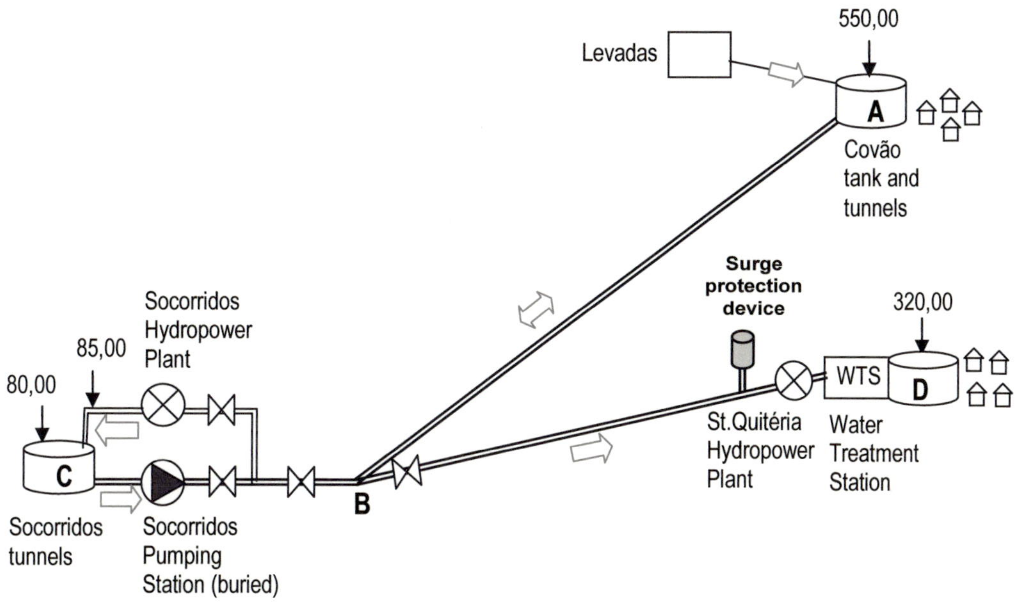

Figure 1. **Multi-purposes scheme of Socorridos system**

The system includes an upper reservoir (Covão) at 540 m, which is used to supply water for the Câmara de Lobos population. In addition to the tunnels, the Covão reservoir is used as storage for the water that flow from the mountains. In Socorridos, there is a tunnel located at 81 m that has the same capacity as the upper one.

Figure 2. Socorridos pumping station – outside view (on the left). Socorridos-St.Quitéria steel pipe (on the centre and on the right)

Figure 3. Socorridos pumping station: centrifugal pumps and control valves

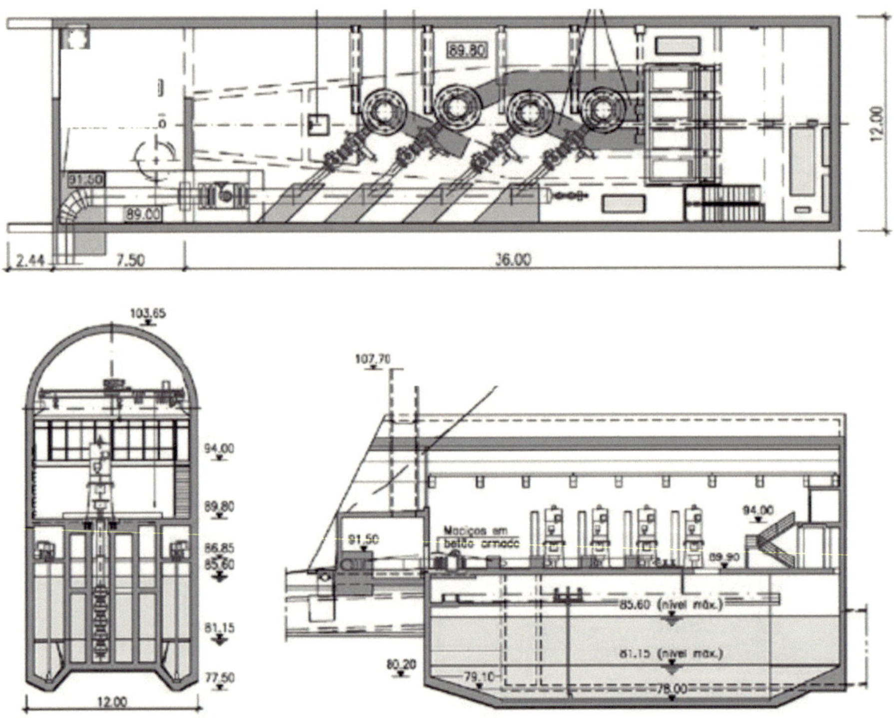

Figure 4. Socorridos pumping station: plant of the ground level (on the top) and transversal and longitudinal views (on the bottom)

St. Quitéria hydropower plant is located at the downstream end of St.Quiteria pipe branch, at immediately upstream a water treatment plant and a storage-tank. This hydropower station has a single Pelton turbine with a nominal flow rate of 1 m3/s and a by-pass to the water treatment plant (Figure 6).

Figure 5. Socorridos storage tank: inlet tunnel (on the left) and centrifugal pump (on the right)

Figure 6. St. Quitéria hydropower station: inside view of Pelton turbine

The pump station is located at level 85 m and has four pumps with an installed power of 3750 kW each. The hydropower station is located at a topographic level of 89 m and has Pelton turbines installed with a nominal power of 8 000 kW and a maximum flow of $2m^3s^{-1}$ each. In Figure 7 the characteristic curves of the Pelton turbines and the pumps are presented.

The penstock between Covão and Socorridos has a total length of 1266.25 m. The characteristics are presented in Table 1.

pipe	L (m)	Φ (mm)	Material
AB	81.25	1000	Steel
BC	132.00	1200	Concrete
CD	303.00	1300	Concrete
DE	440.00	1400	Concrete
EF	310.00	1500	Concrete

Table 1. Covão-Socorridos penstock characteristics.

This pumping station was designed to pump 40000 m³ of water stored in Socorridos reservoir during 6 h, for the electricity low peak hours (from 0 to 6 am). In the remaining hours of the day, the water is discharged from Covão reservoir to Socorridos hydropower station, in reverse flow direction, in order to produce energy. By the end of the day, the total volume of water in the system is in Socorridos reservoir.

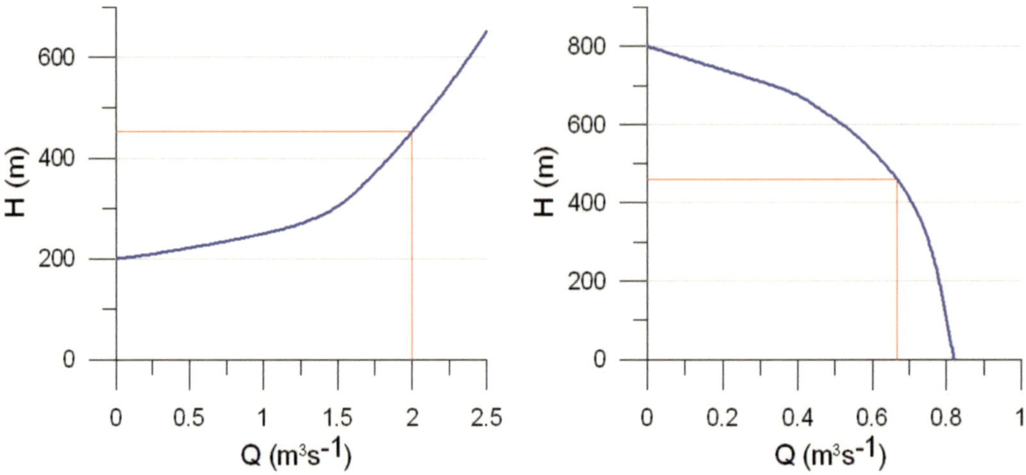

Figure 7. Turbine (Left) and Pump (Right) characteristic curves and operating conditions.

The electricity tariff used in this study, for Socorridos system, is based on the 2006 electricity tariff from Madeira Electricity Company (www.eem.pt), as presented in Figure 8.

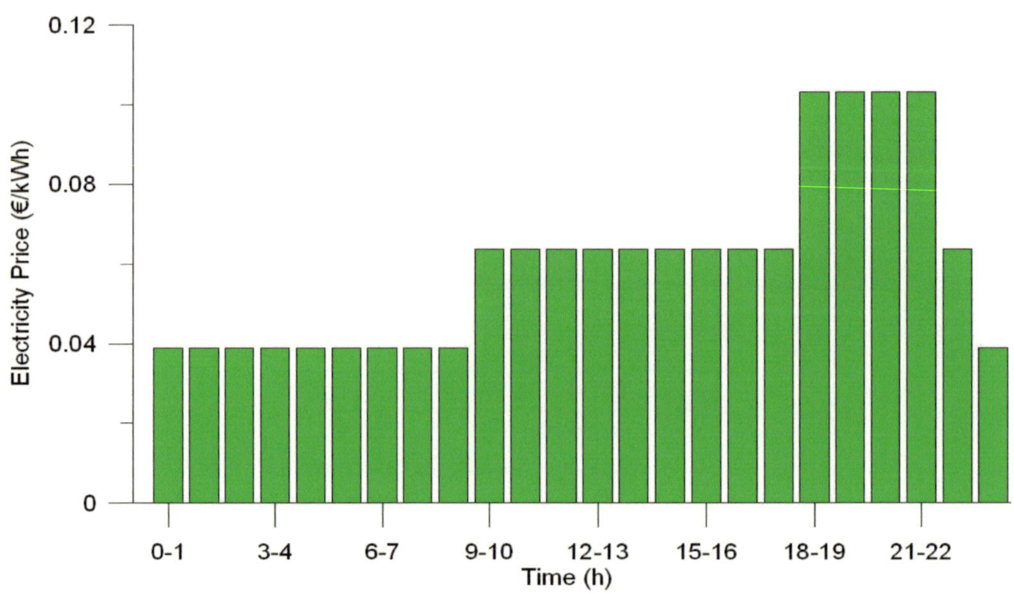

Figure 8. Electricity tariff used in the model (Source: www.eem.pt).

The water consumption in Câmara dos Lobos follows the Manual of Basic Sanitation pattern (DGRN, 1991) and the inflow to Covão reservoir was assumed to have a constant value throughout the day. In Figure 9. the consumption in Câmara dos Lobos and discharge inlet in Covão for one day are presented.

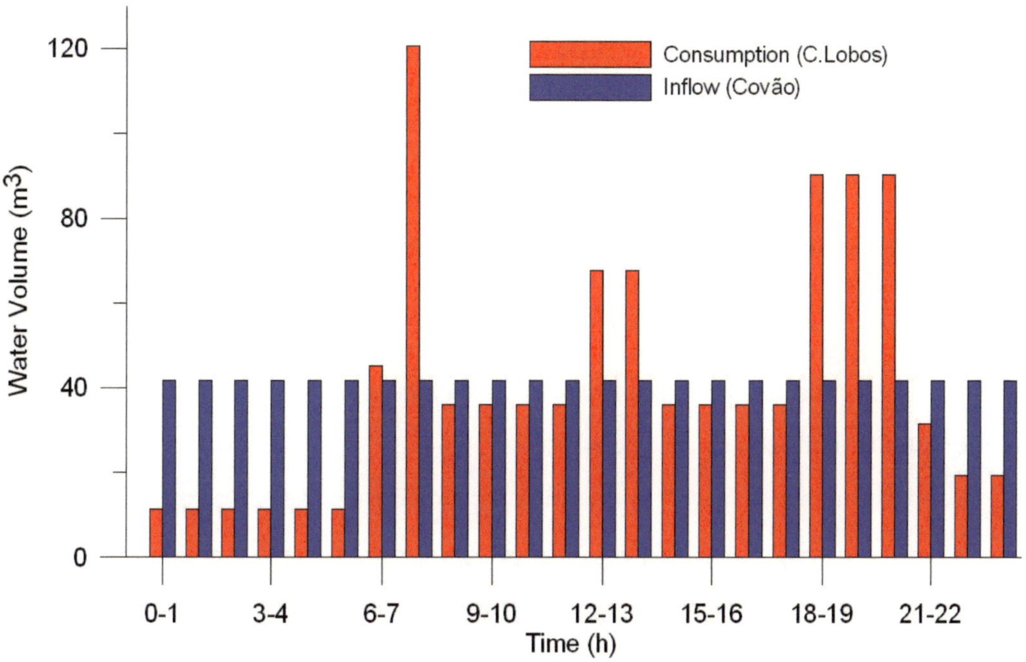

Figure 9. Water volume consumption (Câmara de Lobos) and inlet volume (Covão), in m3, for one day.

For modelling purposes, several assumptions were made to this system:

- the connection to Sta.Quitéria was neglected. This was due to two factors - the energy production is low when compared to the Socorridos hydropower station and the hydropower station does not operate all year;
- the three pumps in the pumping station of Socorridos and the turbines have the same nominal discharge (2 m^3s^{-1});
- it was considered that Socorridos and Covão the reservoirs are cylindrical. Then Socorridos reservoir has a maximum water level of 5 m and a diameter of 101 m. Covão reservoir has a maximum water level of 7 m and a diameter of 85.4 m. The minimum water levels in both reservoirs are 0.5 m.

3. Optimization procedure

Costs associated with the operation of pumping systems represent a significant amount of expenses of a water supply system (Ramos and Covas, 1999). For this reason, it is desirable to optimize the operation of pumps such that all demands are met and, simultaneously, the total pumping cost is minimized.

Typically, pump operation in water supply systems is controlled by water levels at the downstream storage tanks: upper and lower operating water levels are set up in storage tanks such that when these levels are reached, pumps start-up or stop.

The problem consists of defining the hourly operations for the pumps and turbines in the previous presented system, for a period of one day. The goal is to establish a sequence of decisions, for a determined time period, in order to obtain the most economical solution and, at the same time, the better social and environmental solution in terms of guarantee of water supply to populations. The intention is to obtain the pumps and turbines operation time for each hour, so that the maximum benefit from hydropower production and the minimum costs from the pumping station energy consumption are attained.

The problem was solved in terms of water level variation in Covão reservoir. The rules from the optimization model are directly implemented in EPANET model. This procedure is possible since the water level variations are directly proportional to the amount of energy produced or consumed, when the flow and head are considered constants along the simulation time. The time period considered is one day with an hourly time step.

The complexity of the problem is related to:

- the effects of time propagation when an operation is carried, because this will influence the next hours;
- the hydraulic restrictions of the system, like the maximum flow in the pipes, the water level in the reservoirs and the water consumption by the population, have to be completely fulfilled.

An integrated software tool has been developed together with EPANET model (Rossman, 2000) to evaluate the results of the optimization model, in order to verify if the behaviour of the hydraulic components of the system (reservoir levels, flow) is maintained between the desired maximum and minimum limits. To solve the problem several variables are identified, as well as the objective function, which represents the quantity that should be minimized or maximized, the problem constraints associated to the physical capacities of the hydraulic system, and the water demands.

Hence the variables of the optimization process are the following ones:

- Hourly water consumption in Covão (m) - N_{CC}
- Hourly water inlet in Covão (m) - N_{IN}
- Maximum flow in the penstock (m^3s^{-1}) - Q
- Electricity tariff (€) - c
- Initial water level in Covão reservoir (m) - N_{IC}
- Initial water level in Socorridos reservoir (m) - N_{IS}
- Maximum water level in Covão reservoir (m) - N_{MAXC}
- Maximum water level in Socorridos reservoir (m) - N_{MAXS}
- Minimum water level in Covão reservoir (m) - N_{MINC}
- Minimum water level in Socorridos reservoir (m) - N_{MINS}

- Maximum water level rise/decrease for each time step - N_{Qmax}
- Reservoirs diameter (m) - D
- Wind speed curve;
- Wind turbine power curve.

3.1. Pump-storage/hydro

For this situation, two optimization programs were developed: one using linear programming (LP) and the other using a non linear programming (NLP).

For the linear programming case (LP), it is assumed that the water pumping occurs during the first six hours of the day (from 0 to 6 am) and in the remaining hours the system can only produce energy by hydropower. This is to simulate the normal operation mode. The objective function to minimize is the following:

$$f = \sum_{h=1}^{6} \left(\frac{c_{B,h}}{\eta_B} \cdot dN_h \right) + \sum_{h=7}^{24} \left(c_{T,h} \cdot \eta_T \cdot dN_h \right) \tag{1}$$

where c_B represents the electricity tariff for each hour; dN is the water level raise or decrease in Covão reservoir, for each hour; c_T is the produced hydroelectricity selling price for each hour; $\eta_{B,T}$ are the pump and turbine efficiency; and h is the hour of the day.

The above function represents the sum of the water level variation in Covão multiplied by the electricity costs/selling price, throughout one day. Meaning that if there is a raise in Covão reservoir water level ($dN>0$), the pump station is operating and has a cost c_B associated for each hour. If, on the other hand, there is a decrease in Covão reservoir water level ($dN<0$), the system is discharging water from Covão to Socorridos and consequently, producing energy that can be sold at a price c_T. With this function it is possible to provide the electricity costs for the pumping hours and the selling price for the electricity production hours.

The hourly limits relatively to the pipe flow restrictions are presented in Table 2.

Hours	Lower	Upper
0 to 6	0	N_{Qmax}
6 to 24	$-N_{Qmax}$	0

Table 2. Hourly restrictions LP.

From 0 to 6 am it is only possible to pump water since the lower bound is zero. From 6 am forward only turbine operation is allowed.

For the non linear programming case (NLP) there is no need to impose pumping and power hours because the program will choose which solution is the best in order to obtain the major benefits. Then the objective function to minimize is:

$$f = \sum_{h=1}^{24} \left[\frac{c_{B,h}}{\eta_B} \cdot \left(\frac{dN_h + |dN_h|}{2} \right) + c_{Th} \cdot \eta_T \cdot \left(\frac{dN_h - |dN_h|}{2} \right) \right] \tag{2}$$

Hence, if the water level variation in Covão reservoir is positive (pumping situation), the term related to the turbine operation is zero. The opposite is also verified. The hourly limits relatively to the pipe flow restrictions are presented in Table 3.

Hours	Lower	Upper
0 to 24	- N_{Qmax}	N_{Qmax}

Table 3. Hourly restrictions NLP.

This means that, for this case, the only restriction is that the water level variation for each time step cannot be greater than N_{Qmax}.

3.2. Pump-storage/hydro with wind power

For this case the non linear programming (NLP Winter and NLP Summer) was used since the objective function is also non-linear:

$$f = \sum_{h=1}^{24} \left\{ \left[\frac{\left| \frac{Nv_h}{dN_h} - 1 \right| - \left(\frac{Nv_h}{dN_h} - 1 \right)}{2} \right] \cdot \frac{c_{B,h}}{\eta_B} \cdot \left(\frac{dN_h + |dN_h|}{2} \right) + c_{T,h} \cdot \eta_T \cdot \left(\frac{dN_h - |dN_h|}{2} \right) \right\} \tag{3}$$

where N_v is the water level rising to Covão reservoir due to wind power for each hour.

With the former function, the electricity cost for each hour is not the same as the tariff, but it varies according to the contribution of wind available energy. The wind energy is assumed to have a null cost since no energy from the grid is necessary for its generation. Therefore, for each hour, if all of the energy for pumping water is provided by the wind turbines, it has a null cost; if one part of the energy comes from the electrical grid and the other from the wind turbines, the cost is a fraction of the tariff. For example, if one third of the energy for pumping is provided by the wind turbines for one time step, the cost of energy to pump water in this time step is two thirds of the original tariff.

The hourly limits related to the pipe flow restrictions of the system are the same as the previous case (NLP) where the wind component was not considered.

The wind data used for this case is presented in Figure 10. The curves represent typical winter and summer situations for a weather station in Portugal.

The wind turbine chosen was a *Fuhrländer FL2500* (www.friendly-energy.de) with a power curve presented in Figure 11. It was assumed that the system has five of these turbines, so that the maximum power demand by the water pumps in one hour could be provided.

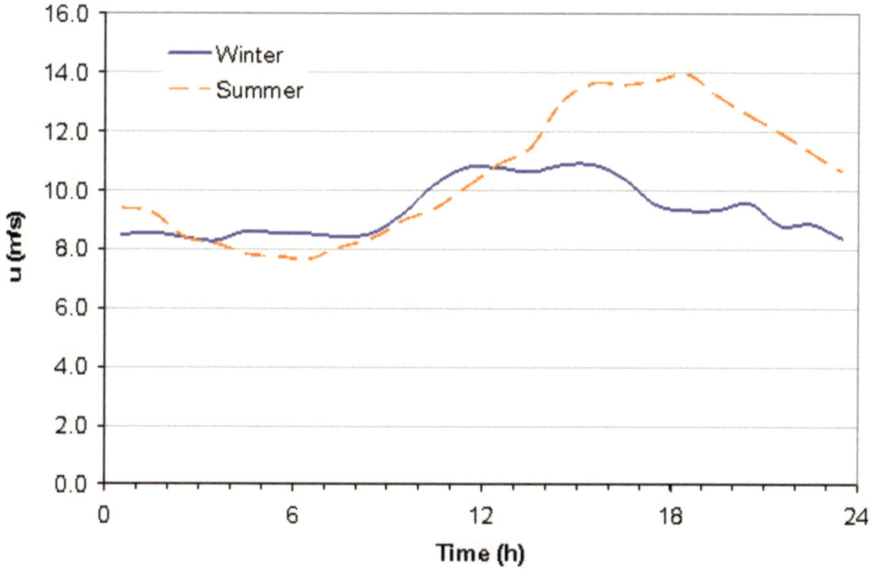

Figure 10. Wind speed average values for typical winter and summer conditions, 100 m above the ground.

The constraints for all of the cases presented, linear and non linear programming, are the following:

1. The guaranty of water supply to the Câmara de Lobos population must be provided, maintaining a minimum water level in both reservoirs;
2. The water in the reservoirs can not exceed the maximum level;
3. The maximum water flow, in each time step, depends on the systems characteristics and the electromechanical equipment.

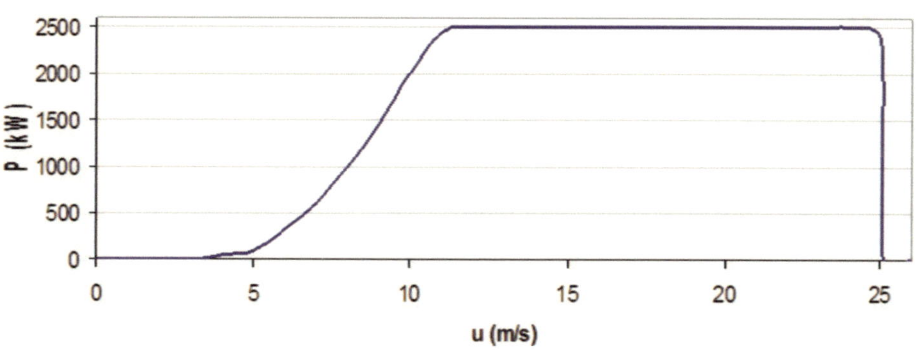

Figure 11. Typical FL2500 power curve.

4. Hydraulic model implementation

An integrated software tool has been developed for determining the optimum pump and turbine schedules and reservoirs water levels, that minimize pumping costs (i.e., maximize off-peak electrical energy consumption) and maximize energy production. This tool incorporates a 'hydraulic simulator' that describes the hydraulic behaviour of the system during 24 hour simulation (EPANET), and an 'optimization solver' based on Linear and Non-linear programming to determine the optimal solution without violating system constraints (e.g., minimum and maximum allowable water levels in the storage reservoirs) and ensuring that downstream demands are satisfied. In Figure 12 the model scheme of Socorridos-Covão system is presented.

As for Covão as for Socorridos the storage reservoirs are tunnels made in rocks. However during modelling implementation these tunnels are approximated by regular cylindrical reservoirs of variable levels with the same volume of the real case. For modelling purposes a discharge control valve, with 2 m³/s as the control parameter, was implemented in order to simulate the hydropower installed at topographic level 89 m. The pump station is installed at 85 m.

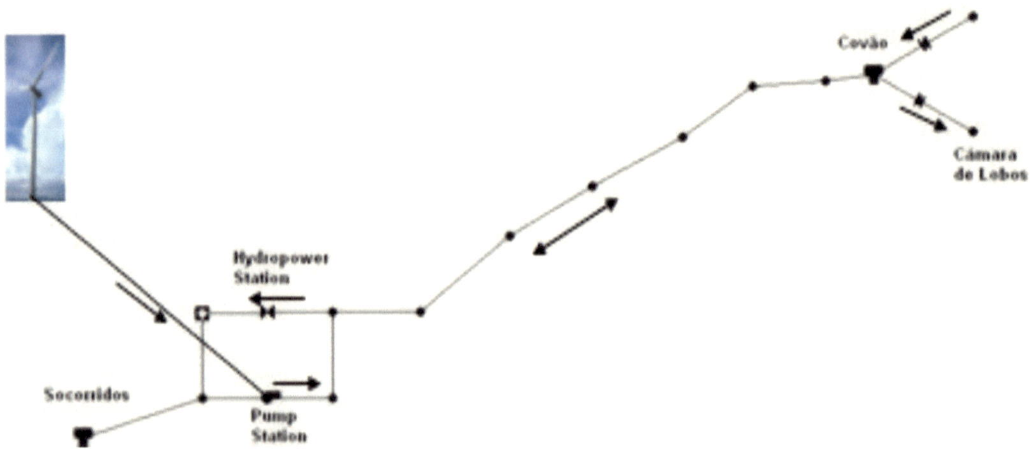

Figure 12. Model scheme of hydraulic system used in the optimization process.

5. Discussion

The four modes of operation (i.e. linear and non-linear programming without wind turbines: LP and NLP; and non-linear programming with wind turbines for summer and winter conditions: NLP Summer and NLP Winter) were compared in terms of water reservoirs levels, pump and turbine operation time and final costs/profits.

In Figure 13, the discharge volumes through the analysis of water levels in Socorridos and Covão reservoirs are presented. It can be seen, that for all modes, the initial and final water levels are imposed to be the same. Then, an adequate comparison in terms of energy production and consumption can be made.

Figure 14 depicts the pumping and turbine operation time for each hour of the day. It can be seen that for the LP mode of operation, the pump station only operates during the first six hours of the day, and in the remaining hours, the hydropower station is working. In the last two hours of the day there is no more water available to power because the Covão reservoir has reached the minimum water level.

For the NLP modes, with and without wind turbines, the behaviour of pump and turbine operation is quite different from each other. The results vary due to the non linearity of the objective function and also according to the wind availability for each hour.

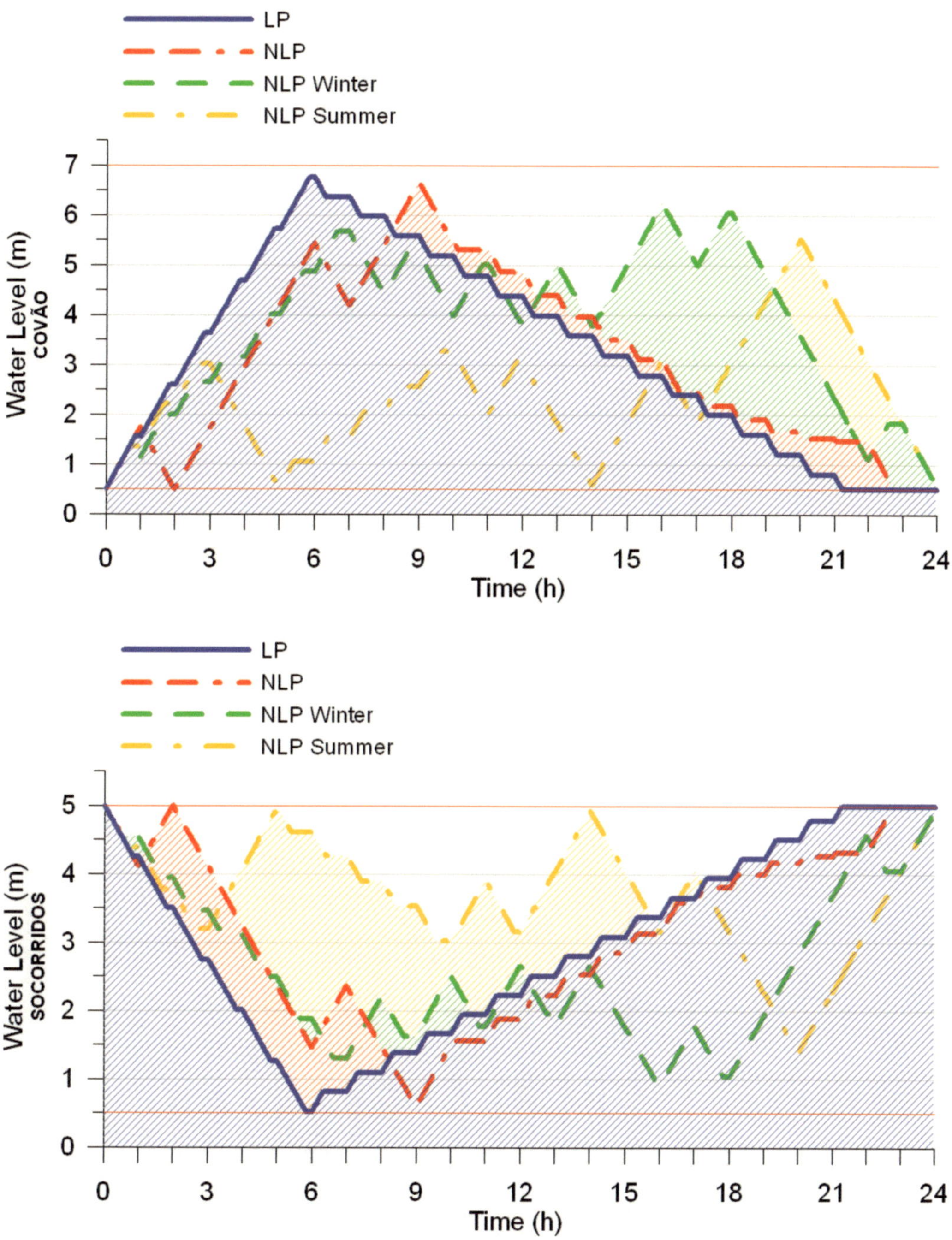

Figure 13. Water level variation in Covão and Socorridos reservoirs.

	E_{Produced} kWh	**E_{Consumed}** kWh	**Cost** €/day	**Benefit** €/day	**Profit** €/day
LP	37247	60107	2342	2754	411
NLP	51110	82747	3224	3734	510
NLP Winter	73344	7012	273	5470	5197
NLP Summer	72933	5009	195	5356	5161

Table 4. Energy produced and consumed, and daily costs, benefits and profits.

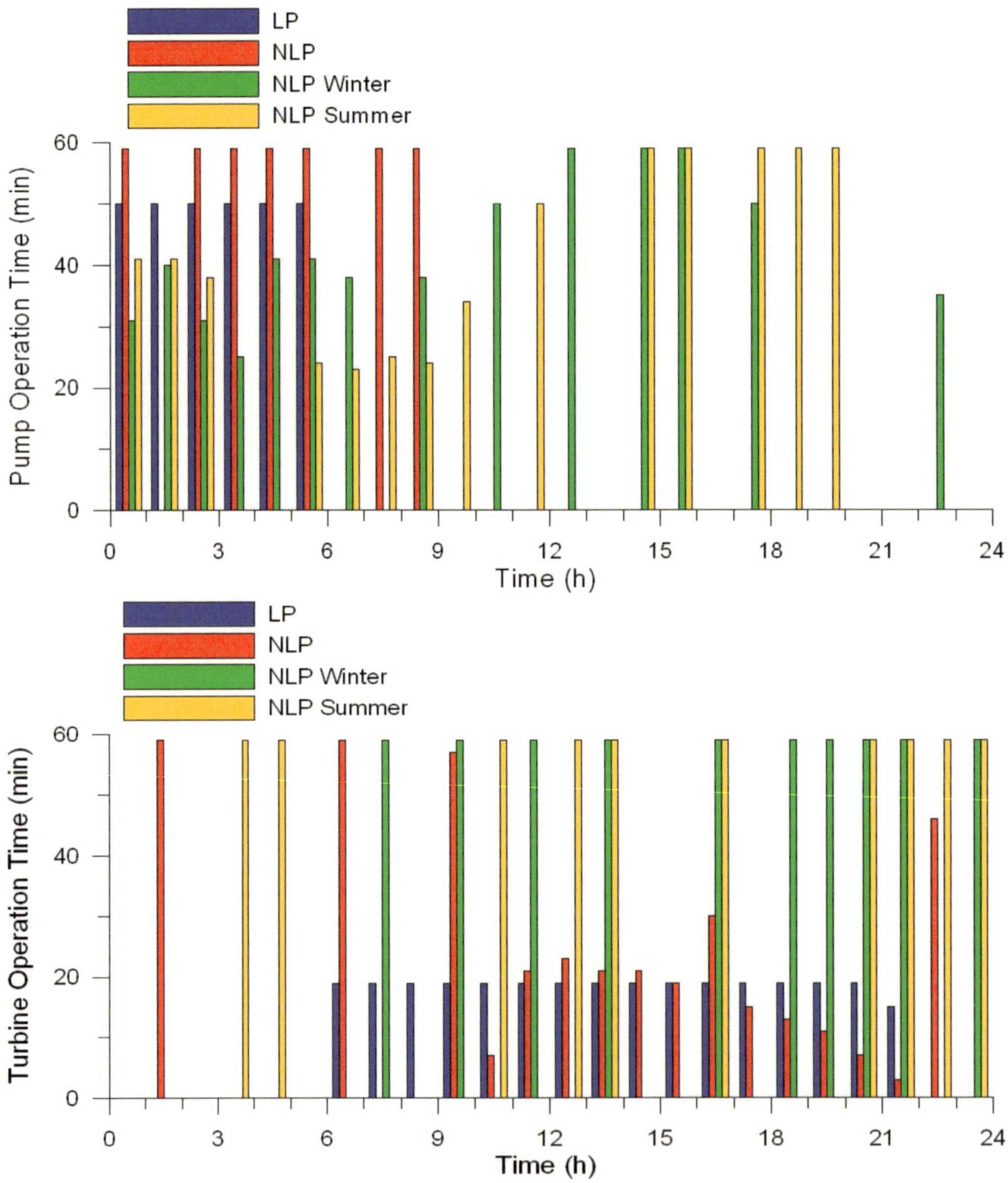

Figure 14. Pump and turbine operation time for the four modes.

In Table 4, a summary of the energy production and consumption, as well as the total costs, benefits and profits are presented.

It can be verified that although, with *NLP* the daily profit is higher than the *LP* case, there is more energy consumption but also more energy production. The profits are approximately 100 €/day higher for the *NLP*.

When considering the wind park as energy supplier for the pump station, the energy consumption from the electrical grid is much lower and the energy production higher. There is not a relevant difference in the profits between summer and winter wind conditions. The costs of installation, operation and maintenance of the wind park are not considered here since it was a component out of the purpose of this work.

6. Conclusions

In the last decades, the managers of water distribution systems have been concerned with the reduction of energy consumption, and the strong influence of climate changes on water patterns. The subsequent increase in oil prices has increased the search for alternatives to generate energy using renewable sources and creating hybrid energy solutions, in particular associated to the water consumption. Renewable energy includes hydro, wind, solar and many others resources. To avoid problems caused by weather and environment uncertainties that hinder the reliability of a continuous production of energy from renewable sources, when only one source production system model is considered, the possibility of integrating various sources, creating hybrid energy solutions, can greatly reduce the intermittences and uncertainties of energy production bringing a new perspective for the future. These hybrid solutions are feasible applications for water distribution systems that need to decrease their costs with the electrical component. These solutions, when installed in water systems, take the advantage of power production based on its own available flow energy, as well as on local available renewable sources, saving on the purchase of energy produced by fossil sources and contributing for the reduction of the greenhouse effect.

An optimization model for determining the best pump and turbine hourly operation for one day was developed. The model was applied to the "Multi-purposes Socorridos" system located in Madeira Island, Portugal, which is a pumped storage system with water consumption and hydropower production.

The model is very flexible in terms of input data: wind speed, water consumption, reservoirs volume, maximum flow and electricity tariff, and the numerical computations take less than a minute. The results can immediately be introduced in EPANET hydraulic simulator in order to verify the system behaviour.

With non linear programming, the results showed that a saving of nearly 100 €/day can be achieved when compared to the normal operation mode, maintaining the hydraulic restrictions and water delivery to the population. When a wind park is added to the system, the profits are much higher, approximately 5200 €/day, for winter and summer wind conditions.

Author details

H.M. Ramos

Department of Civil Engineering, Instituto Superior Técnico (IST) from Technical University of Lisbon, Lisbon, Portugal

Acknowledgement

The authors wish to thank to FCT through the researchers projects PTDC / ECM / 65731 / 2006 and PTDC/ECM/64821/2006 (partly supported by funding under the European Commission – FEDER), as well as the project FP7 HYLOW - 212423.

7. References

Bose, S., Liu, Y., Talya, S., Vyas, P., Videhult, S., Bjerke, M.,Boerresen, B., "A methodology for sizing and cost optimization of wind power with pumped-hydro storage", *Proceedings,* International Conference in Larnaca, Cyprus, "RES and RUE for islands – Sustainable Energy Solutions, 30-31 August 2004.

Castronuovo, E.D.; Lopes, J.A.P., 2004, *On the optimization of the daily operation of a wind-hydro power plant,* IEEE Transactions on Power Systems, Volume 19, Issue 3, Page(s): 1599 – 1606, August 2004.

Direcção-geral dos recursos naturais, "Manual de Saneamento Básico", Volumes 1 e 2 (Ministério do Ambiente e dos Recursos Naturais), 1991.

Empresa Electricidade da Madeira, www.eem.pt, accessed on December 2006.

Firmino, M., Albuquerque, A., Curi, W., Silva, N., "Método de eficiência energética no bombeamento de água via programação linear e inteira", *Proceeedings,* VI SEREA – Seminário Iberoamericano sobre sistemas de abastecimento urbano de água, João Pessoa, Brazil, 5th-7th June 2006.

Gonçalves, F. V., Costa, L. H. e Ramos, H. M. (2011). Best economical hybrid energy solution: Model development and case study of a WDS in Portugal. Energy Policy, v.39, n.6, pp. 3361-3369, 2011.

Furhlander, www.friendly-energy.de, accessed on January 2007.

Papathanassiou, S.A., Tziantzi, M., Papadopoulos, M.P., Tentzerakis, S.T., Vionis, P.S., "Possible benefits from the combined operation of wind parks and pumped storage stations", *Proceedings,* EWEC'03, Madrid, May 2003.

Ramos, H., Covas, D. "The Economical and Environmental Benefit due to Renewable Energy Production in Water Supply Systems" (in Portuguese), *Proceedings,* IV Silusba, 24-26 May, 1999.

Refocus, www.re-focus.net, accessed on November 2006

Rossman, L.A., "EPANET 2.0 User's manual", Ed. Drinking Water Research Division, Risk Reduction Engineering Laboratory Environmental Protection Agency, 2000.

Section 3

Energy Efficiency – Measurement and Analysis

Chapter 9

Energy Measurement Techniques for Energy Efficiency Programs

Luís F. C. Duarte, Elnatan C. Ferreira and José A. Siqueira Dias

Additional information is available at the end of the chapter

http://dx.doi.org/10.5772/47791

1. Introduction

The reduction of energy consumption and elimination of energy waste are among the main goals of all countries that recognize that the world climate is changing due to these problems. Reducing greenhouse gas (GHG) emissions is mandatory for preserving a healthy biosphere and all earth's ecological systems.

The main source of greenhouse gases originated by human activities is the burning of fossil fuels [21]. Studies show how it is possible to generate electricity from renewable energy sources other than burning fossil fuel. Unfortunately, those technologies are not yet widely spread or haven't reached a development stage where the production costs are adequate for the economic scenarios [18, 20, 23].

BP Statistical Review of World Energy 2009 [3] showed that in 2008, the total worldwide energy consumption was 474 exajoules (474×10^{18} J), with 80 to 90 % derived from the combustion of fossil fuels. If this scenario does not change, it can lead us to an unpleasant future with several environmental issues, from global warming caused by fossil fuel combustion and deforestation due to acid rain and destruction of the ozone layer.

In order to change this scenario, utilities and governmental agencies all over the world are implementing energy efficiency programs. These programs are designed to implement solutions which help customers to manage the energy use and save money on energy bills. However, the first step in achieving an energy-efficient houses is the understanding of where the energy is going. Is has been shown that the effectiveness of an energy efficiency program depends strongly on the feedback that the consumers receive about their energy use.

A good knowledge of where the energy is going is fundamental for the customers to decide how it is possible to reduce energy waste and maximize energy bill savings. There are different approaches for implementing energy efficiency programs, and basically two types of actions that can be taken: by changing the behavior and habits of the customers in what concerns the use of home appliances or by making investments in energy-efficient technologies. Many

INTECH
open science | open minds

© 2012 The Author(s). Licensee InTech. This chapter is distributed under the terms of the Creative Commons Attribution License http://creativecommons.org/licenses/by/3.0), which permits unrestricted use, distribution, and reproduction in any medium, provided the original work is properly cited.

times, in order to achieve a significant improvement in energy savings, these two actions must be implemented simultaneously.

Some programs provide direct payments or subsidies (rebates, discounts and loans) to customers who decide to purchase or install a specific energy efficiency home appliance. Other programs address non-financial incentives such as information and technical services. Non-financial incentives/services may be bundled with direct or indirect incentives, or may be offered on a stand-alone basis. Depending on the customer type and market characteristics that a given program targets, an effective energy efficiency program design may include any one of these incentive types, or may bundle them together in various ways [17].

It has been shown that the reduction of energy consumption by customers or the adequacy of their consumption behavior is strongly related with the quality of feedback on where (an what for) the energy spent was used. This understanding leads to changes in behavior that can both reduce energy consumption and shift the energy use from peak periods to off-peak periods. The more detailed is the feedback information that the customer receives, the more efficient and substantial are the energy savings.

The results from surveys conducted by the American Council for an Energy-Efficient Economy in [8] are shown in Figure 1, where it can be noticed that Real-Time Plus Feedback (real-time information down to the appliance level) offered to households can provide energy savings of up to 12%.

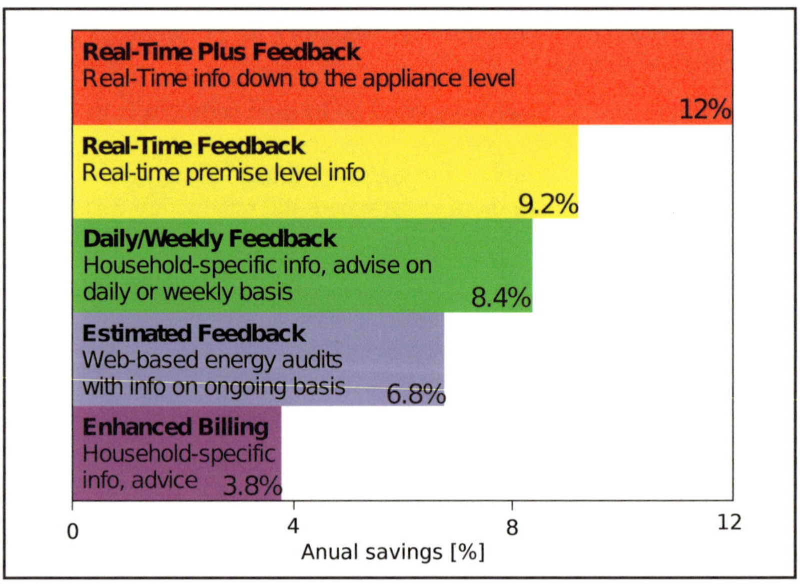

Figure 1. Average household electricity savings, based on 36 studies carried out between 1995-2010

It has been observed that technical and physical improvements in housing are not enough to guarantee reduced energy consumption since consumption in identical homes can easily differ by a factor of two or more depending on the behavior of the inhabitants. Thus, smart metering (or advanced metering) was proposed as a promising way of developing the energy market and contributing to social, environmental and security-of-supply objectives [4] because this technology allows households to have a feedback on the energy consumption of each home appliance, contributing to the knowledge of the relative demand of different appliances and also showing bad habits of the inhabitants.

This chapter will present and discuss several energy measurement methods and technologies of advanced metering initiatives (AMI) for in-house use (devices which are installed after the electric panel) which can be used to distinguish the power consumption of each electrical appliance in a building and provide a detailed breakdown of the energy bill.

2. Electrical energy measurement basics

Power is, by definition, a work done per unit time. Measured in Watts (W), electrical power is commonly acquired indirectly, by measuring the voltage and the current of the circuit. In alternating current (AC) circuits, instantaneous electrical power is calculated by multiplying the instantaneous values of the voltage and current:

$$P_i = V_i \times I_i \tag{1}$$

Electrical energy is obtained by accumulating instantaneous electrical power measurements over a period of time, and is written as:

$$E = \int_0^t V(t) \times I(t)\, dt \tag{2}$$

In energy meters integrated circuits, the signals of voltage and current are discretized in time by an analog to digital converter (ADC), and the electrical energy is then written as:

$$E = \sum_{i=0}^{i=T} V_i \times I_i \tag{3}$$

In the energy meters integrated circuits (ICs) the voltage is is usually obtained using a resistive voltage divider connected directly to the ac line, and the current can be acquired by measuring the voltage drop across a shunt resistor or across a burden resistor connected to a current transformer.

Figure 2 shows the basic schematic diagram of an electrical energy meter IC. A current transformer with a burden resistor is used to provide the current-to-voltage conversion needed by the current channel ADC, and a simple resistive divider network attenuates the line voltage which will be fed into the voltage input channel ADC.

In order to achieve high accuracy, modern electrical energy metering ICs perform the signal processing, such as multiplication and filtering, in the digital domain. This approach provides superior stability and accuracy over time even in extreme environmental conditions. Their operation is based on high precision Sigma-Delta Analog to Digital Converters with resolution between 16 - 24 bits, and signal data processing integrated in hardware level.

To achieve high resolution and reduce noise, the Sigma-Delta Analog to Digital Converters in the IC convert the signals from the current and voltage channels using oversampling. The signals are sampled at a frequency that is many times higher than the bandwidth of interest and this spreads the quantization noise over a wider bandwidth. With the noise spread over a wider bandwidth, the quantization noise within the band of interest is lowered [2].

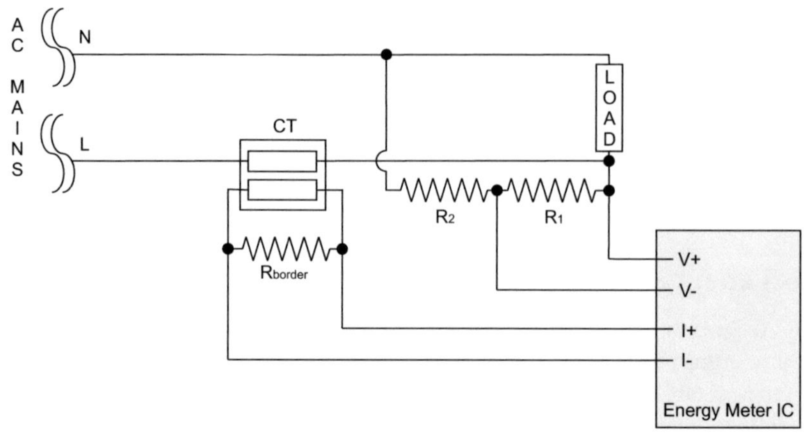

Figure 2. Energy meter IC connected to a resistive voltage divider and current transformer with burden resistor.

This analog input structure greatly simplifies sensor interfacing by providing a wide dynamic range for direct connection to the sensor and also simplifies the antialiasing filter design. A high-pass filter in the current channel removes any dc component from the current signal, eliminating inaccuracies in the real power calculation which may appear due to offsets in the voltage or current signals [1].

The real power calculation is derived from the instantaneous power signal, which is generated by a direct multiplication of the current and voltage signals. To extract the real power component (that is, the DC component), the instantaneous power signal is low-pass filtered. Figure 3 presents a graph of the instantaneous real power signal and shows how the real power information is extracted by low-pass filtering of the instantaneous power signal. This scheme calculates real power for sinusoidal current and voltage waveforms at all power factors.

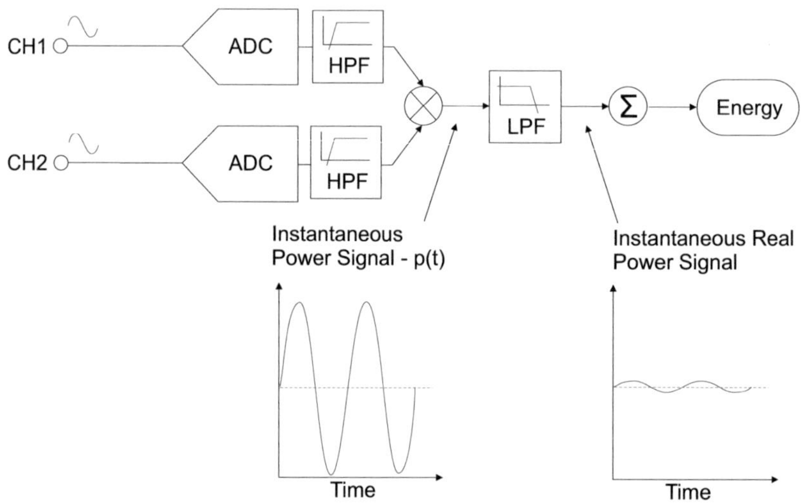

Figure 3. Signal Processing Block Diagram

Practically every current waveform in real appliances have some harmonic content. Using the Fourier transform, the instantaneous voltage can be expressed in terms of their harmonic content as:

$$v(t) = V_0 + \sqrt{2} \times \sum_{h \neq 0}^{\infty} V_h \times \sin(h\omega t + \alpha_h) \tag{4}$$

where:
$v(t)$ is the instantaneous voltage.
V_0 is the average value.
V_h is the RMS value of Voltage Harmonic h.
α_h is the phase angle of the voltage harmonic.

and the instantaneous current is given by:

$$i(t) = I_0 + \sqrt{2} \times \sum_{h \neq 0}^{\infty} I_h \times \sin(h\omega t + \beta_h) \tag{5}$$

where:
$i(t)$ is the instantaneous current.
I_0 is the dc component.
I_h is the RMS value of Current Harmonic h.
β_h is the phase angle of the current harmonic.

Thus, using Equations 1, 4 and 5, the real power P can be expressed in terms of its fundamental real power (P_1) and the harmonic real power (P_h) as:

$$P = P_1 + P_h \tag{6}$$

where P_1 is given by:

$$\begin{aligned} P_1 &= V_1 \times I_1 \cos(\phi_1) \\ \phi_1 &= \alpha_1 - \beta_1 \end{aligned} \tag{7}$$

and P_h is:

$$\begin{aligned} P_h &= \sum_{h \neq 1}^{\infty} V_h \times I_h \cos(\phi_h) \\ \phi_h &= \alpha_h - \beta_h \end{aligned} \tag{8}$$

Some energy meter ICs stores in registers the information processed in the digital domain. According to the energy meter IC, all that information (or part of it) can be retrieved via serial communication (like SPI, I^2C or RS-232). Access to that information is a key element to the development of centralized systems that can identify different appliances on a household installation, as it will be shown in the further sections.

Another technique to implement precision low-cost electrical energy meters is to perform all the signal processing at software level. This can be done by using microcontrollers ICs with integrated ADCs. If a hardware multiplier is integrated in the microcontroller IC, the multiplications that are constantly executed during the energy metering process can be executed very quickly, and measuring energy with these general purpose microcontroller ICs can be very competitive when compared to the energy meter ICs.

This technique is similar to the one implemented in hardware level, using the same acquisition methods and digital processing principles. However, using this approach, the system can be customized to operate in ultra-low power, as in the technique developed in [15] to design a battery powered energy meter using the microcontroler MSP430AFE2xx [16]. The system was configured to operate in a 60 Hz AC line and during every period of one second, make the measurements only during 3 cycles and stay sleeping for the others 57 cycles. The system calculates the RMS value of the product $V \times I$ and adds this value a register. Assuming that there is no significant change in the current and voltage during each period of less then one second, the accumulated value in the register is equal to the energy.

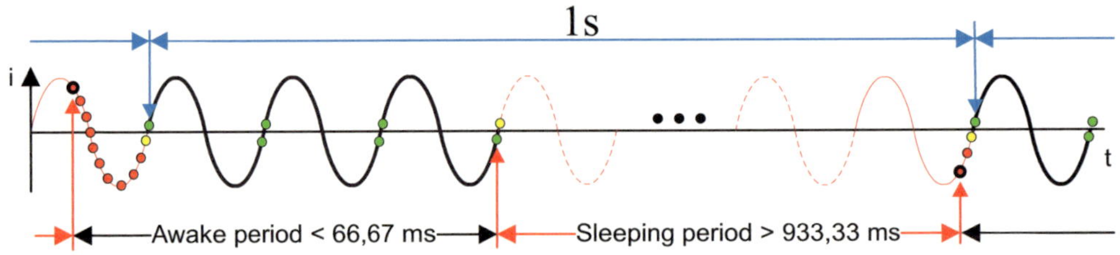

Figure 4. Awake and sleeping periods

3. Decentralized measurement methods

Establishing mesh networks using wireless and Power Line Communications modems (PLC) concentrators/routers together with smart metering sensors which are connected to the home appliances can provide both accumulated and real-time information of the energy spent in every appliance in a house. Adding actuators to these smart sensors, it is possible to manage remotely the status (on-off) of a few important loads (like clothes dryer, air conditioning) in order to avoid that these loads be turned-on during a peak period. In this case the household has to buy the system, since the information has to be continuously monitored and the energy savings will be dependent on this monitoring.

Another possibility, which does not involve the costs associated with the ownership of the measurement system, are offered as a service by the utilities or an ESCO (Energy Service Company). The energy monitoring is performed through the installation of low-cost smart metering sensors to the home appliances, and after a sampling period (typically 2 weeks), the meters are read and a diagnosis report of the energy use is prepared by the company that is offering the service. When there is no need for getting real time information, the installation is extremely easy, since the devices do not need to form a mesh network.

An example of energy monitoring system of the typical appliances in a home was presented by [19]. This proposed monitoring system presents a limited performance because only electrical appliances which were connected to the outlets could be measured, since the monitoring system could not access the lighting equipment, which represents an important part of the energy spent in many households.

4. Experimental results of a pilot project

4.1. Project overview

Recently a pilot project, supported by ANEEL (the Brazilian Electrical Energy Regulating Agency) and AES Eletropaulo (the major power electric utility in the State of São Paulo) was developed by the Department of Electronics and Microelectronics of the School of Electrical and Computer Engineering, at the University of Campinas [7]. The objective was to develop and test a prototype of a hybrid (wireless and PLC) intelligent sensor network system which should be able to perform the breakdown of the electricity bill of the customers. The premise was that with detailed information in hands, consumers could understand better how much they spend the electrical energy in every single electrical device in a house and, this would lead to changes in habits and substantial savings in electrical energy.

The developed system was planned to be offered as a service by the ESCOs, so that both customers who want to know better how they use the electrical energy and energy-efficiency programs were potential candidates for using the service. In order to be practical, the developed system had to be simple to install and easy to be deployed in a residence or small business, without requiring any changes in the original electrical wiring. Although the electrical appliances which are connected to the mains outlet could be read using a small module containing an energy meter IC, it was mandatory that the energy spent on lighting should be monitored, and special sensors had to be designed for this application.

The core of the system was designed around a ZigBee wireless sensor network, and is composed by five types of modules: a coordinator, a displaying-processing unit (DPU) and three types of smart energy meters. A PLC module was also implemented, for special cases where the mesh wireless network is difficult to be implemented.

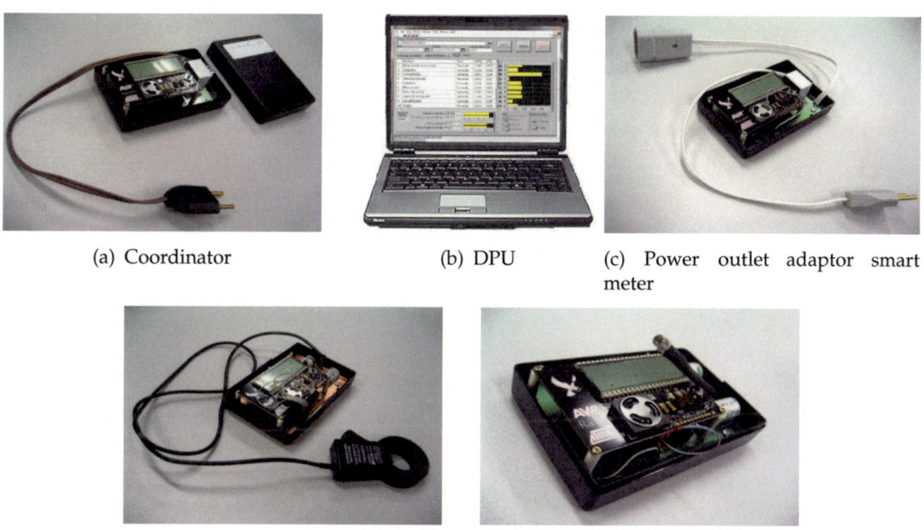

(a) Coordinator

(b) DPU

(c) Power outlet adaptor smart meter

(d) Clamp smart meter

(e) Light smart meter

Figure 5. System modules

The coordinator is the main element, and only one module is required per each mesh network. The coordinator is responsible for storing the information sent by the smart energy meters, and also to sustain the ZigBee mesh network. It also measures and stores the AC voltage to accurately calculate the energy from clamp smart meters. Sometimes, due to the characteristics of the local being monitored or due to the building structure/construction, it is difficult to establish the communications between some end devices and the coordinator. If the devices in the network are too far away from each other, a second coordinator can be added to create and maintain a second ZigBee mesh network. The coordinators will communicate via Power Line Communications (PLC) and a master coordinator will concatenate the information of every smart energy meter.

The hybrid communication technology used (ZigBee + PLC) made the system very robust in terms of coverage. However, the PLC is used only to transfer the information from a coordinator to another and since reading a coordinator is a simple and fast process, two (or more) coordinators can be read with the same DPU, eliminating the PLC module and resulting in a lower cost system that can be easily installed.

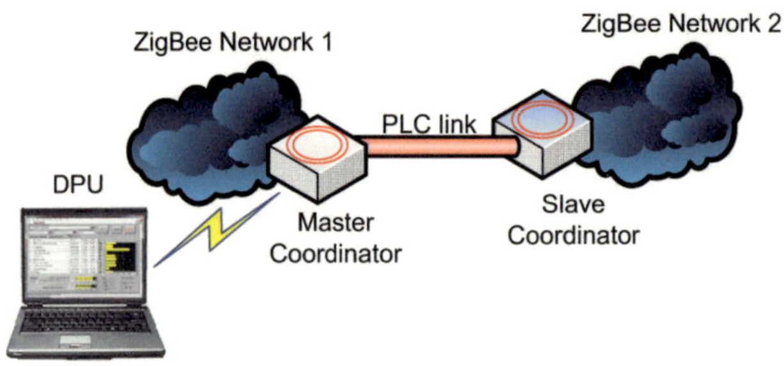

Figure 6. PLC communication connecting two wireless networks.

The DPU is basically a PC that connects to the wireless network to retrieve the information stored in the coordinator. It is also responsible for processing the information according to the configuration file created during the set-up phase(by the person who is installing the system), linking each of the appliances to the IDs of the smart energy meters. After processing, the DPU displays the report in an user-friendly screen, showing the energy spent of items in a user-friendly way, like John's notebook, lighting of dining room, air conditioning of Rose's bedroom, etc.

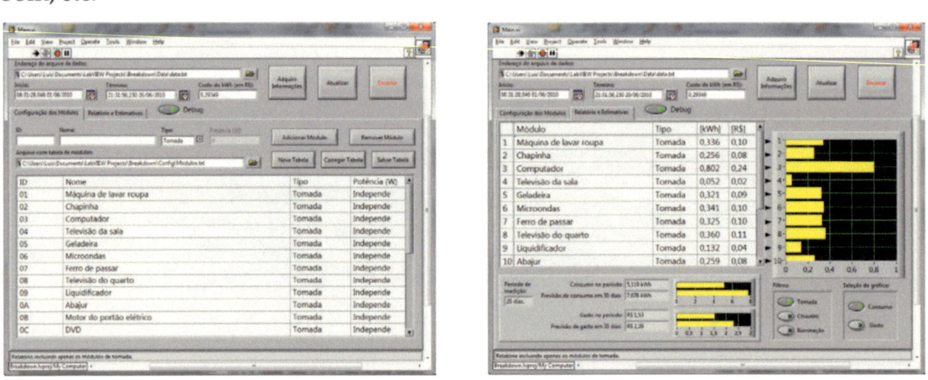

(a) First tab - Configuration file

(b) Second tab - Report

Figure 7. DPU running the reporting application.

The report has a filter that can show only one type of smart meter, two, or all tree of them. A horizontal graphic bar shows the energy consumption of each electrical device. This graphic shows the energy spent (in KWh) or the energy cost (in this software version in R$ - Brazilian Real, but it can easily be converted to any currency) for each appliance.

Three types of smart energy meters were developed to measure and monitor the energy consumption of all electric devices in a house, without requiring any changes in the original electrical wiring installation. The first one, the power outlet adapter smart meter, is employed to measure and monitor the energy consumption of every device that can be plugged to a power outlet. The home appliance power cord is simply plugged to the smart energy meter module and the module is connected to the AC power outlet, as shown in Figure 8.

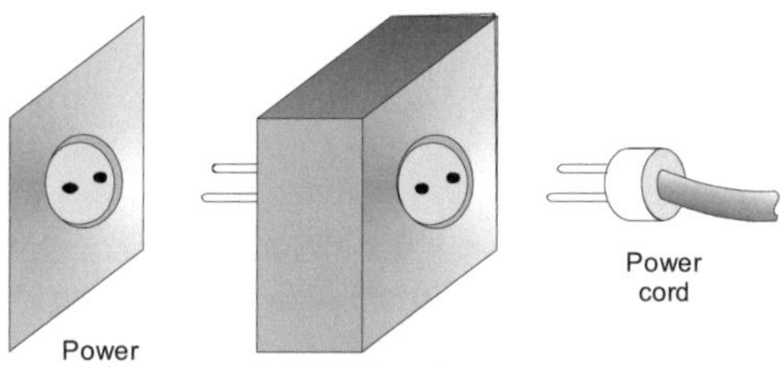

Figure 8. Power outlet adapter smart meter.

The outlet adapter meter measures the amount of energy consumed by the appliance it is monitoring, and send this information to the coordinator every five minutes. Because it is connected to the AC mains, it does not need batteries, even in the presence of a long power supply shutdown, since the data is stored in an internal flash memory. It also performs the function of a router in the wireless network, forwarding the information from the others smart meters to the coordinator, if the smart meters cannot talk to the coordinator directly.

The second smart meter is a clamp smart meter, and it runs on batteries. It was designed to monitor resistive loads that demand currents over 15 A, as it is the case of electrical showers, heaters and other high current loads. The current clamp transforms the ac current in ac voltage proportional to the amplitude of the current, and the circuit of the meter calculates the time integral of the electrical current (in Ah). It is important to remember that only the current is measured and, therefore, the energy has to be calculated. This energy calculation is performed by the network coordinator, that multiplies the data (in Ah) sent by the smart clamp meter by the average value of voltage measured by the coordinator's circuit.

The lighting smart meter developed also runs on batteries. It is employed to measure and monitor the energy consumption of lighting devices which are not accessible by the clamp meter (for example, fluorescent lights in ceiling mounting fixtures). It uses a photodiode to detect the light status (on or off). When the photodiode detects that a light (or a set of lighting fixtures) was turned on, the microcontroller starts to integrate the amount of time that the light is kept on. The system uses one light smart meter for each lighting set that is controlled by one switch. During the system installation, the nominal power of all lights that are monitored by one photodiode is stored in the microcontroller's memory, and the energy consumed by the set of lights is then calculated by the software in the DPU.

The system is designed to monitor a house for a period of one month, but it can also be used to monitor a fraction of that time and estimate the monthly cost of the electricity bill based on the acquisition period, for households with very stable and repetitive habits. A period of less than two weeks is not accurate, because some important energy hungry events may be missed, like when there is no need for using the cloth drier machine because the weather was very good in that week.

4.2. Field test results

The system was tested in the field, and the results showed that most of the consumers waste a lot of energy due to habits that could be easily changed without any impact on the comfort or the quality of life.

Table 1 shows the results measured in a very small apartment, with one person who leaves the apartment early in the morning and returns only at 7:00 PM. It can be seen that this household spends 76% of the electricity with only two appliances: a Computer and a VCR. The PC is a high load in this household, with 64% of the monitored energy consumption. After the report on how the energy is used was presented, the resident reported that the PC is a high end device with powerful video cards and was kept on 24/7 just to eliminate the slow boot time. Concerning the VCR (which is responsible for about 12% of the consumption), it was informed that it is a very old model, that is never turned-off and is used only to change the channels in the TV, because the TV's remote control was not working. In this case, the energy savings could reach a level of almost 80% if the computer is turned off when it is not in use and a new TV remote control is bought.

Appliance	**Energy %**
TV	17.93%
Cell Phone + Charger	1.00%
VCR	12.71%
Refrigerator	3.78%
PC	64.00%
Light bulb 20W (kitchen)	0.58%
TOTAL	100.00%

Table 1. Results measured in a small apartment with 1 people

The results from another household, a medium size apartment with 3 people, is presented in table 2.

This test found that the resident has an old digital alarm clock/radio (which, obviously, was kept turned on 24/7) that presents a huge energy consumption. If this alarm clock was replace with a new and more economic model, it would pay itself back after a few months and energy savings in the order of 12% would be achieved.

The results measured in another interesting case, a large house with three people, are presented in table 3. This household shows a high energy consumption by computers and televisions. The residents reported that the treadmill was kept in stand by during all the acquisition period, and a TV set on a bedroom is always kept on during the whole night. The PCs are also left on continuously to avoid the slow boot time.

Appliance	Energy %
Notebook	0.03%
Hair Dryer	0.53%
Washer/Dryer Machine	0.72%
TV (living room)	11.26%
Refrigerator 1	10.37%
Microwave	1.04%
Alarm clock/radio	12.20%
Lighting 50W (over the piano)	0.03%
Lighting 50W (bathroom)	0.00%
Lighting 50W (kitchen)	22.55%
Lighting 50W (office - over the PC)	7.47%
Lighting 50W (office)	2.37%
Refrigerator 2	15.48%
Cordless telephone	12.23%
TV (bedroom)	3.72%
TOTAL	**100.00%**

Table 2. Results measured in a medium size apartment with 3 people

Appliance	Energy %
PC with amplified speakers	10.83%
Home Theater	0.00%
Electric rice cooker	1.58%
Satellite TV receptor	2.57%
Treadmill	2.24%
PC with wireless dongle	19.40%
PC + Wireless router + notebook	18.48%
Cordless telephone	0.89%
TV LCD 43" (bedroom)	16.71%
Microwave	0.56%
Lighting 120W (TV room)	0.05%
Refrigerator + freezer (side-by-side)	16.36%
TV LCD 53" (TV room)	9.00%
Lighting 160W (kitchen)	1.30%
Lighting 50W (closet)	0.02%
TOTAL	**100.00%**

Table 3. Results measured in a large house with 3 people

From the results of the field tests it was observed that habits and behavior are responsible for a lot of energy waste. From the results measured in these households it becomes clear that a detailed report with the breakdown of the electricity bill can result in significant electrical energy savings just by changing habits.

In the case presented in table 3, the electricity bill was monitored during the next six months after the results were presented. The family reported that, after the report was presented, all three PCs were turned-off during the night and the timer of the bedroom TV was set to turn-off the TV set at midnight.In the case of the The monthly savings in the electricity bills during these six months period were over 20%, when compared to the same month of the previous year. This result is a clear evidence that if a customer is informed of a habit that is wasting energy and this habit can be easily changed, a lot of energy can be saved and and this savings tend to be permanent.

The measurement system was designed to get the information from every appliance in the house in real time, but the customers reported that they were interested only in the two weeks final result. Thus, the ZigBee network could be removed from the smart meters and a simple peer-to-peer communication (from each smart meter to the DPU) could be used, reducing the power consumption of the measurement system and increasing the batteries' life.

5. Centralized measurement methods

With the main objective of minimizing the number of sensors needed to monitor all appliances and also reducing the complexity of the installation service, many researchers are proposing centralized measurement approaches. There are several proposals of different techniques to identify the appliances connected in the same circuit.

In [22] the authors presented the design and implementation of a wireless monitoring system for residential spaces, based on a a multi-layer decision architecture called TinyEARS. This system was able to recognize the appliances by deploying one acoustic sensor node per room, that will identify the appliances' acoustic signatures. Combining this information with the data acquired by a real-time power meter installed at the main electric panel and with relatively simple processing algorithms the system can recognize the appliances with an overall success rate of 94%.

Another technique widely adopted is the identification based on load signatures. Load Signature is an electrical expression that an appliance distinctly possesses regarding its electrical consumption behavior. It can be measured in various forms. From power consumption levels to waveforms of electrical quantities such as voltage and current. Almost every electrical measurement can be treated as a load signature. It can be represented in the frequency domain [6], in the time domain [11] and can also be represented mathematically in terms of wavelets, eigenvalues, or components of the Singular Value Decomposition (SVD) [12].

Methodologies which use signal processing techniques and estimation algorithms for signal load recognition based on load signatures allows the use of a single intelligent device in only one point of the installation (in the electric panel). The detailed energy breakdown of the whole installation is then calculated by sophisticated algorithms.

One of the earliest works (1980's) in nonintrusive monitoring was developed at MIT and had its origins in load monitoring for residential buildings [9]. In the developed technique the operating schedules of individual loads are determined by identifying times at which electrical power measurements change from one steady-state value to another. These steady-state changes, known as events, correspond to the load either being turning on (or turning off), and can be characterized by the magnitude and sign in real and reactive power values. Figure 9 shows the power consumption of a refrigerator and a microwave oven, where two different-sized step changes are clearly present, providing characteristic signatures of the refrigerator and the microwave oven. Knowing the time of each on and off event, it is possible to determine the total energy consumption of the refrigerator and the microwave oven.

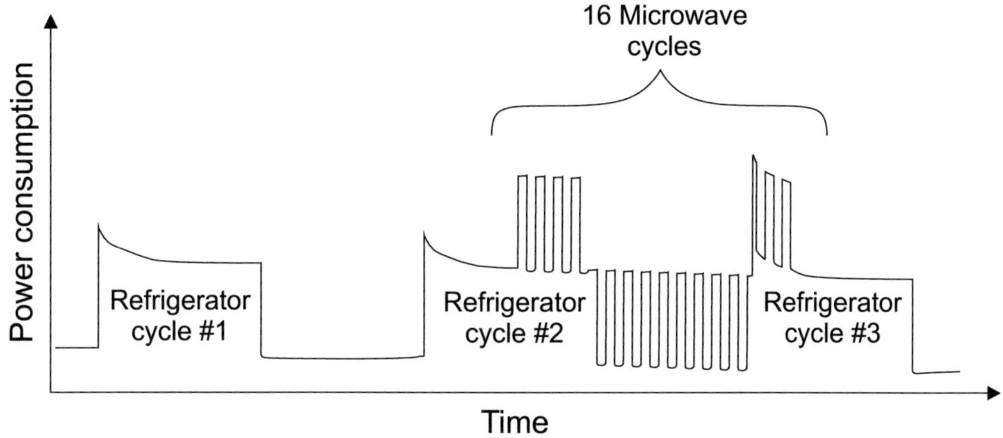

Figure 9. Characteristic signatures of a refrigerator and a microwave oven sensed on the same circuit

In [12] the authors highlighted the fact that the complex electrical loads of today have signatures that vary with time, depending on their state and mode of use and that common appliances can have non-linear load characteristics. They propose a conceptual modeling to characterize an appliance based on three sets of signatures that are extracted from the appliance: steady state, transient state and operational pattern and therefore construct a taxonomy for the appliances.

A method to construct taxonomy of electrical appliances based on load signatures is presented in [10]. In this work the authors suggests a 2-dimensional form of load signature denominated voltage-current (V-I) trajectory to characterize typical household appliances. The V-I trajectory load signatures consists in acquiring the steady-state voltage and current in one-cycle long, normalize them to eliminate the effect of the current magnitude in the size of V-I trajectory, and then plot the V-I trajectory. After creating the trajectories for the appliances, the shapes of the trajectories of the appliances can be analyzed.

The proposed methodology for constructing the load taxonomy is summarized as: (1) the voltage and current waveforms of the household appliances are measured; (2) load signatures in the form of V-I trajectory are constructed; (3) shape features are extracted from the V-I trajectories; (4) hierarchical clustering method is applied to cluster the appliances; (5) the load taxonomy is constructed according to the clustering results.

In [13] the authors proposed a methodology of using load signatures and Genetic Algorithms (GA) to identify electrical appliances from a composite load signal. They introduced a classification method to group the appliances and how to disaggregate the composite load signals by a GA identification process from a generated random combinations of load signatures from the groups of appliances.

The methodology consists of defining a signature for each appliance by averaging 50 consecutive one-cycle steady-state current waveforms. Then the current waveforms are grouped by the ratio of their fundamental (50Hz) component verses their Root-Mean-Square (RMS) total, after a Fast Fourier Transform (FFT) calculation. That means that the higher the

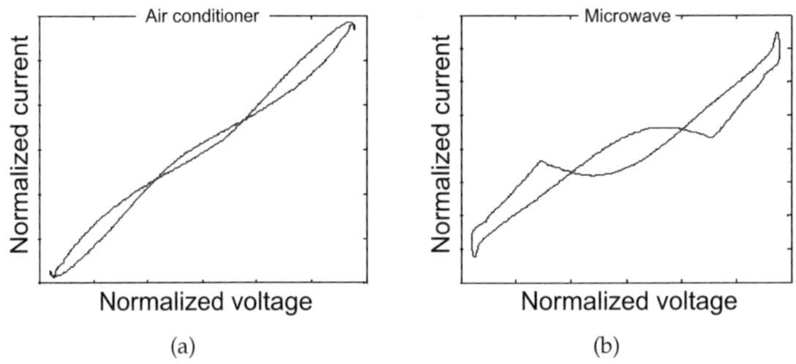

Figure 10. Examples of V-I trajectories of (a) an air conditioner and (b) a microwave oven

ratio, the more sinusoidal shape of the signature is. They considered a sampling rate of 200 points per cycle (50Hz), which is sufficient for steady-state assessment in the time-domain. However, higher sampling rate is desirable if transients will be also analyzed.

The appliances identification can be done by making use of Genetic Algorithms, which are stochastic global search methods based on the principle of the fittest survives. A fitness function, which calculates the least sum squared error between the proposed aggregated signal and the measured signal, is firstly defined. A group of potential solutions are also defined as the initial population. Different variables/attributes, termed the genes, which would affect the fitness function, are allowed to cross-over and mutate to form new generation of potential solutions. In each generation, those matches having the highest fitness value would be retained for further reproduction. The process repeated itself until the best fit is found or the generation limit is reached.

The identification accuracy of the Genetic Algorithm technique is very high for a small number of appliances, but it decreases as the number of aggregated appliances increases. Also, the identification accuracy for sinusoid and quasi-sinusoid waveforms is lower than those of non-sinusoid signatures.

A different approach based on centralized measurement is presented in [14]. The proposed methodology detects the state changes of appliances and acquire energy information simultaneously. The appliance identification is performed using power meters installed in the circuit-level of the electrical panel, and measures the total electrical consumption every 5 seconds, assuming that all appliances and the states of theses appliances in the circuit are known.

In addition, user behavior is taken into account, as the algorithms assume that there are some patterns for using appliances. For example, when using the computer, first the user may switch on the light in the room, then start up the power of the computer and finally turn on the monitor. If the user has a regular lifestyle, the pattern is likely to be regular. Based on that assumption, the temporal character is taken into consideration, therefore, the Dynamic Bayesian Network (DBN) is applicable.

A recent proposal [5] proposes the use of several energy meters ICs, one for each circuit breaker in the electrical panel. This approach aims to minimize the number of appliances sensed in the same circuit breaker. If equal loads, which would present almost perfectly matched load signatures (for example, two air conditioning systems) are connected to different circuit breakers, the identification will be very simple, since it is performed per circuit breaker. The technique under development presents a methodology to disaggregate the power consumption information of each appliance based on synchrophasors as load signatures.

6. Conclusions

A compilation of several state-of-the-art methods that can be used to obtain detailed information about the energy consumed by the appliances in a household has been presented. The decentralized systems, where it is necessary to install one meter per home appliance, are more suitable to be employed in energy efficiency programs where the objective is to present a report of the breakdown of the electricity bill to the family. It has been observed that, based on the reports, many customers implement important (and sometimes simple) changes in habits that lead to effective energy savings.

When it is desired to permanently monitor the household appliances in real time, for example to control the demand in small business installations or to shift the use of energy from peak periods to low-cost tariffs non-peak periods, usually the centralized systems are more efficient. However, these systems are not yet capable of performing a complete identification of all the appliances installed, and further investigations are required in these type of systems.

Advanced metering initiatives are fundamental for providing accurate information and understanding on how and where the energy is being used in a household, and also to evaluate the actual results of implemented energy efficiency programs. Educating people by showing how they can make a better use of the electrical energy seems to be the key element to promote permanent energy savings.

7. Acknowledgements

This research is partially supported by CAPES.

Author details

Luís F. C. Duarte, Elnatan C. Ferreira and José A. Siqueira Dias
Department of Electronics and Microelectronics. School of Electrical and Computer Engineering of University of Campinas. Campinas, SP - Brazil

8. References

[1] *AD71056 Data Sheet - Energy Metering IC with Integrated Oscillator and Reverse Polarity Indication* [2006]. Analog Devices.

[2] *ADE7953 Data Sheet - Single Phase, Multifunction Metering IC with Neutral Current Measurement* [2011]. Analog Devices.

[3] BP [2009]. Statistical review of world energy 2009.

[4] Darby, S. [2006]. *The effectiveness of feedback on energy consumption. A review for DEFRA of the literature on metering, billing and direct displays*, Environmental Change Institute, University of Oxford.

[5] DEMIC/FEEC/UNICAMP [2012]. Monitoring power consumpion of in-house appliances by measuring the distribution board in circuit-level. Pre-proposal for a corporate-sponsored research agreement.

[6] Duan, J., Czarkowski, D. & Zabar, Z. [2004]. Neural network approach for estimation of load composition, *IEEE International Symposium on Circuit and Systems (ISCAS)*, pp. 988–991.

[7] Duarte, L. F. C., Zambianco, J. D., Airoldi, D., Ferreira, E. C. & Dias, J. A. S. [2011]. Characterization and breakdown of the electricity bill using custom smart meters: A tool for energy-efficiency programs, *International Journal of Circuits, Systems and Signal Processing* 5(2): 116–123.

[8] Ehrhardt-Martinez, K., Donnelly, K. A. & Laitner, J. A. [2010]. Advanced metering initiatives and residential feedback programs: A meta-review for household electricity-saving opportunities, *Technical report*, ACEEE.

[9] Hart, G. W. [1992]. Nonintrusive appliance load monitoring, *Proceedings of the IEEE* 80(12): 1870–1891.

[10] Lam, H. Y., Fung, G. S. K. & Lee, W. K. [2007]. A novel method to construct taxonomy of electrical appliances based on load signatures, 53(2): 653–660.

[11] Laughman, C., Lee, K., Cox, R., Shaw, S., Leeb, S., Norford, L. & Armstrong, P. [2003]. Power signature analysis.

[12] Lee, W. K., Fung, G. S. K., Lam, H. Y. & Chan, F. H. Y. [2004]. Exploration on load signatures, *Proceedings of International Conference on Electrical Engineering (ICEE)*.

[13] Leung, J. S. K., Ng, K. S. H. & Cheng, J. W. M. [2007]. Identifying appliances using load signatures and genetic algorithms, *Proceedings of International Conference on Electrical Engineering (ICEE)*.

[14] Lin, G., Lee, S., Hsu, J. Y. & Jih, W. [2010]. Applying power meters for appliance recognition on the electric panel, *Proceedings of the 5th IEEE Conference on Industrial Electronics and Applications (ICIEA)*, pp. 2254–2259.

[15] Morais, F. J. O. [2011]. *Low power electronic equipment with wireless communication to aid in electrical power theft detection*, Master's thesis, University of Campinas.

[16] *MSP430AFE253 Data Sheet - Mixed Signal Microcontroller* [2011]. Texas Instruments.

[17] Prindle, W. [2010]. *Customer incentives for energy efficiency through program offerings*, National Action Plan for Energy Efficiency, ICF International, Inc.

[18] Rosca, M. [2008]. Thermal energy supply strategy for the city of Oradea, Romania, *4th IASME/WSEAS International Conference on ENERGY, ENVIRONMENT, ECOSYSTEMS and SUSTAINABLE DEVELOPMENT (EEESD'08)*.

[19] Serra, H., Correia, J., Gano, A. J., de Campos, A. M. & Teixeira, I. [2005]. Domestic power consumption measurement and automatic home appliance detection, *Procedings of 2005 IEEE International Workshop Intelligent Signal Processing*, pp. 128–132.

[20] Shanming, W. & Schaefer, U. [2007]. Design and optimization of 8MW directly driven surface mounted permanent magnet wind generators, *Proceedings of the Seventh IASTED International Conference on Power and Energy Systems*.

[21] Solomon, S., Qin, D., Manning, M., Chen, Z., Marquis, M., Averyt, K. B., Tignor, M. & Miller, H. L. [2007]. In: *Climate Change 2007: The physical science basis. Contribution of working group I to the fourth assessment report of the intergovernmental panel on climate change.* IPCC, 2007: Summary for Policymakers. Cambridge University Press, Cambridge, United Kingdom and New York, NY, USA.

[22] Taysi, Z. C., Guvensan, M. A. & Melodia, T. [2010]. TinyEARS: Spying on house appliances with audio sensor nodes, *Proceedings of the 2nd ACM Workshop on Embedded Sensing Systems for Energy-Efficiency in Building (BuildSys '10)*.

[23] Vojcinak, P., Vrtek, M., Hajovsky, R. & Koziorek, J. [2010]. Evaluation and monitoring of effectiveness of heat pumps via COP parameter, *Proceedings of the International Conference on Circuits, System ans Signals*.

Chapter 10

Comparing the Dynamic Analysis of Energy Efficiency in China with Other Countries

Chenchen Yang, Feng Yang, Liang Liang and Xiping Xu

Additional information is available at the end of the chapter

http://dx.doi.org/10.5772/48589

1. Introduction

The development of world economy is closely related with energy consumption. According to the economics research by Department of Agriculture of the U.S., since 2000, the energy consumption quantity begins to rise sharply at an average annual growth rate of 2.5% which approximates with the global real GDP (gross domestic product) growth. In China, the second largest economy in the world, this index increases by 15.1% in 2004. Excessive energy consumption, together with the environmental pollution, has become a huge threat to the sustainable development of human beings. Thus, the continuous pursuing for higher energy utilization has drawn the attention of many researchers.

There are three categories of indices to evaluate energy utilization summarized by Ang[1] which are thermodynamic indicators, physical-based indicators, and monetary-based indicators. Different with the first two indicators, outputs in monetary-based indicators are measured in form of currency. This causes monetary-based indicators popularly used in measuring energy efficiency of various levels, not only the common production process at the micro-level but also the comparison between countries at the macro-level.

Ang [1] introduces some key indices belonging to the category of monetary-based indicators. Energy intensity (EI), which is defined as the quotient of total energy consumption divided by total output (GDP or GNP), is used to estimate one's energy efficiency roughly[1]. Energy coefficient is another index referring to the quotient of growth rate of total energy consumption divided by growth rate of total output, which is usually applied in comparison among various countries or regions [2]. However, the stability of energy coefficient is very poor, especially when the growth rate of one country's GDP approaches to 0. Benefit for its clear definition, simple calculation and easily improvement, EI becomes the most frequently-used index in energy efficiency evaluation from both points of practice and research.

INTECH
open science | open minds

© 2012 The Author(s). Licensee InTech. This chapter is distributed under the terms of the Creative Commons Attribution License http://creativecommons.org/licenses/by/3.0), which permits unrestricted use, distribution, and reproduction in any medium, provided the original work is properly cited.

Most of literatures studying energy efficiency adopt energy intensity to analyze energy utilization efficiency, for instance, Howarth et al. [3] and Greening et al. [4], both of which are quoted frequently by other researchers. However, for simplicity, total energy consumption used in EI calculation only considers the sum of all kinds of energy consumption. EI neglects the structure of energy consumption, that's why the index may estimate the energy efficiency inaccurately. Different energy storage capacities and consumption habits make energy consumption structure to be an indispensable influence factor in evaluation. In order to deal with this problem, Xu and Liang [5] introduced a weighted energy intensity model based on data envelopment analysis to evaluate the energy efficiency considering energy consumption structure.

Data envelopment analysis (DEA), a popular approach to evaluate the relative efficiency of homogenous decision making units (DMU) with multiple inputs and multiple outputs[6], has been widely used in the energy efficiency analysis and gained a lot of research achievement [7]. For example, in recent literatures, Mohammadi et al. [8] used DEA approach to evaluate energy efficiency of kiwifruit production in Iran. Rao et al.[9] developed an improved DEA model to analyze energy efficiency and energy savings potential in China. Bian and Yang [10] summarized several DEA models for measuring the energy efficiency and proposed an extended Shannon-DEA method to define a comprehensive concept of energy efficiency.

However, EI index based on DEA concentrates on the transforming degree of energy consumption to GDP or other economic statistical data, and ignores the function of non resource inputs such as labor and capital stock which also play an essential role during the production process. Boyd and Pang [11] introduced the concept of total factor energy efficiency (TFEE) and proposed a model to estimate the linkage between energy efficiency and productivity of the glass industry. References [12] and [13, 14] developed a series of models in estimating total factor energy efficiencies of 29 regions of China and Japan.

Except for using DEA model to analyze the energy efficiency at a given time, this chapter intends to investigate the dynamic change of energy efficiency over periods by adopting Malmquist production index (MPI) technique. First applied to study on the consumers' behavior, after improved for many years, MPI approach deserves high praise in input-output analysis for the reason as follows: (1) no need for the price of input or output; (2) no need for the assumption of behavior pattern; (3) to get more intensive result of dynamic change easily[15]. MPI divides the total production growth rate into two parts, catch-up effect and frontier-shift effect, from which the cause of the change in energy efficiency can be clarified[16].

The current chapter tries to compare the total factor energy efficiencies of 48 countries all over the world in 2003 and analyze the dynamic change in total factor energy efficiencies of provinces of China over the period of 2000-2003 by the proposed model. The rest of this chapter is organized as follows. In Section 2, we introduce several methods for measuring the total factor energy efficiency and the dynamic change based on DEA and MPI technique. Section 3 shows how to use the proposed approach in analyzing the energy efficiency of 48 countries in 2003 and section 4 presents a dynamic example of total factor energy efficiency estimation of 30 provinces in china. Section 5 concludes this chapter.

2. Methodology

2.1. Energy efficiency considering energy structure based on DEA model

Suppose that there are n homogenous decision making units (DMU) to be evaluated, denoted by DMUj (j = 1, 2, …, n). Each DMU consumes m type of energy inputs xij (i = 1, 2, …, m) to produce s types of outputs yrj (r = 1, 2, …, s).

Xu and Liang introduced weighted energy intensity model (WEI) based on DEA to evaluate the energy efficiency considering energy structure. Energy efficiency of DMU_0 is obtained by the following fractional programming:

$$
\begin{aligned}
\max\ & h_0 = \frac{\sum_{r=1}^{s}\mu_r y_{r0}}{\sum_{i=1}^{m} v_i x_{i0}} \\
s.t.\ & \frac{\sum_{r=1}^{s}\mu_r y_{rj}}{\sum_{i=1}^{m} v_i x_{ij}} \le 1, \quad j=1,2,\ldots,n \\
& \mu_r, v_i \ge 0, \quad r=1,2,\ldots,s;\ i=1,2,\ldots,m.
\end{aligned} \tag{1}
$$

In the empirical example, xij stand for all kinds of energy consumption like crude oil, natural gas, coal and so on while yrj are outputs. The vector of vi stands for the weights of the energy consumption xij which represents the energy structure. In addition, the vector of μr is the weight of the output yrj. According to DEA technique, DMU_0 is efficient if there is a parameter bundle (vi, μr) making the target value equal to 1. The production frontier constituted by all of the efficient DMUs suggests an improvement direction to the non-efficient DMUs.

Halkos & Tzeremes have noticed that the scale of countries has influence on the energy efficiency especially when estimating the various countries and regions[17]. Some small countries could be efficient under the condition of variable return-to-scale (VRS) as there is less restrictive[18]. Banker et al.[19] improved an extension based on the variable return-to-scale assumption by adding a convexity constraint.

Here we transform Programming (1) into an integral linear programming and add the VRS assumption. Then we obtain the following program:

$$
\begin{aligned}
\min\ & \theta \\
s.t.\ & \sum_{j=1}^{n}\lambda_j x_{ij} \le \theta x_{i0}, \quad i=1,2,\ldots,m. \\
& \sum_{j=1}^{n}\lambda_j y_{rj} \ge y_{r0}, \quad r=1,2,\ldots,s. \\
& \sum_{j=1}^{n}\lambda_j = 1, \\
& \lambda_j \ge 0, \quad j=1,2,\ldots,n.
\end{aligned} \tag{2}
$$

2.2. Total factor energy efficiency based on DEA model

The concept of total factor energy efficiency investigates deeply into the energy consumption and production procedure and takes the non-resource inputs into account. As some representative examples, capital stock and labor are usually included. Following program is used to evaluate the total factor energy efficiency:

$$
\begin{aligned}
&\min \quad \theta \\
&s.t. \quad \textstyle\sum_{j=1}^{n} \lambda_j x_{ij} \le \theta x_{i0}, \quad i = 1,2,\ldots,m. \\
&\qquad \textstyle\sum_{j=1}^{n} \lambda_j z_{tj} \le \theta z_{t0}, \quad t = 1,2,\ldots,p. \\
&\qquad \textstyle\sum_{j=1}^{n} \lambda_j y_{rj} \ge y_{r0}, \quad r = 1,2,\ldots,s. \\
&\qquad \lambda_j \ge 0, \qquad\qquad j = 1,2,\ldots,n.
\end{aligned}
\tag{3}
$$

Here ztj (t = 1, 2, …, p) stands for the non-resource inputs of DMUj. Adding the VRS assumption turns Model (3) into the following linear programming:

$$
\begin{aligned}
&\min \quad \theta \\
&s.t. \quad \textstyle\sum_{j=1}^{n} \lambda_j x_{ij} \le \theta x_{i0}, \quad i = 1,2,\ldots,m. \\
&\qquad \textstyle\sum_{j=1}^{n} \lambda_j z_{tj} \le \theta z_{t0}, \quad t = 1,2,\ldots,p. \\
&\qquad \textstyle\sum_{j=1}^{n} \lambda_j y_{rj} \ge y_{r0}, \quad r = 1,2,\ldots,s. \\
&\qquad \textstyle\sum_{j=1}^{n} \lambda_j = 1, \\
&\qquad \lambda_j \ge 0, \qquad\qquad j = 1,2,\ldots,n.
\end{aligned}
\tag{4}
$$

2.3. Total factor energy efficiency based on Malmquist production index

The above sections discuss the efficiency evaluation at a given time while this section presents the efficiency evaluating model during a period. Malmquist production index (MPI) is widely applied in measuring the dynamic variation trend of input-output efficiency by dividing the total efficiency into two parts, catch-up effect and frontier-shift effect [20]. Catch-up effect detects whether the efficiency of DMU makes progress during the period. If the numerical value of catch-up effect is more than 1, then we can make sure that the technical efficiency of DMU gets improvement and DMU is closer to the production frontier. Frontier-shift effect is used to assess the technique advancement which is measured by the transform degree of production frontier at different time-points. If the numerical value of frontier-shift effect is more than 1, it means the production technique of the latter is better than that of the former.

We assume that the production possibility set at time t, denoted by St, includes all of the feasible production bundles, input xt and output yt. For each time-point t, we have

$$S^t = \{(x^t, y^t) : x^t \text{ can produce } y^t\}$$

And the input distance function at time t is

$$D_i^t(x^t, y^t) = \sup\{\delta : (x^t / \delta, y^t) \in S^t\} = \frac{1}{\inf\{\delta : (\delta x^t, y^t) \in S^t\}}$$

Following Färe et al. [21] and Boussemart et al. [22], the catch-up effect can be defined as

$$catch-up = \frac{D_i^{t+1}(x^{t+1}, y^{t+1})}{D_i^t(x^t, y^t)}$$

Where $D_i^{t+1}(x^{t+1}, y^{t+1})$ means the efficiency of DMU $\left(x^{t+1}, y^{t+1}\right)$ at time $t+1$ and $D_i^t(x^t, y^t)$ means the efficiency of DMU $\left(x^t, y^t\right)$ at time t.

The frontier-shift effect is defined as

$$frontier-shift = \sqrt{\frac{D_i^t(x^t, y^t)}{D_i^{t+1}(x^t, y^t)} \times \frac{D_i^t(x^{t+1}, y^{t+1})}{D_i^{t+1}(x^{t+1}, y^{t+1})}}$$

Where $D_i^t(x^{t+1}, y^{t+1})$ means the efficiency of DMU $\left(x^{t+1}, y^{t+1}\right)$ at time t and $D_i^{t+1}(x^t, y^t)$ means the efficiency of DMU $\left(x^t, y^t\right)$ at time $t+1$.

The Malmquist production index can be measured as follows:

$$MPI = catch\text{-}up \times frontier\text{-}shift$$

We notice that there need four efficiencies to obtain the MPI and two of which can be obtained by the linear program (3). The other two efficiencies, $D_i^t(x^{t+1}, y^{t+1})$ and $D_i^{t+1}(x^t, y^t)$, can be measured by the following two models.

$$\begin{aligned} &\min\ \theta \\ &s.t.\quad \lambda_j x_j^{t+1} \le \theta x_0^t, \\ &\qquad \lambda_j z_j^{t+1} \le \theta z_0^t, \\ &\qquad \lambda_j y_j^{t+1} \ge y_0^t, \\ &\qquad \textstyle\sum_{j=1}^{n} \lambda_j = 1, \\ &\qquad \lambda_j \ge 0,\ j = 1, 2, \ldots, n. \end{aligned} \tag{5}$$

$$\begin{aligned} &\min\ \theta \\ &s.t.\quad \lambda_j x_j^t \le \theta x_0^{t+1}, \\ &\qquad \lambda_j z_j^t \le \theta z_0^{t+1}, \\ &\qquad \lambda_j y_j^t \ge y_0^{t+1}, \\ &\qquad \sum_{j=1}^{n} \lambda_j = 1, \\ &\qquad \lambda_j \ge 0,\ j = 1,2,\ldots,n. \end{aligned} \tag{6}$$

3. A comparative analysis of energy efficiency of 48 countries

In this chapter, energy efficiency analysis of 48 countries in 2003 is illustrated. The major countries and regions all over the world are included in our consideration such as the United States, China, Russia, Japan and so on. Primary energy consumption includes oil, natural gas, coal, nuclear energy and hydropower. We incorporate oil and natural gas consumption as the first part of energy input. Nuclear power and hydropower are incorporated as the second part of energy input. Coal is the third input. Labor and capital stock are adopted as the non-resource input. Gross Domestic Product (GDP) is the only output.

The data on energy input are collected from World Petroleum Yearbook (2004). GDP and labor are obtained from the World Development Indicators database (2003). Due to the unavailability on the data of capital stock of some countries, we use the index of adjust savings after consumption of fixed capital as a substitute. The data is available from the website of World Bank. All of the data collected are summarized in Table 1.

Category	Indicators	Max	Min	Mean
Energy inputs	Oil & natural gas	1481.1	5.3	105.62
	Nuclear power & hydropower	799.7	0.1	51
	Coal	242.8	0.2	22.27
Non-energy inputs	Labor	129483.9	362.1	10208.44
	Capital stock	13012	8.1	933.06
Output	GDP	109486	99	6828.9

* The units of data on energy inputs are all million tones oil equivalents. Labor is expressed in units of 10-thousand persons. Capital stock is stated in units of 100-million USD. GDP is described in units of 100-million USD.

Table 1. Summary of inputs and output

Table 2 shows the results of energy efficiency considering energy structure measured by model (2). Countries in column 2 are ranked by GDP. The third column represents energy efficiency considering energy structure. Results indicate that: (1) there are 10 efficient DMUs including US, Japan, Italy and so on; (2) 21 countries' energy efficiency scores lie on the interval of 0.5-1 and the typical countries are Britain, Germany, Mexico, etc; (3) the energy efficiency scores of the rest 17 countries are at very low level, less than 0.5; (4) the return-to-scale situation of most developed countries is in decreasing stage while in contrast many developing countries behave increasing returns to scale.

No.	Country	WEI-VRS	Rank	RTS
1	United States	1.0000	1	D
2	Japan	1.0000	1	D
3	Germany	0.9086	12	D
4	Britain	0.9969	11	D
5	France	1.0000	1	D
6	Italy	1.0000	1	D
7	China	0.3391	40	D
8	Canada	0.3471	38	D
9	Mexico	0.7755	18	D
10	Korea	0.3359	41	D
11	India	0.3526	36	D
12	Australia	0.8422	16	D
13	Netherlands	1.0000	1	D
14	Brazil	0.2938	42	D
15	Russian Federation	0.0707	48	D
16	Switzerland	1.0000	1	C
17	Sweden	0.8560	15	I
18	Austria	0.7669	19	D
19	Turkey	0.3469	39	D
20	Norway	0.8601	14	I
21	Poland	0.6479	22	D
22	Indonesia	0.2256	45	D
23	Greece	0.5793	26	D
24	Finland	0.7047	21	I
25	South Africa	0.4902	32	I
26	Ireland	1.0000	1	C
27	Portugal	0.5838	24	I
28	Thailand	0.1894	46	I
29	Iran	1.0000	1	C
30	Argentina	0.4893	33	I
31	Malaysia	0.2928	43	D
32	Czech Republic	0.5073	30	I
33	Hungary	0.5485	27	I
34	Egypt	0.5417	28	I
35	Pakistan	0.2818	44	I
36	Philippines	0.5805	25	I

37	New Zealand	0.7114	20	I
38	Columbia	0.5025	31	I
39	Chile	0.4777	34	I
40	Peru	0.8972	13	D
41	Romania	0.3488	37	I
42	Bangladesh	1.0000	1	C
43	Ukraine	0.1096	47	I
44	Slovakia	0.6364	23	I
45	Kazakhstan	0.5090	29	I
46	Bulgaria	0.8084	17	I
47	Lithuania	1.0000	1	D
48	Uzbekistan	0.3637	35	I

* D, I and C indicate decreasing, increasing and constant return-to-scale respectively.

Table 2. Energy efficiencies of 48 countries considering energy structure

It is particularly pointed out that the energy efficiency of China is only 0.3394 which is the worst among the top 10 countries ranked by GDP. The information of the input/output shown in Table 3 release that there are two reasons for that. First, the technical efficiency of energy consumption of china is lower, compared with Italy for example which has approximate output. Second, by comparison with 10 efficient countries, China has an improper construction of energy consumption that mainly relies on coal resource. Considering the heavy environmental pollution with coal's burning, adjusting the structure of energy consumption is imperative.

No.	Country	GDP	Oil & Gas	Nuclear & hydropower	Coal
1	Japan	43009	317.6	112.2	75
2	Ireland	1537	12.1	1.6	0.2
3	Bangladesh	519	15.2	0.4	0.2
4	Netherlands	5115	79.9	9.2	0.9
5	France	17576	133.6	12.4	114.6
6	Italy	14683	156.6	15.3	10
7	Iran	1371	126.4	0.7	2
8	Switzerland	3201	14.7	0.1	14.5
9	Lithuania	182	5.3	0.1	3.7
10	United States	109486	1481.1	573.9	242.8
11	China	14170	304.7	799.7	73.8

* The units of data on energy inputs are all million tones oil equivalents. GDP is described in units of 100-million USD.

Table 3. Input/output of 10 efficient countries and China

Table 4 represents total factor energy efficiency calculated by model (3) & (4). Countries in column 2 are ranked by GDP. Column 3 and 5 show two kinds of results due to the different setting-ups of return-to-scale. Column 3 indicates the total factor energy efficiency based on constant return-to-scale which can be viewed as pure technical efficiency. Column 5 indicates the total factor energy efficiency based on variable return-to-scale. The last column shows the status of each one's return-to-scale. It is noticed that only five countries are efficient both in CRS and VRS. Quantity of efficient country in column 5 is more than that in column 3. Notice that the return-to-scale effect of top 14 countries in the table is in decreasing stage while most of the last 18 countries are in increasing stage.

It is interesting to analyze the situation of china. It can be observed from table 4 that China is in stage of decreasing return-to-scale effect and TFEE is ranked 30, still lower than all of the developed countries and most of the developing countries. This is mainly caused by lower technical efficiency shown in column 3. Therefore, there are at least 3 ways to enhance the total factor energy efficiency of China, including (1) improving the output of GDP, (2) re-arranging the allocation of energy inputs and non-resource inputs and (3) improving the technical efficiency of production.

No.	Country	TFEE-CRS	Rank	TFEE-VRS	Rank	RTS
1	United States	0.8654	8	1.0000	1	D
2	Japan	0.9177	6	1.0000	1	D
3	Germany	0.8179	10	1.0000	1	D
4	Britain	0.8016	11	1.0000	1	D
5	France	0.6395	17	1.0000	1	D
6	Italy	0.8892	7	1.0000	1	D
7	China	0.3139	32	0.8663	30	D
8	Canada	0.5787	18	0.8417	31	D
9	Mexico	0.7008	14	1.0000	1	D
10	Korea	0.3154	31	0.7827	37	D
11	India	0.3086	33	0.8897	28	D
12	Australia	0.6449	16	0.9020	26	D
13	Netherlands	0.7462	12	1.0000	1	D
14	Brazil	0.2581	37	0.8341	33	D
15	Russian Federation	0.0696	47	1.0000	1	C
16	Switzerland	1.0000	1	1.0000	1	C
17	Sweden	0.8394	9	1.0000	1	C
18	Austria	0.7431	13	0.8304	34	D
19	Turkey	0.3375	27	1.0000	1	C
20	Norway	1.0000	1	1.0000	1	C
21	Poland	0.5168	22	0.8130	35	D
22	Indonesia	0.1895	40	0.8941	27	D
23	Greece	0.5743	19	0.7196	42	D
24	Finland	0.6992	15	0.7779	38	I
25	South Africa	0.4727	25	0.7215	41	C

26	Ireland	1.0000	1	1.0000	1	C
27	Portugal	0.5362	21	0.6063	46	I
28	Thailand	0.1784	42	0.6815	44	D
29	Iran	1.0000	1	1.0000	1	C
30	Argentina	0.4820	23	1.0000	1	C
31	Malaysia	0.3039	34	0.8014	36	I
32	Czech Republic	0.3341	28	0.5315	48	I
33	Hungary	0.3335	29	0.6999	43	I
34	Egypt	0.5630	20	1.0000	1	C
35	Pakistan	0.2114	38	0.8803	29	I
36	Philippines	0.3321	30	1.0000	1	I
37	New Zealand	0.4625	26	0.9426	25	I
38	Columbia	0.2926	35	0.7719	39	I
39	Chile	0.2814	36	0.7570	40	I
40	Peru	0.4730	24	1.0000	1	C
41	Romania	0.1478	44	0.8352	32	I
42	Bangladesh	1.0000	1	1.0000	1	C
43	Ukraine	0.0442	48	0.5624	47	I
44	Slovakia	0.1883	41	0.6789	45	I
45	Kazakhstan	0.1065	45	1.0000	1	I
46	Bulgaria	0.1558	43	1.0000	1	I
47	Lithuania	0.2046	39	1.0000	1	I
48	Uzbekistan	0.0710	46	1.0000	1	I

*D, I and C indicate decreasing, increasing and constant return-to-scale respectively.

Table 4. TFEE of 48 countries

4. A dynamic analysis of energy efficiencies of 30 Chinese provinces during 2000-2003

This section aims to investigate the total factor energy efficiency of main areas in china using the time-series data from 2000 to 2003. These areas shown in table 5 include 12 provinces in the east area, 10 provinces in the central area and 8 provinces in the west area. Consisting of fast-developing regions like Beijing, Shanghai, Guangdong etc., the east area owns GDP output around half of the country. The central area contains inland provinces such as Shanxi, Jilin, Henan etc. This area has a large population and tremendous potential. Compared with the other areas, the west area is the least developed region in China, containing provinces of Sichuan, Guizhou, Yunnan etc. In our study, Tibet, which is also a province in the west area, is missing due to the unavailability of data. Similar as the analysis in the above section, GDP is the only output and non-resource inputs are capital stock and labor while energy inputs are represented as crud oil, coal and electric power. The data on energy input are collected from China Energy Statistical Year Book (2004). GDP and labor data are collected from the Statistical Year Book of China published by National Bureau of Statistics during 2000-2003. The data on capital stock comes from Jun et al. [23].

Areas	Num.	Provinces
East area	12	Beijing, Tianjin, Hebei, Liaoning, Shanghai, Jiangsu, Zhejiang, Fujian, Shandong, Guangdong, Guangxi, Hainan
Central area	10	Shanxi, Inner Mongolia, Jilin, Heilongjiang, Anhui, Jiangxi, Henan, Hubei, Hunan, Chongqing
West area	8	Sichuan, Guizhou, Yunnan, Shaanxi, Gansu, Qinghai, Ningxia, Xinjiang

Table 5. Chinese provinces in different areas

Curves in Figure 1 show the difference among the average TFEE scores of the provinces in the east, central and west areas using model (4). Obviously the east area is the most efficient and the west area is worst in any year. Meanwhile, it is shown that energy efficiencies for all areas are gradually improving. The detailed results are listed in Table 6. It can be easily observed from the table that most of efficient provinces are in the east area. TFEE scores of Liaoning, Shanghai, Jiangsu, Guangdong, Guangxi, Hainan, Fujian are all at a high level. Provinces in the central area are not as good as the provinces in the east area except Anhui which is adjacent to the east area. Another province in the central area, Shanxi, for specially, has very low TFEE scores during the four years and makes little progress. The situation in the west area is even worse other than Sichuan, Yunnan, Qinghai and Ningxia.

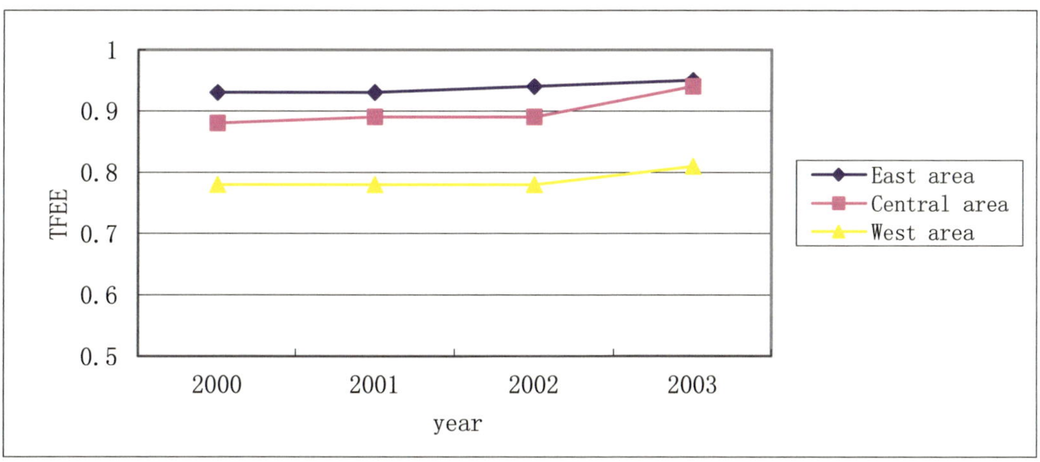

Figure 1. TFEE of 30 provinces during 2000-2003

Table 7 is used to clarify which part makes the energy efficiency get improvement. During 2000 to 2001, the average value of Malmquist production index (MPI) for all provinces is 1.13 which means the efficiency in 2001 is better than 2000. Catch-up effect (CE) and frontier-shift effect (FE) are two parameters to distinguish which part is functioned. The data on the last row show that the average value of CE is 1.00 and FE is 1.13. That is to say, the technical efficiency in 2001 is almost the same as that in 2000, while the production frontier gets improvement during the two years. Improvement is also achieved in next two years, but there is something different that both CE and FE are working.

In order to compare the trends of 3 areas, we make a summary in table 8 using the average data. It is clear that the central area has the greatest achievement and the west area is following. FE is always doing better than CE.

No.	Province	Area	2000	2001	2002	2003
1	Beijing	E	0.8644	0.9557	0.9634	1.0000
2	Tianjin	E	0.9911	1.0000	1.0000	1.0000
3	Hebei	E	0.7702	0.8028	0.7848	0.7663
4	Liaoning	E	1.0000	1.0000	1.0000	1.0000
5	Shanghai	E	1.0000	1.0000	1.0000	1.0000
6	Jiangsu	E	1.0000	1.0000	1.0000	1.0000
7	Zhejiang	E	0.9949	1.0000	1.0000	1.0000
8	Shandong	E	1.0000	0.8836	0.9994	1.0000
9	Guangdong	E	1.0000	1.0000	1.0000	1.0000
10	Guangxi	E	1.0000	1.0000	1.0000	0.9406
11	Hainan	E	1.0000	1.0000	1.0000	1.0000
12	Fujian	E	1.0000	1.0000	1.0000	1.0000
13	Shanxi	C	0.4510	0.4469	0.5150	0.5847
14	Inner M	C	0.6664	0.6929	0.7194	0.8298
15	Jilin	C	0.6657	0.6826	0.6890	0.7548
16	Heilongjiang	C	0.8323	0.7791	0.8795	0.9683
17	Anhui	C	1.0000	1.0000	1.0000	1.0000
18	Jiangxi	C	0.9508	0.9800	1.0000	1.0000
19	Henan	C	0.9075	0.9044	0.8967	1.0000
20	Hubei	C	1.0000	1.0000	0.9077	0.9541
21	Hunan	C	0.9284	0.9305	0.9138	0.9342
22	Chongqing	C	1.0000	1.0000	1.0000	1.0000
23	Sichuan	W	1.0000	1.0000	1.0000	1.0000
24	Guizhou	W	0.5258	0.5232	0.5203	0.6556
25	Yunnan	W	1.0000	1.0000	1.0000	1.0000
26	Shaanxi	W	0.5191	0.5307	0.5701	0.6067
27	Gansu	W	0.4206	0.4229	0.4153	0.4126
28	Qinghai	W	1.0000	1.0000	1.0000	1.0000
29	Ningxia	W	1.0000	1.0000	1.0000	1.0000
30	Xinjiang	W	0.7356	0.7431	0.7497	0.8371

*E, C and W indicate east area, central area and west area respectively.

Table 6. TFEE of 30 provinces during 2000-2003

No.	Province	2000-2001			2001-2002			2002-2003		
		CE	FE	MPI	CE	FE	MPI	CE	FE	MPI
1	Beijing	1.11	1.05	1.17	1.01	1.05	1.06	1.03	1.07	1.11
2	Tianjin	1.01	1.03	1.04	1.00	1.02	1.02	1.00	1.07	1.07
3	Hebei	1.04	1.16	1.20	0.98	1.13	1.11	0.98	1.11	1.09
4	Liaoning	1.00	1.04	1.04	1.00	1.00	1.00	1.00	1.09	1.09
5	Shanghai	1.00	1.16	1.16	1.00	1.14	1.14	1.00	1.19	1.19
6	Jiangsu	1.00	1.20	1.20	1.00	1.16	1.16	1.00	1.10	1.10
7	Zhejiang	1.00	1.08	1.08	1.00	1.09	1.09	1.00	1.13	1.13
8	Shandong	0.88	1.10	0.97	1.13	1.10	1.24	1.00	1.15	1.15
9	Guangdong	1.00	1.18	1.18	1.00	1.03	1.03	1.00	1.06	1.06
10	Guangxi	1.00	1.08	1.08	1.00	1.00	1.00	0.94	1.07	1.01
11	Hainan	1.00	0.23	0.23	1.00	1.01	1.01	1.00	0.72	0.72
12	Fujian	1.00	1.08	1.08	1.00	1.10	1.10	1.00	1.07	1.07
13	Shanxi	0.99	1.04	1.03	1.16	1.06	1.23	1.14	1.05	1.19
14	Inner M	1.04	1.04	1.08	1.04	1.12	1.16	1.15	1.04	1.20
15	Jilin	1.03	1.01	1.04	1.01	1.02	1.03	1.10	0.98	1.08
16	Heilongjiang	0.94	1.01	0.95	1.13	0.98	1.11	1.10	0.96	1.05
17	Anhui	1.00	1.06	1.06	1.00	1.03	1.03	1.00	1.07	1.07
18	Jiangxi	1.03	1.00	1.03	1.02	1.00	1.02	1.00	0.96	0.96
19	Henan	1.00	1.16	1.16	0.99	1.14	1.13	1.11	4.19	4.65
20	Hubei	1.00	1.02	1.02	0.91	1.03	0.94	1.05	0.99	1.04
21	Hunan	1.00	1.00	1.00	0.98	0.98	0.96	1.02	0.96	0.98
22	Chongqing	1.00	3.58	3.58	1.00	3.82	3.82	1.00	4.27	4.27
23	Sichuan	1.00	1.53	1.53	1.00	1.68	1.68	1.00	1.59	1.59
24	Guizhou	1.00	0.95	0.95	0.99	1.02	1.01	1.26	0.82	1.04
25	Yunnan	1.00	1.09	1.09	1.00	1.12	1.12	1.00	1.16	1.16
26	Shaanxi	1.02	1.00	1.02	1.07	0.99	1.06	1.06	0.96	1.02
27	Gansu	1.00	1.02	1.02	0.98	1.03	1.01	0.99	1.04	1.03
28	Qinghai	1.00	0.94	0.94	1.00	0.94	0.94	1.00	0.94	0.94
29	Ningxia	1.00	0.91	0.91	1.00	0.91	0.91	1.00	0.90	0.90
30	Xinjiang	1.01	1.01	1.02	1.01	1.01	1.02	1.12	0.97	1.09
	Average	1.00	1.13	1.13	1.01	1.16	1.17	1.04	1.26	1.30

Table 7. Changes of 30 provinces during 2000-2003

Areas	2000-2001			2001-2002			2002-2003		
	CE	FE	MPI	CE	FE	MPI	CE	FE	MPI
East area	1.00	1.03	1.04	1.02	1.07	1.09	1.01	1.07	1.08
Central area	1.00	1.32	1.32	1.01	1.35	1.36	1.06	1.71	1.81
West area	1.00	1.06	1.06	1.01	1.09	1.09	1.05	1.05	1.10

Table 8. Average data of areas during 2000-2003

It is interesting to investigate the individual province. Here are some examples. First, we make a comparison between Shanghai and Hainan both of which are efficient during the periods. However, detailed data indicate that Shanghai keeps making frontier forward gradually while Hainan are opposite except year 2002. This could be explained by that the location of Hainan on the frontier is on the edge. Second, take Beijing for example. Beijing is non-efficient from 2000 to 2002 and finally becomes efficient at 2003 by making efforts on improving technical efficiency and putting frontier forward. Third, energy efficiency of Shandong province suffers a decline and is back to the normal level later. MPI during first period is decreasing mainly caused by declining CE. In the next two years, some parameters get recovery which makes MPI increasing.

5. Conclusion

This chapter reviews the development process of the evaluation technique of energy efficiency and focuses on introducing the concept of energy intensity. However, missing the structure of energy consumption causes the energy efficiency estimated inaccurately. Thus, the current chapter introduces a weighed energy efficiency model based on DEA to fix it. Energy cannot produce production without non-energy inputs such as labor and capital. That's why we extend the method to the total factor energy efficiency model. Energy efficiency of China and other 47 countries in 2003 are employed to illustrate the models. Results show that unbalance of energy efficiency exists. For china, specially, it needs to adjust energy consumption structure as its poor energy efficiency and improve GDP since its total factor energy efficiency is at a lower level than some developed countries.

As a key part, the chapter adopts Malmquist production index technique to analyze the dynamic change in energy efficiency of Chinese provinces which can further explore the reason for the variation of energy efficiency deeply. The chapter uses the proposed models to investigate the changes in energy efficiency of provinces in china during the period of 2000 to 2003. We find that the east area has better energy efficiency than the central and west area but lower improving rapid. In addition, it is interesting to find that energy efficiency of most provinces improves due to the extending frontier. Although our work mainly focuses on estimating energy efficiency at the macro-level, it can provide guidance to managers and manufacturers at the micro level.

Author details

Chenchen Yang, Feng Yang, Liang Liang* and Xiping Xu
School of Management, University of Science and Technology of China, Hefei, P. R. China

* Corresponding Author

Acknowledgement

We would like to thank the financial support by Program for New Century Excellent Talents in University, Ministry of Education of China and National Natural Science Foundation of China (Grant No. 70801056, 71121061, 71090401/71090400 and 71110107024).

6. References

[1] Ang, B., Monitoring changes in economy-wide energy efficiency: From energy-GDP ratio to composite efficiency index. Energy Policy, 2006. 34(5): p. 574-582.

[2] Ang, B., A statistical analysis of energy coefficients. Energy economics, 1991. 13(2): p. 93-110.

[3] Howarth, R.B., et al., Manufacturing energy use in eight OECD countries: Decomposing the impacts of changes in output, industry structure and energy intensity. Energy economics, 1991. 13(2): p. 135-142.

[4] A Greening, L., D.L. Greene, and C. Difiglio, Energy efficiency and consumption—the rebound effect—a survey. Energy Policy, 2000. 28(6): p. 389-401.

[5] Xu Xiping, liang Liang. Energy Efficiency and Productivity of China: Compared with other Countries. ICCS 2007, Part III, LNCS 4489, pp.988-991, 2007, Beijing, China.

[6] Charnes, A., W.W. Cooper, and E. Rhodes, Measuring the efficiency of decision making units. European Journal of Operational Research, 1978. 2(6): p. 429-444.

[7] Zhou, P., B.W. Ang, and K.L. Poh, A survey of data envelopment analysis in energy and environmental studies. European Journal of Operational Research, 2008. 189(1): p. 1-18.

[8] Mohammadi, A., et al., Energy efficiency improvement and input cost saving in kiwifruit production using Data Envelopment Analysis approach. Renewable Energy, 2011. 36(9): p. 2573-2579.

[9] Rao, X., et al., Energy efficiency and energy saving potential in China: An analysis based on slacks-based measure model. Computers & Industrial Engineering, 2011(0).

[10] Bian, Y. and F. Yang, Resource and environment efficiency analysis of provinces in China: A DEA approach based on Shannon's entropy. Energy Policy, 2010. 38(4): p. 1909-1917.

[11] Boyd, G.A. and J.X. Pang, Estimating the linkage between energy efficiency and productivity. Energy Policy, 2000. 28(5): p. 289-296.

[12] Hu, J.L. and S.C. Wang, Total-factor energy efficiency of regions in China. Energy Policy, 2006. 34(17): p. 3206-3217.

[13] Honma, S. and J.L. Hu, Total-factor energy efficiency of regions in Japan. Energy Policy, 2008. 36(2): p. 821-833.

[14] Honma, S. and J.L. Hu, Total-factor energy productivity growth of regions in Japan. Energy Policy, 2009. 37(10): p. 3941-3950.

[15] Odeck, J., Assessing the relative efficiency and productivity growth of vehicle inspection services: An application of DEA and Malmquist indices. European Journal of Operational Research, 2000. 126(3): p. 501-514.

[16] Tone, K., Malmquist productivity index. Handbook on data envelopment analysis, 2004: p. 203-227.

[17] Halkos, G.E. and N.G. Tzeremes, Exploring the existence of Kuznets curve in countries' environmental efficiency using DEA window analysis. Ecological Economics, 2009. 68(7): p. 2168-2176.

[18] Zhang, X.-P., et al., Total-factor energy efficiency in developing countries. Energy Policy, 2011. 39(2): p. 644-650.

[19] Banker, R.D., A. Charnes, and W.W. Cooper, Some models for estimating technical and scale inefficiencies in data envelopment analysis. Management science, 1984: p. 1078-1092.

[20] Cooper, W.W., L.M. Seiford, and J. Zhu, Handbook on data envelopment analysis. 2004, Springer.

[21] Färe, R., S. Grosskopf, and C.A.K. Lovell, The measurement of efficiency of production. Vol. 6. 1985: Springer.

[22] Boussemart, J.P., et al., Luenberger and Malmquist productivity indices: theoretical comparisons and empirical illustration. Bulletin of Economic Research, 2003. 55(4): p. 391-405.

[23] Jun, Z., W. Guiying, and Z. Jipeng, The Estimation of China's provincial capital stock: 1952—2000. Economic Research Journal, 2004. 10: p. 35-44.

Chapter 11

The Reliability Design and Its Direct Effect on the Energy Efficiency

Seong-woo Woo, Jungwan Park, Jongyun Yoon and HongGyu Jeon

Additional information is available at the end of the chapter

http://dx.doi.org/10.5772/48790

1. Introduction

Reliability refers to the ability of system or component to perform a required function under stated environmental and operational conditions for a specified period of time. Traditionally, the reliability over the product life can be illustrated by a bathtub curve that has three regions: a decreasing rate of failure, a constant rate of failure, and an increasing rate of failure, as shown in Figure 1(a). As the reliability of a product (or part) improves, failure of the part becomes less frequent in the field. The bathtub curve may change into a straight line with the slope angle β. In a straight line there are two variables to be measured: product life L_B (or mean time between failures) and failure rate λ, as shown in Eq. (1):

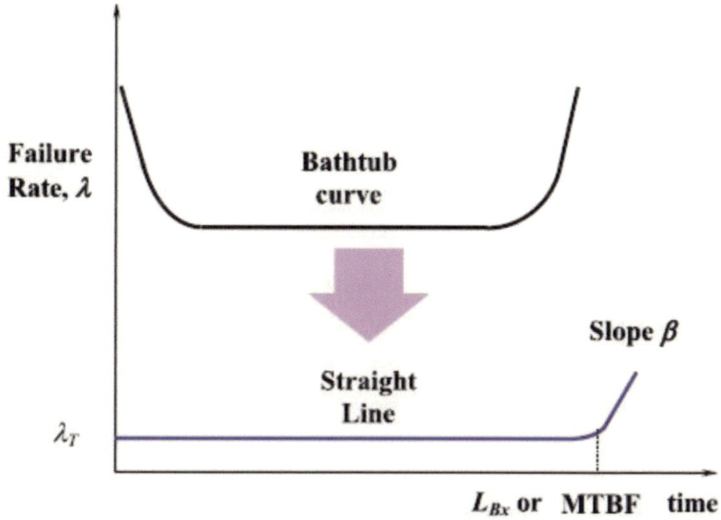

(a) The bathtub curve and straight line with slope β

INTECH
open science | open minds

© 2012 The Author(s). Licensee InTech. This chapter is distributed under the terms of the Creative Commons Attribution License http://creativecommons.org/licenses/by/3.0), which permits unrestricted use, distribution, and reproduction in any medium, provided the original work is properly cited.

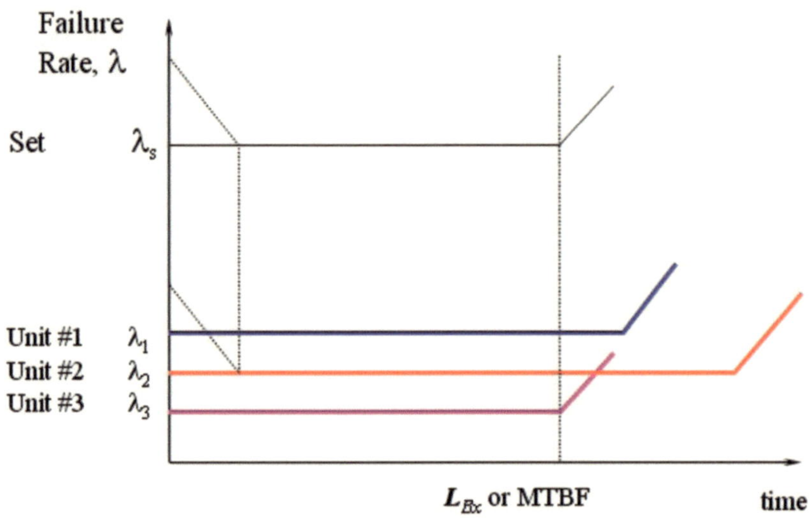

(b) System life and failure rate consisting of unit #1, unit #2 and unit #3

Figure 1. System life and failure rate

$$\mathrm{R}(\mathrm{L_B}) = e^{-\lambda L_B} \cong 1 - \lambda L_B \qquad (1)$$

We can thus establish the reliability growth plan of parts with a constant failure rate.

A company generally designs its new products to (1) minimize initial failures, (2) reduce random failures during the expected product working period, and (3) lengthen product life. Such aims are met through the use of robust design techniques, including statistical design of experiment (SDE) and the Taguchi methods [1]. The Taguchi methods describe the robustness of a system for evaluation and design improvement, which is also known as quality engineering [2-3] or robust engineering [4]. Robust design processes include concept design, parameter design, and tolerance design [5]. Taguchi's robust design methods place a design in an optimum position where random "noise" does not cause failure, which then and helps in determining the proper design parameters [6].

However, for a simple mechanical structure, the Taguchi methods' robust design processes need to consider a large number of design parameters. They also have difficulty in predicting the product life, L_B (or MTBF).

In this study we present a new method for the reliability design of mechanical systems. This new method takes into account the fact that products with missing or improper design parameters can result in recalls and loss of brand name value. Based on the analysis of a failed refrigerator drawer and handle systems, we demonstrated our new reliability design method. The new method uses ALT; the new concept of product life, L_B; and sample size, as a novel means of determining proper design parameters [7-14].

1.1. Targeting the refrigerator *Bx* life and failure rate λ

The multi-unit refrigerator used as a case study for this method consists of a compressor, a drawer, a door, a cabinet, and other units. For the drawer, the B_1 life of the new design is targeted to be over 10 years with a yearly failure rate of 0.1%. The entire refrigerator's *Bx* life can be obtained by summing up the failure rates of each refrigerator unit. The refrigerator's B_{12} life with the new design is targeted to be over 10 years with a yearly failure rate of 1.2 % (Table 1) [19].

No	Units	Market Data Failure Rate	Market Data *Bx* Life	Design	Conversion	Expected Failure Rate	Target *Bx* Life	*Bx* Life	Based *Bx*
1	Compressor	0.34	5.3	New	x5	1.70	0.10	10	B1.0
2	Door	0.35	5.1	Given	x1	0.35	0.15	10	B1.5
3	Cabinet	0.25	4.8	Modified	x2	0.50	0.10	10	B1.0
4	Drawer	0.20	6.0	New	x2	0.40	0.10	10	B1.0
5	Heat exchanger	0.15	8.0	Given	x1	0.15	0.10	10	B1.0
6	etc	0.50	12.0	Given	x1	0.50	0.50	10	B6.0
Sum	R-Set	1.79	7.4	-	-	3.60	1.10	10	B12.0

Table 1. Total parametric ALT plan of refrigerator

1.2. Analysis of the problems identified in field samples (loads analysis)

In the field, certain components in these refrigerators had been failing or making noise, causing consumers to replace their refrigerators. Data from the failed products in the field showed how common used the refrigerators under common usage conditions. Refrigerator reliability problems in the field occur when the parts cannot endure repetitive stresses due to internal or external forces over a specified period of time. The energy flow in a refrigerator (or other mechanical) system can generally be expressed as efforts and flows (Table 2) [15]. Thus, the stresses come from the efforts.

Refrigerator Units (or Parts)	Effort, $e(t)$	Flow, $f(t)$
Mechanical translation (draws, dispenser lever)	Force component, $F(t)$	Velocity component, $V(t)$
Mechanical rotation (door, cooling fan)	Torque component, $\tau(t)$	Angular velocity component, $V(t)$
Compressor	Pressure difference, $\Delta P(t)$	Volume flow rate, $Q(t)$
Electric (PCB, condenser)	Voltage, $V(t)$	Current, $i(t)$

Table 2. Effort and flow in the multi-port system

For a mechanical system, the time-to-failure approach employs a generalized life model (LS model) [16], such as:

$$T_f = A(S)^{-n} \exp\frac{E_a}{kT} = A(e)^{-n} \exp\frac{E_a}{kT} \tag{2}$$

Repetitive stress can be expressed as the duty effect that carries the on/off cycles and shortens part life [17]. Under accelerated stress conditions, the acceleration factor (AF) can be described as:

$$AF = \left(\frac{S_1}{S_0}\right)^n \left[\frac{E_a}{k}\left(\frac{1}{T_0} - \frac{1}{T_1}\right)\right] = \left(\frac{e_1}{e_0}\right)^n \left[\frac{E_a}{k}\left(\frac{1}{T_0} - \frac{1}{T_1}\right)\right] \tag{3}$$

And n can be determined by multiple testings with different stress levels.

1.3. Parametric ALT with *B*x life and sample size

Traditionally, the characteristic life is defined as:

$$\eta^\beta \equiv \frac{\sum t_i^\beta}{r} \cong \frac{n \cdot h^\beta}{r} \tag{4}$$

As the reliability of a product (or part) improves, failures of the product become less frequent in laboratory tests. Thus, it becomes more difficult to evaluate the characteristic life using Equation (4). The distribution of failed samples should follow the Poisson distribution for small samples [18]. For a 60% confidence level, the characteristic life can be redefined as

$$\eta^\beta \cong \frac{1}{r+1} \cdot n \cdot h^\beta \tag{5}$$

In order to introduce the *B*x life in the Weibull distribution, the characteristic life can be modified as

$$L_B^\beta \cong x \cdot \eta^\beta = \frac{x}{r+1} \cdot n \cdot h^\beta \tag{6}$$

where L_B = *B*x life and $x = 0.01X$, on the condition that $x \le 0.2$.

*B*x is the time by which X % of the drawer and handle system installed in a particular population of refrigerators will have failed. In order to assess the *B*x life with about a 60% confidence level, the number of test samples is derived in Eq. (7). That is,

$$n \cong \frac{1}{x} \cdot (r+1) \cdot \left(\frac{1}{h^*}\right)^\beta \tag{7}$$

with the condition that the durability target is defined as follows,

$$h^* = (AF \cdot h)/L_B \ge 1 \tag{8}$$

Based on the customer usage conditions, the normal range of operating conditions and cycles of the product (or parts) are determined. Under the worst case, the objective number of cycles and the number of required test cycles can be obtained from Eq. (7). ALT equipment can then be conducted on the basis of load analysis. Using ALT we can find the missing or improper parameters in the design phase.

1.4. Refrigerator unit L_{Bx} life and failure rate, λ, with the improved designs

The parameter design criterion of the newly designed samples can be more than the target life of B_X = 10 years. From the field data and from a sample under ALT with a corrective action plans, we can obtain the missing or improper parameters of parts and their levels in the design phase.

With the improved design parameters, we can derive the expected L_{Bx} life of the final design samples using Equation (6).

$$L_B{}^{\beta} \cong x \cdot \frac{n \cdot (h \cdot AF)^{\beta}}{r+1} \tag{9}$$

Let $x = \lambda \cdot L_B$ in Equation (9). The failure rate of the final design samples is derived in Equation (10)

$$\lambda \cong \frac{1}{L_B} \cdot (r+1) \cdot \frac{L_B^{\beta}}{n \cdot (h \cdot AF)^{\beta}} \tag{10}$$

2. Case study: Reliability design of a refrigerator drawer and handle system

Figure 2 shows a refrigerator with the newly designed drawer and handle system and its parts. In the field, the refrigerator drawer and handle system had been failing, causing consumers to replace their refrigerators (Figure 3). The specific causes of failures of the refrigerator drawers during operation were repetitive stress and/or the consumer improper usage. Field data indicated that the damaged products had structural design flaws, including sharp corner angles and weak ribs that resulted in stress risers in high stress areas.

A consumer stores food in a refrigerator to have convenient access to fresh food. Putting food in the refrigerator drawer involves opening the drawer to store or takeout food, closing the drawer by force. Depending on the consumer usage conditions, the drawer and handle parts receive repetitive mechanical loads when the consumer opens and closes the drawer.

Figure 4 shows the functional design concept of the drawer and handle system. The stress due to the weight load of the food is concentrated on the handle and support slide rail of the drawer. Thus, the drawer must be designed to endure these repetitive stresses.

The force balance around the drawer and handle system cans be expressed as:

$$F_{draw} = \mu W_{load} \tag{11}$$

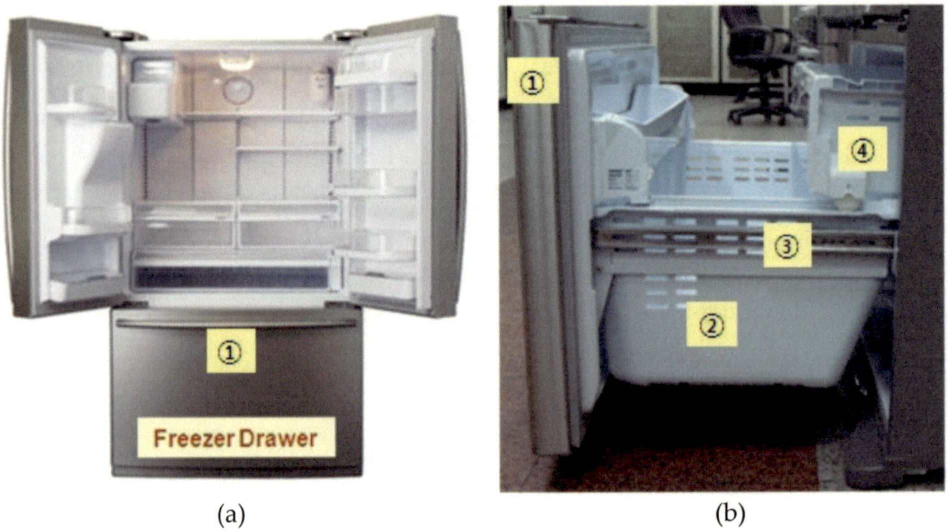

Figure 2. Refrigerator and drawer assembly. (a) French refrigerator (b) Mechanical parts of the drawer: handle ①, drawer ②, slide rail ③, and pocket box ④

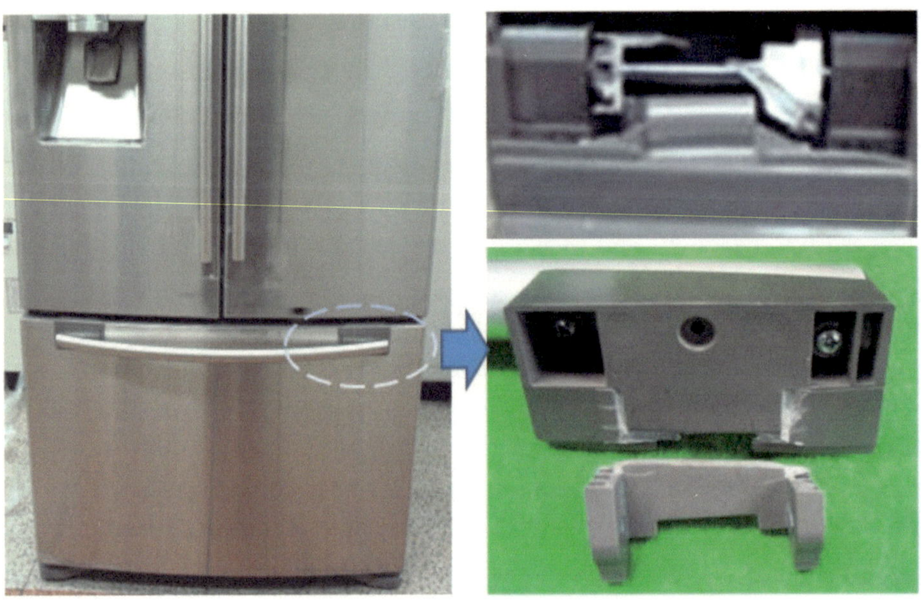

Figure 3. A damaged product after use

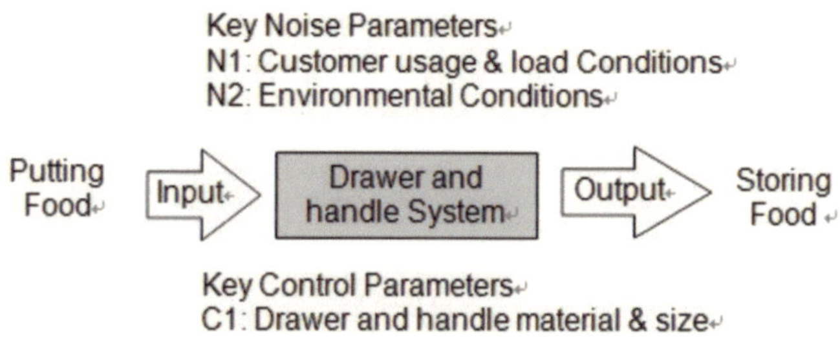

(a) Parameter diagram of drawer and handle system

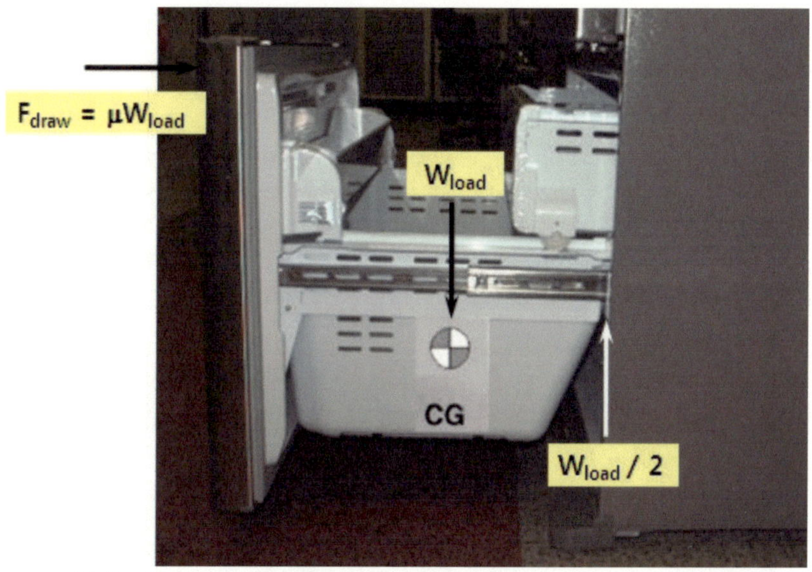

(b) Design concept of mechanical drawer and handle system

Figure 4. Functional design concept of the drawer and handle system

Because the stress of the drawer and handle system depends on the food weight, the life-stress model (LS model) can be modified as follows:

$$T_f = A(S)^{-n} = A(F_{draw})^{-n} = A(\mu W_{load})^{-n} \tag{12}$$

where A is constant. Thus, the acceleration factor (*AF*) can be derived as

$$AF = \left(\frac{S_1}{S_0}\right)^n = \left(\frac{F_1}{F_0}\right)^2 = \left(\frac{\mu W_1}{\mu W_0}\right)^2 = \left(\frac{W_1}{W_0}\right)^2 \tag{13}$$

3. Laboratory experiments

The normal ranges of the operating conditions for the drawer system and handle were 0 to 50°C ambient temperature, 0 to 85% relative humidity and 0.2 to 0.24G vibration. The normal number of operating cycles for one day was approximately 5; the worst case was 9. Under the worst case, the objective drawer open/close cycles for ten years would be 32,850 cycles (Table 3).

Item	Number of operations (times)			
	1 day		10 years	
	Normal	Worst	Normal	Worst
Drawer	5	9	18,250	32,850

Table 3. Operating number of a drawer

For the worst case, the food weight force on the handle of the drawer was 0.34 *kN*. The applied food weight force for the ALT was 0.68 *kN*. With a quotient, *n*, of 2, the total *AF* was approximately 4.0 using equation (13).

The parameter design criterion of the newly designed drawer can be more than the target life of B_1 = 10 years. Assuming the shape parameter β was 2.0 and x was 0.01, the test cycles and test sample numbers calculated in Equation (7) were 67,000 cycles and 3 units, respectively. The ALT was designed to ensure a B_1 life of 10 years with about a 60% level of confidence that it would fail less than once during 67,000 cycles.

(a) ALT equipment and controller

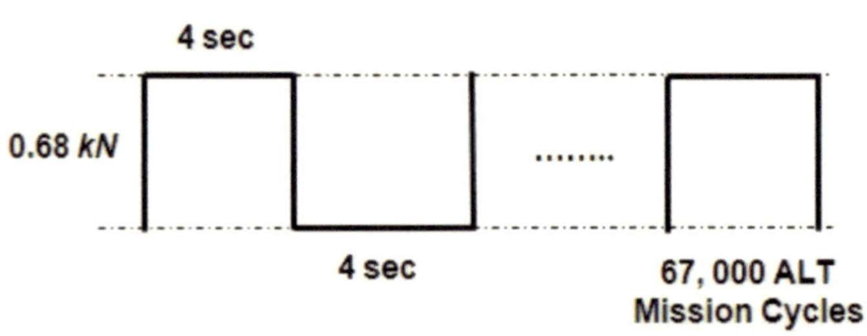

(b) Duty cycles of repetitive food weight force on the drawer

Figure 5. ALT equipment and duty cycles.

Figure 5 shows ALT equipment and duty cycles for the repetitive food weight force, F_{draw}. For the ALT experiments, the control panel on top of the testing equipment started and stopped the drawer during the mission cycles. The food load, F, was controlled by the accelerated weight load in the drawer storage. When a button on the control panel was pushed, mechanical arms and hands pushed and pulled the drawer.

4. Parametric ALTs with corrective action plans

Figures 6(a) and 6(b) show the failed product from the field and the 1st accelerated life testing, respectively. The failure sites in the field and the first ALT occurred at the drawer handle as a result of high concentrated stress. Figure 7 shows a graphical analysis of the ALT results and field data on a Weibull plot. For the shape parameter, the estimated value on the chart was 2.0. For the final design, the shape parameter was determined to be 3.1. These methodologies were valid for pinpointing the weak design responsible for failures in the field and 1st ALT.

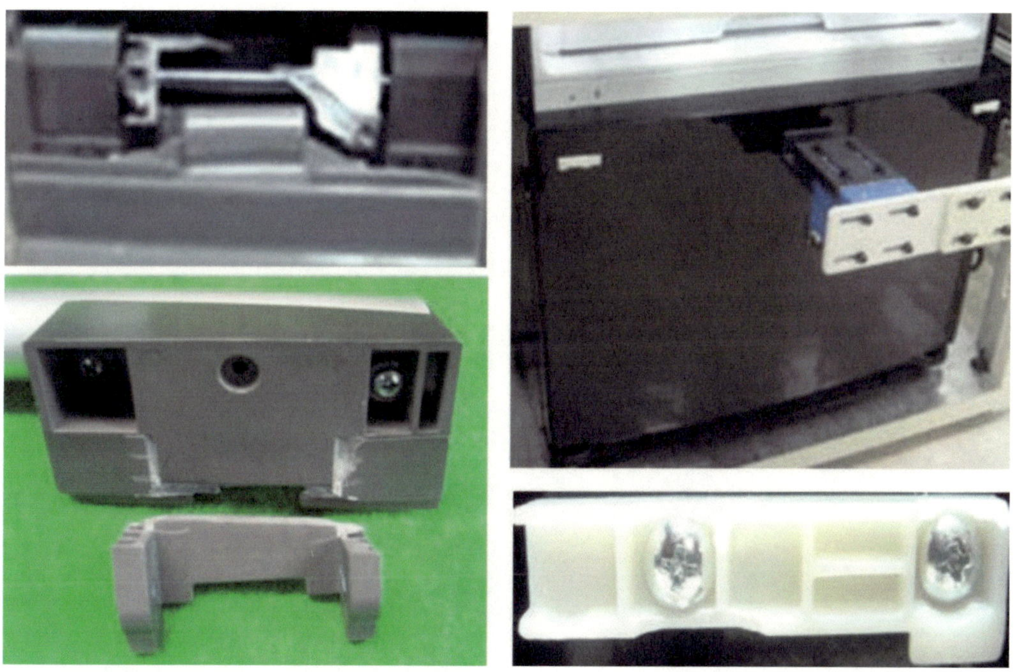

(a) Failed product in field (b) Failed sample in first accelerated life testing

Figure 6. Failed products in field and first ALT

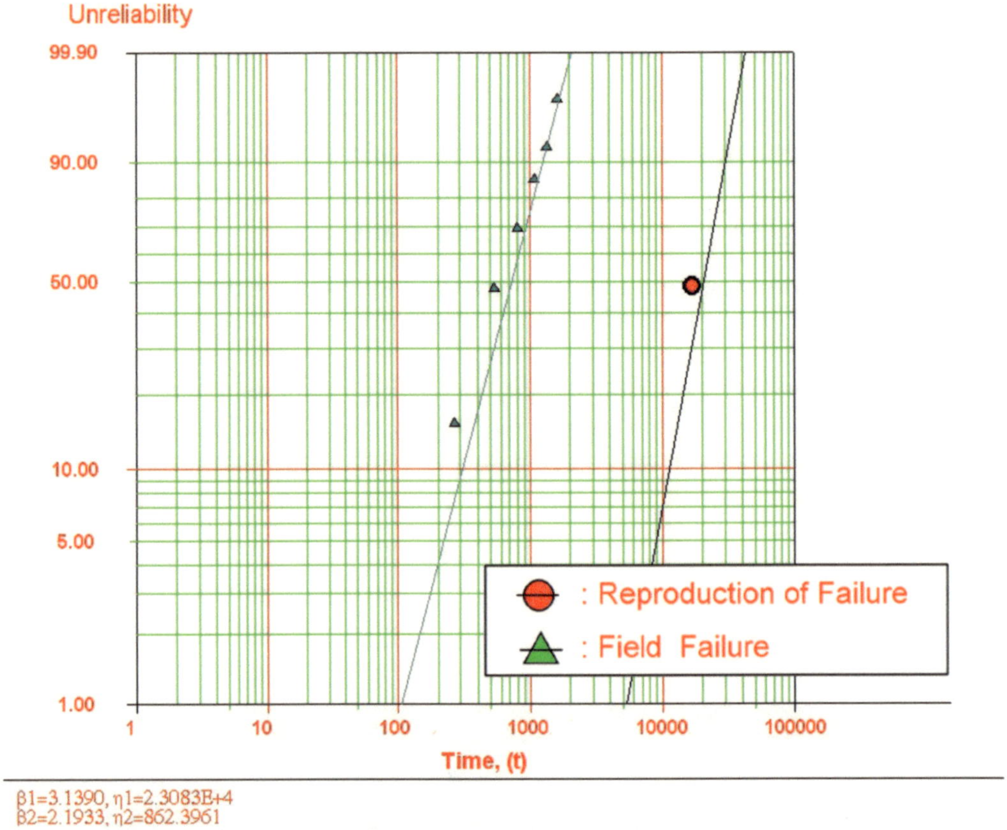

Figure 7. Field data and results of 1st ALT on Weibull chart.

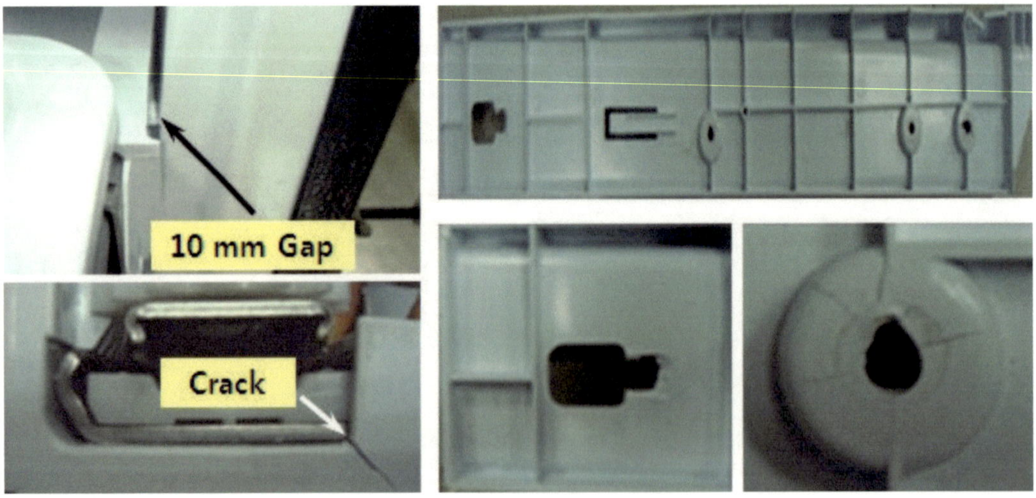

Figure 8. Failed slide rails in second ALT

The fracture of the drawer in the first and second ALTs occurred in the handle and slide rails (Figure 6(b) and Figure 8). The missing or improper parameters of the handle and slide rails in the design phase are listed in Table 4. These design flaws can result in a fracture when the repetitive food load is applied.

CTQ			Parameters	Unit
Fracture	KNP	N1	Consumer food loading	kN
	KCP	C1	Reinforced handle width	–
		C2	Handle hooker width	–
		C3	Fastening screw number	–
		C4	Slide rail chamber	mm
		C5	Slide rail boss thickness	mm
		C6	New added rib	–

Table 4. Vital parameters based on ALTs

To prevent the fracture problem and release the repetitive stresses, the handle and slide rails were redesigned. The corrective action plan for the design parameters included: (1) increasing the width of the reinforced handle, C1, from 90mm to 122mm; (2) increasing the handle hooker size, C2, from 8mm to 19mm; (3) increasing the rail fastening screw number, C3, from 1 to 2; (4) adding an inner chamber and plastic material, C4, from HIPS to ABS; (5) thickening the boss, C5, from 2.0mm to 3.0mm; (6) adding a new support rib, C6 (Table 5).

The parameter design criterion of the newly designed samples was more than the target life, B_1, of ten years. The confirmed value, β, on the Weibull chart was 3.1. For the second ALT, the recalculated test cycles and sample size in Equation (7) were 32,000 and 3 units, respectively. In the third ALT, no problems were found with the drawer after 32,000 cycles and 65,000 cycles. We therefore concluded that the modified design parameters were effective.

Table 6 provides a summary of the ALT results. Figure 9 shows the results of the 1st ALT and 3rd ALT plotted in a Weibull chart. With the improved design parameters, the B_1 life of the samples in the third ALT was lengthened to more than 10.0 years.

Handle	Right/left slide rail
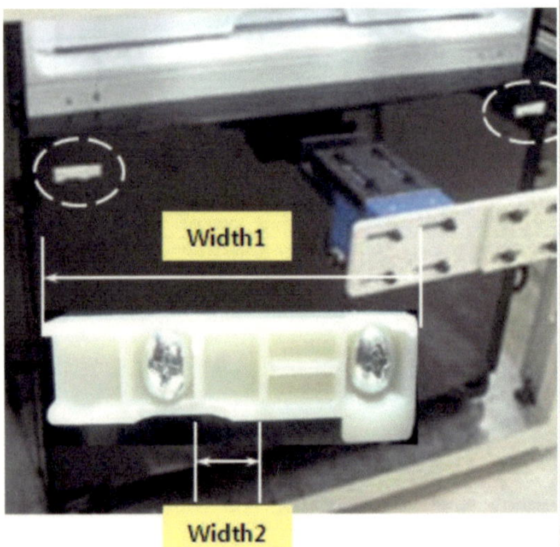	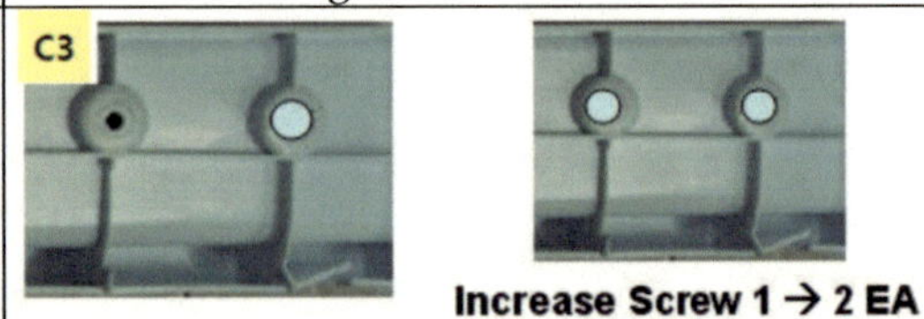 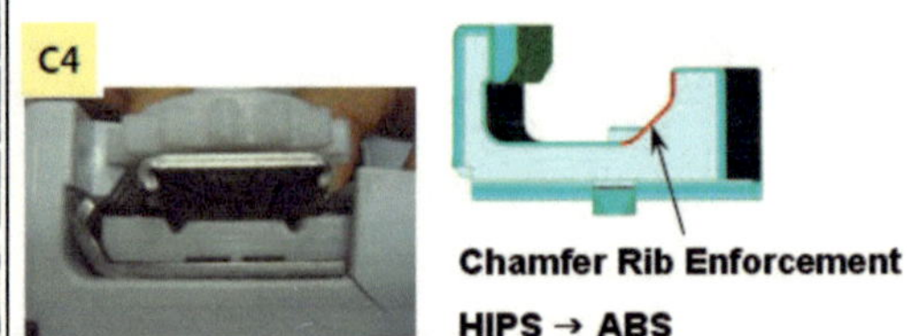 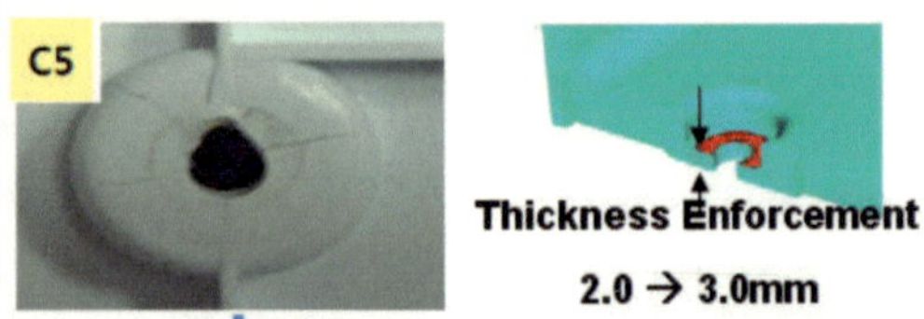 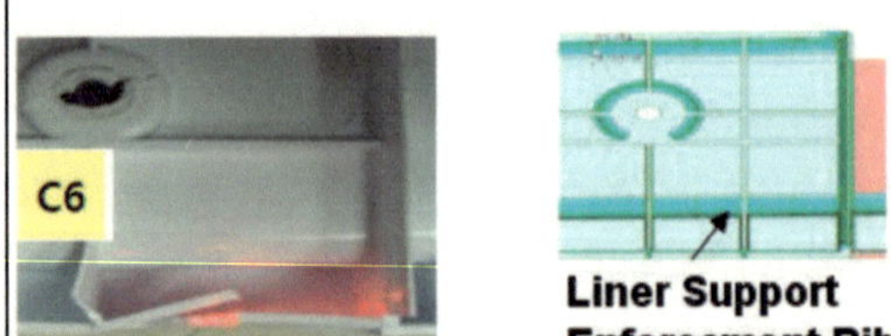
C1: Width L90mm → L122mm (1st ALT) C2: Width L8mm → L19mm (1st ALT)	C3: Rail fastening screw number 1→2 (2nd ALT) C4: Chamfer: Corner chamfer Plastic material HIPS → ABS (2nd ALT) C5: Boss thickness 2.0 → 3.0 mm (2nd ALT) C6: New support rib (2nd ALT)

Table 5. Redesigned handle and right/left slide rail

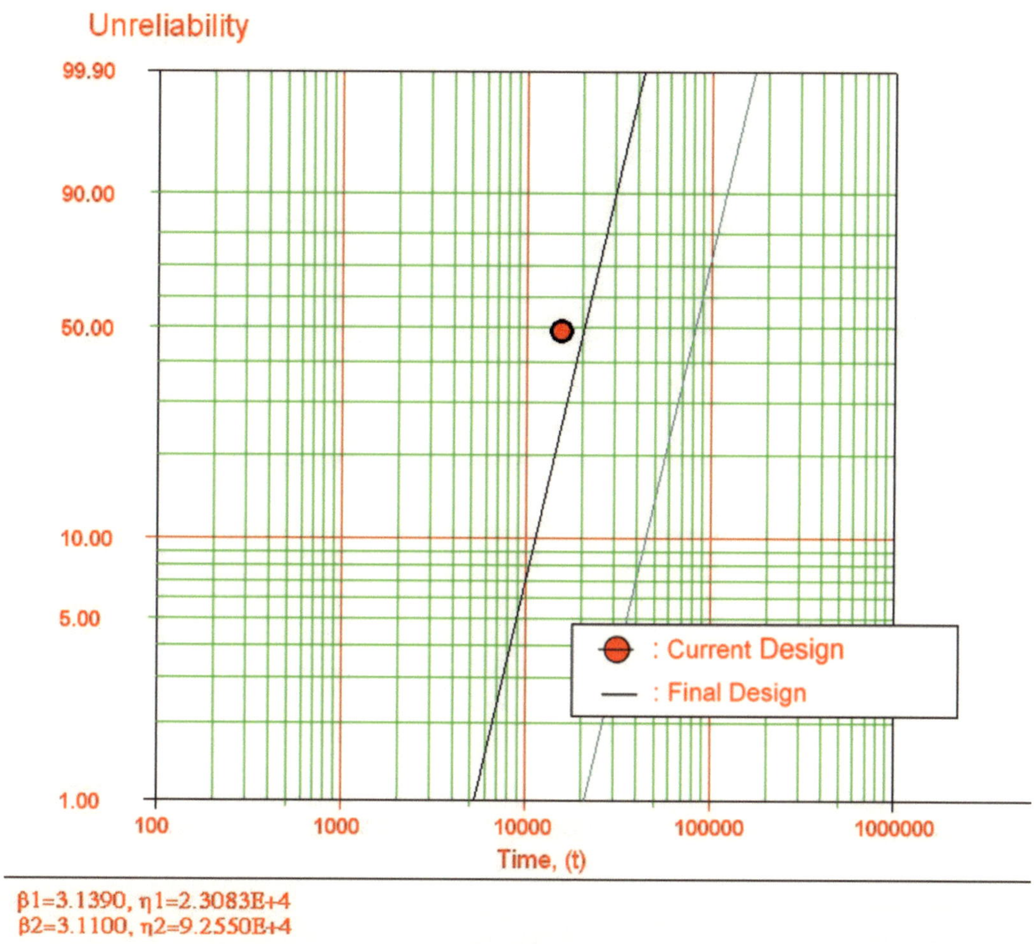

Figure 9. Results of 1st ALT and 3rd ALT plotted in Weibull chart

	First ALT	Second ALT	Third ALT
	Initial design	First design iteration	Final design
In 32,000 cycles, Fracturing is less than 1	7,500 cycles: 2/3 Fail 12,000 cycles: 1/3 OK λ = 26.6 %/year B_1 = 3.4 year	16,000 cycles: 2/3 Fail λ= 2.46 %/year B_1 = 7.3 year	32,000 cycles: 3/3 OK 65,000 cycles: 3/3 OK λ = 0.1 %/year B_1 = 10 year
Drawer structure	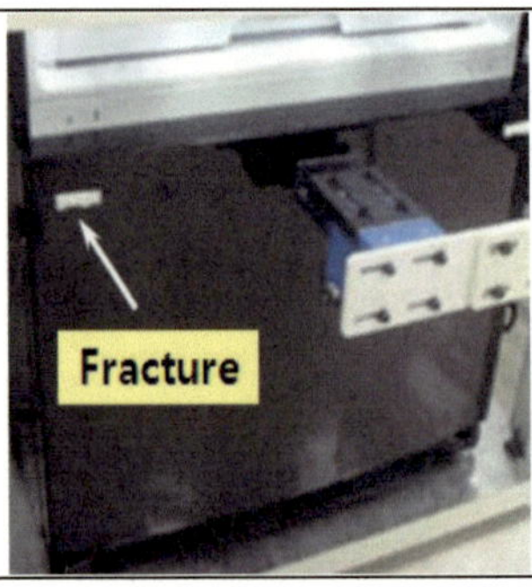	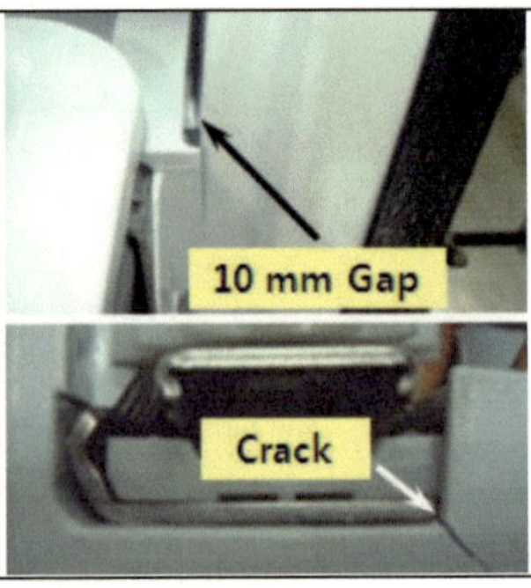	
Material & Spec.	Width1: L90 →L122 Width2: L8→L19.0	Rib1: new support rib boss: 2.0 → 3.0 mm Chamfer1: Corner Material: HIPS →ABS	

Table 6. Results of ALTs

6. Conclusions

We developed a new reliability design method based on a study of a defective refrigerator drawer and handle system that was failing under field use conditions. The failure modes and mechanisms for the drawer in the field and in the ALTs were identified. Important design parameters were studied and improvements were evaluated using ALTs.

Based on the products returned from the field and the results of the first ALT, we found that the handles were fracturing because of design flaws. The handle design was corrected by increasing the handle width. During the second ALT, the slide rails fractured because they did not have enough strength to endure the repetitive food storage loads. The slide rails were corrected by providing additional reinforced ribs, reinforced boss, and an inner chamber. As a result these modified design parameters, there were no problems in the third ALT. We therefore concluded that the values for the design parameters were effective to meet the life cycle requirements. The yearly failure rate and B_1 life of the redesigned drawer and handle system, based on the results of ALT, were under 0.1% and more than 10 years, respectively. The study of the missing or improper design parameters in the design phase, through the inspection of failed products in the field, load analysis, and ALTs was very effective in redesigning more reliable parts with significantly longer life.

The case study focused on a mechanical structure consisting of several parts subjected to repetitive stresses under consumer usage conditions. The same principles developed for the new reliability design methodology could be applied to other mechanical systems, including construction equipment, automobile gear trains and engines, forklifts, washing machines, vacuum cleaners, and motor fan systems. We recommend that the missing or improper controllable design parameters on these systems also be studied for reliability design. These parameter studies would also include failure analysis, load analysis, and a tailored series of accelerated life tests. These methodologies could then predict part life quantitatively through accelerated factors and exact sample size.

7. Nomenclature

AF	acceleration factor
B_X	durability index
$C1$	width of reinforced handle, mm
$C2$	width of handle hooker, mm
$C3$	back rib of slide rail
$C4$	screw boss height of slide rail, mm
$C5$	inner chamber of slide rail
$C6$	material of slide rail
$C7$	screw number of slide number
e	effort
e_0	effort under normal stress conditions
e_1	effort under accelerated stress conditions
E_a	activation energy
f	flow
$F(t)$	unreliability
F_{draw}	open/close force of the freezer drawer system, kN
F_1	weight force under accelerated stress conditions, kN
F_0	weight force under normal conditions, kN
h	testing time (or cycles)
h^*	non-dimensional testing cycles, $h^* = h/L_B \geq 1$
i	current, A
k	Boltzmann's constant, 8.62 x 10^{-5} eV/deg
KCP	Key Control Parameter
KNP	Key Noise Parameter
L_B	the target B_X life and $x = 0.01X$, on the condition that $x \leq 0.2$
n	the number of test samples
$N1$	consumer freezer door drawer open/close force, kN
ΔP	pressure difference, MPa
r	failed numbers

R	reliability function
S	stress
S_0	mechanical stress under normal stress conditions
S_1	mechanical stress under accelerated stress conditions
t_i	test time for each sample
T	absolute temperature, *K*
T_1	absolute temperature under accelerated stress conditions, *K*
T_0	absolute temperature under normal stress conditions, *K*
T_f	time to failure
V	velocity, *m/s*
V	voltage, *volt*
W_1	food weight force under accelerated stress conditions, *kN*
W_0	food weight force under normal stress conditions, *kN*
W_{load}	total food weight force in the freezer door drawer, *kN*
X	accumulated failure rate, %
x	$x = 0.01 \cdot X$, on condition that $x \le 0.2$.

Greek symbols

η	characteristic life
λ	failure rate
μ	friction coefficient

Superscripts

β	shape parameter in a Weibull distribution
n	stress dependence, $n = -\left[\frac{\partial \ln(T_f)}{\partial \ln(S)}\right]_T$

Subscripts

0	normal stress conditions
1	accelerated stress conditions

Author details

Seong-woo Woo, Jungwan Park,
Jongyun Yoon and HongGyu Jeon
STX Institute of Technology, Korea

8. References

[1] Taguchi, G., 1978, Off-line and On-line Quality Control Systems, Proceedings of the International Conference on Quality Control, Tokyo, Japan.

[2] Taguchi, G., Shih-Chung, T, 1992, Introduction to Quality Engineering: Bringing Quality Engineering Upstream, American Society of Mechanical Engineering, New York.

[3] Ashley, S., 1992, Applying Taguchi's Quality Engineering to Technology Development, Mechanical Engineering.

[4] Wilkins, J., 2000, Putting Taguchi Methods to Work to Solve Design Flaws, Quality Progress, 33(5), 55-59.

[5] Phadke, M., 1989, Quality Engineering Using Robust Design, Englewood Cliffs, NJ: Prentice Hall.

[6] Byrne, D., Taguchi, S., 1987, Taguchi Approach to Parameter Design, Quality Progress, 20(12), 19-26.

[7] Woo, S., Pecht, M., 2008, Failure Analysis and Redesign of a Helix Upper Dispenser, Engineering Failure Analysis, 15 (4), 642–653.

[8] Woo, S., O'Neal, D., Pecht, M., 2009, Improving the Reliability of a Water Dispenser Lever in a Refrigerator Subjected to Repetitive Stresses, Engineering Failure Analysis, 16 (5), 1597–1606.

[9] Woo, S., O'Neal, D., Pecht, M., 2009, Design of a Hinge Kit System in a Kimchi Refrigerator Receiving Repetitive Stresses, Engineering Failure Analysis, 16 (5), 1655–1665.

[10] Woo, S., Ryu, D., Pecht, M., 2009, Design Evaluation of a French Refrigerator Drawer System Subjected to Repeated Food Storage Loads, Engineering Failure Analysis, 16 (7), 2224–2234.

[11] Woo, S., O'Neal, D., Pecht, M., 2010, Failure Analysis and Redesign of the Evaporator Tubing in a Kimchi Refrigerator, Engineering Failure Analysis, 2010, 17(2), 369-379.

[12] Woo, S., O'Neal, D., Pecht, M., 2010, Reliability design of a reciprocating compressor suction reed valve in a common refrigerator subjected to repetitive pressure loads, Engineering Failure Analysis, 2010, 17(4), 979-991.

[13] Woo, S., Pecht, M., O'Neal, D., 2009, Reliability Design and Case Study of a Refrigerator Compressor Subjected to Repetitive Loads, International Journal of Refrigeration, 32 (3), 478-486

[14] Woo S, O'Neal D, Pecht M. Reliability design of residential sized refrigerators subjected to repetitive random vibration loads during rail transport, Engineering Failure Analysis 2011, 18(5), 1322–1332.

[15] Karnopp, D., Margolis, D., Rosenberg, R., 2000, System Dynamics: Modeling and Simulation of Mechatronic Systems, third ed. John Wiley & Sons, Inc, New York.

[16] McPherson, J., 1989, Accelerated Testing, Packaging, Electronic Materials Handbook, ASM International 1, 887-894.

[17] Ajiki, T., Sugimoto, M., Higuchi, H., 1979. A new cyclic biased THB power dissipating ICs. In: 17th Annual Proceedings Reliability Physics. pp. 118–126.

[18] Ryu, D., Chang, S., 2005, Novel Concept for Reliability Technology, Microelectronics Reliability, 45 (3), 611-622.

[19] Ryu, D., 2012, Improving Reliability and Quality for Product Success, CRC Press

Chapter 12

Data Processing Approaches for the Measurements of Steam Pipe Networks in Iron and Steel Enterprises

Luo Xianxi, Yuan Mingzhe, Wang Hong and Li Yuezhong

Additional information is available at the end of the chapter

http://dx.doi.org/10.5772/47781

1. Introduction

Steam is important secondary energy resource media in iron& steel enterprises, amounting to nearly 10% of the whole energy consumption. When an enterprise is running, if all the produced steam can meet all the demands and no steam is bled, the overall energy efficiency can be effectively improve. Thus the complex networks of steam pipes and steam production scheduling systems were set up. Obviously, steam scheduling has to depend on the real time measured data from the steam pipe networks. Accurately measuring the variables of pressure, temperature and flow rate is essential to secure the safety and economic efficiency. It's also necessary for accumulating the amount of steam production or the consumption to calculate the energy cost for each working procedure.

With the help of Energy Management System (EMS), all the data collected from the distributed instruments. In practice, the measurements of pressure and temperature are usually accurate enough for the application except that the sensors or the transducers fail. However, the measurements of mass flow rate are not so accurate because of the complex nature of the steam itself, lacking of high precision measuring instruments, the impact of interference and information transmitting network failures and other reasons. The reliability of the measurement of mass flow rate is poor.

When the steam mass flow measurement values deviate from the actual values to a certain extent, the automatic control system may largely deviate from the process requirements substantially. Even worse steam bleeding or accident could happen [1]. Therefore, it is not a satisfactory to decide or adjust the production process according to the data from the flow rate meters [2]. In energy management, the accumulation differences between production and consumption make it difficult to calculate the energy costs, analyze the segments of

INTECH
open science | open minds

© 2012 The Author(s). Licensee InTech. This chapter is distributed under the terms of the Creative Commons Attribution License http://creativecommons.org/licenses/by/3.0), which permits unrestricted use, distribution, and reproduction in any medium, provided the original work is properly cited.

irrational use of steam, and find the weakness in the management links. Therefore, improving the reliability the steam flow rate measurement data is essential for the normal production and energy conservation in iron& steel enterprises.

The objective of the work may be depicted by figure 1. The real time data (mainly referring to the mass flow rate variables) would be mainly processed by three approaches. By fault data detection and reconstruction, the fault data would be picked out and the real value would be reconstructed or estimated. By gross error detection, the data with gross error would be discovered and re-estimated. By data reconciliation, the random error would be decreased and the quality of the data would be further improved.

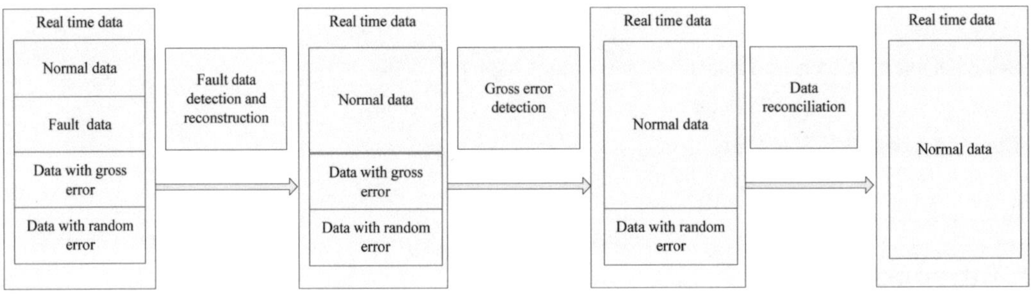

Figure 1. The objective of the work

For fault data detection and reconstruction, the reasons of the low accuracy of steam flow rate data are introduced. By applying the statistical process control theory[3] to determine univariate and multivariate process control limit to monitor the abnormal data online. The approach to calculate the real data (mass flow rates) through the thermal and hydraulic mathematical models of the steam pipe networks is proposed.

For the section of gross error detection, the definition of the problem is introduced. And the two basic gross error detection approaches, the Measurement Test (MT) method and the Method of pseudonodes (MP), are demonstrated.

For the section of data reconciliation, the constrained least-squares problems stated in the section of gross error detection is discussed in detail, including the assumptions for the application,the constraint equations and the selection of the weighted parameter matrix.

The presented approaches of data processing can be programmed for computers to determine the abnormality and improve the precision of the mass flow rate data.

2. Fault data detection and reconstruction

2.1. The reasons of the poor accuracy of steam flow rate measurement

There are many reasons for the poor accuracy of steam measurement data. The most important reasons can be concluded as follows:

1. The working conditions of the instrument deviating from the designed conditions

At present, the mass flow rates are mostly deduced by the volume flow rates and the density. However, the changes of temperature and pressure in the transmission process would lead to the density of steam deviate from the original designed value [3]. The measurement errors would be very large [1]. Any more, some superheated steam would change into a vapor-liquid two-phase medium, which making precision worse.

2. The Complexity of Steam Characteristics
With the ambient temperature changing, the total amount of the condensate water in the transmission process will be different. That makes difference between the amount of the production and the consumption of steam. In addition, the steam pipe leakage will add to the difference. Therefore, the accumulated readings are always doubtful.

3. The Occurrence of wearing or damage to the Key Components
As the orifice differential pressure flow rate meter being in use for long time, the size of aperture would differ from the original size because of the adhered foreign bodies, or erosion by the durative high temperature steam flow. As the parameters can not be adjusted on time, and it is hard to calibrate the instrument, the measurement errors will accumulate.

4. The external Interference or the Failure of Data Transmission Channel
The instrument interfered by the disturbances or failure happening in data transmission channels will make significant errors to the data derived in the control center.

2.2. Determining control limits for fault data detecting

The abnormal data from certain sensors (including temperature, pressure and flow rate sensors) may be characterized as fluctuation quickly with a large magnitude (induced by the poor contact in the instruments), keeping a certain value without any varying (induced by the failure of the data sampling systems) or being outside the normal variable range. The first two cases are not discussed here because they are easy to be discovered. As for the last case, the statistical process control approaches to determine the control limits for data monitoring is applied. When the values of the controlled variables (or functions) exceed the limits, it shows there are abnormal data except that the process is actually abnormal. The statistical process control includes two types--univariate and multivariate control.

2.2.1. Determining the control limits for single variable

The statistical process control (SPC) and control chats were first proposed by Shewhart in quality prediction. Traditional SPC mainly treated single variables. When the value locates outside the normal range, the system would output an alarming signal to notify the operators to check and make sure whether the state is abnormal. In the work, if the process state is actually normally operating, the values of the variables can be judged as fault data. Reasonably determined control limits can reduce the probability of false alarm. Many researches focus on the problem [4-6]. In the work, empirical distribution function combined with the principle of "3σ"[3]is applied to determine the control limits of different variables.

Denote $X \in R^{n\times 1}$ as one of the measurement data matrix, n is the number of the samples, each row of X is a measured sample.If $X = (x_1, x_2, ..., x_n)^T$, the range of the data matrix is:

$$R(X) = \max(x_i) - \min(x_i), (i \le i \le n) \tag{1}$$

Divide the range into N intervals with the same length, each length of the interval is:

$$c = \frac{R(X\ )}{N} \tag{2}$$

The corresponding intervals are marked as $c_1, c_2, ..., c_N$. So each element in X belongs to one of the intervals and these data are divided into N groups. The probability of the sampled variable value locates in each interval can be estimated as:

$$f_r = \frac{v_r}{N} \tag{3}$$

$r = 1,2,...N, v_i$ is the number of the data locate in the ith interval. As the data are independent to each other (usually the hypothesis is reasonable). The probability defined:

$$P_r\{X_i \in c_r\} \approx f_r \tag{4}$$

The average probability density of each interval can be written as:

$$p_r\{X_i \in c_r\} \approx \frac{f_r}{R(X_i)/N} = \frac{Nf_r}{R(X_i)} \tag{5}$$

The empirical distribution of X can be got with the lengths of intervals as bottom and p_r as the height. When n and N are large enough, the bar graph would be more similar to the global distribution of X. According to the central limit theorem, when the system has only one stable state, the measured data would follow normal distribution as the sample size grows. So the global distribution function of X can be written as normal distribution:

$$f(x) = \frac{1}{\sqrt{2\pi\sigma}} e^{-\frac{(x-\mu)^2}{2\sigma^2}} \tag{6}$$

Figure 2 shows the distribution graph. The probability of data locating in the range of μ±3σis 99.73%. When the value of X is outside the range, it's reasonable to suspect the value being abnormal (otherwise, the process doesn't normally function). That's the principle of "3σ". In practice, the outside of "3σ"occupying about 0.25% probability is named "red zone", and the range inside μ±2σis named "green zone"(about 95% probability). The intervals between them are "yellow zone". Four control limits are determined by the method.

As the expectation and the deviation of the X global distribution are unknown, and the estimated values with the sample are not accurate especially when the empirical distribution isn't similar to normal distribution, the principle of "3σ" can not be applied directly.

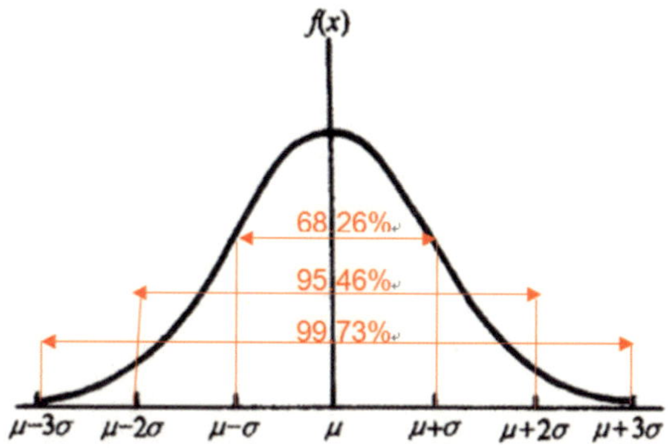

Figure 2. The probability density distribution graph of normal distribution

However, according the principle of "3σ" and the empirical distribution of X, the control limits can be deduced reversely by searching the respective probability of limits.

Two kinds of special instances have to be noticed:

1. No crossing point between empirical distribution and the testing level.
   It shows the corresponding sample value is not appeared in the significance test level range. To determine the limits, the empirical distribution can be linearly extended to outer range or estimated with experience by comparing with the real process performance state.

2. Two more crossing points between empirical distribution and the testing level at one side.
   It shows there are disturbances in the data, which makes the empirical distribution fluctuate at one lateral side. According to principle of "3σ" , find the highest probability interval and search from it to the two laterals for the points first amounting to the probability $(1-\alpha)/2$.

Figure 3-5 show the control limits (4 lines) and monitoring result in a plant. The interval between the two green lines is the "green zone" showing normally working. The intervals outside the red lines are the "red zones" indicating the fault data or process abnormality. The other two intervals are the "yellow zones" which reminds the operators to notice the changes of the data. Figure 4 and 5, in the right side the red and green line are coincident. It can be explained that the outer equipment constraints the variable freely moving.

By determining the limits of single variable, the wild values considered as fault data can easily be discovered. However, two kinds of inherent errors can not be avoided.

The approach of Univariate Statistical Process Control (USPC) can roughly judge fault data. The approach is also appropriate to monitor the temperature and pressure variables.

2.2.2. Determining the control limits of multivariate system based on PCA

The shortage of USPC is the operators being prone to "information overload"when the number of the variables is large. Anymore, it's too rough to judge the normalty of data just by monitor alarming. In practice, the variables maybe constrained by certain inherent functions. It shows the existence of fault data when the functions are unsatisfied. Principal Component Analysis (PCA) is one of the approaches to discover the relationship among the variables by means of statistics. By determining the control limits of two guidelines, Hotelling's T^2 and SPE (Squared Predictive Error, Q), the multivariate process can be monitored conveniently.

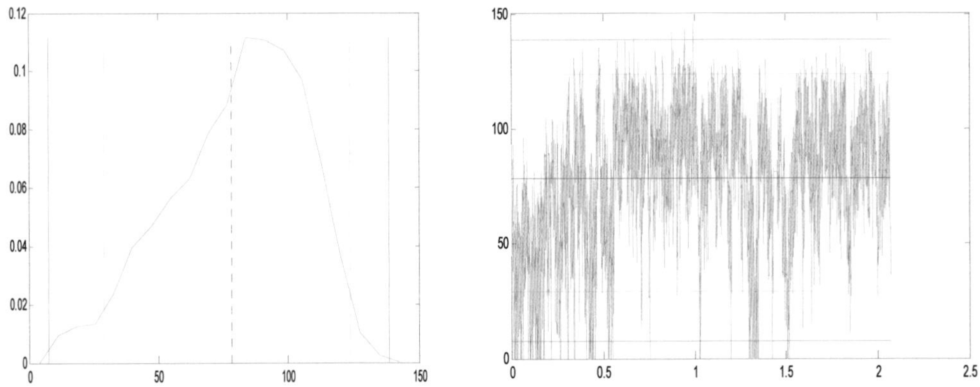

Figure 3. Monitoring the Steam Flow Rate of the Steel Making Process

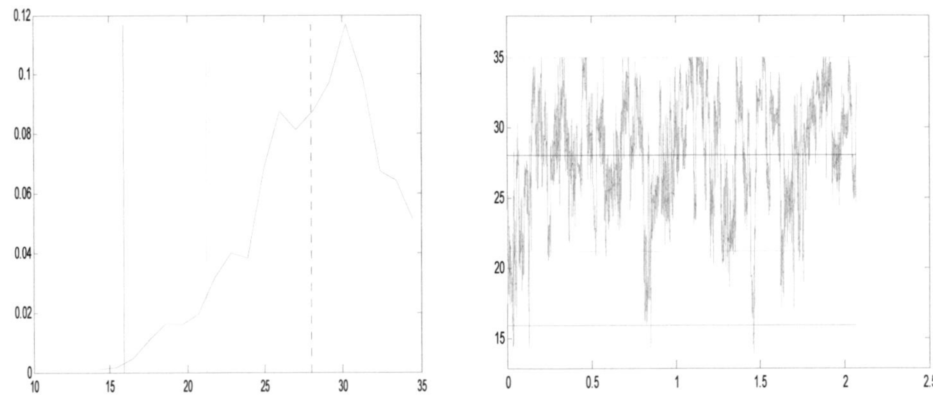

Figure 4. Monitoring the Flow Rate of Start Steam Boiler

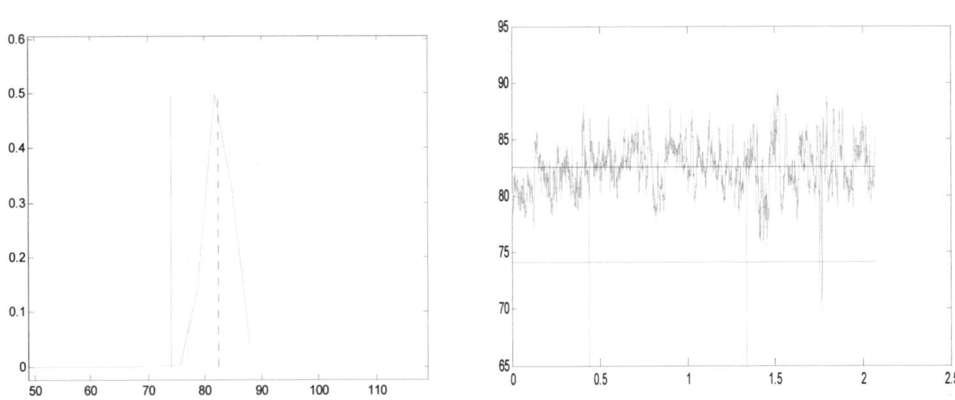

Figure 5. Monitoring the Steam Flow Rata of CDQ (Coke Dry Quenching)

The proposed approach is to distinguish different stable states of the process by empirical distribution function, and determine the control limits for each stable state.

i. Differentiate the stable states and group the samples

Denote the measurement matrix as $X \in R^{n \times m}$, n is the size of samples, m is the number of the monitored variables. The rows of X are the serial samples in time order.

Denote $X_i = (x_{1i}, x_{2i}, ..., x_{ni})$ (i=1,2,...,m) as the ith column. The empirical distribution of X_i can be derived as just mentioned. When the process has two or more stable states, the distribution bar graph would characterize as multi peaks.

Figure 6 is the curve of flow rate and the distribution bar graph of a chemical process in a period. The two peaks represent different demands for steam in different states.

Without loss of generality, the two peaks distribution is to be discussed. If the control limits are determined without considering the different states, the sensitivity of detecting fault data will be too low for the multivariate process. Apply two normal distributions to fit the two peak sections. If the averages and deviations are signed as $X_{i.s1}, \sigma_{i.s1}, X_{i.s2}, \sigma_{i.s2}$, the rule to divide the samples into different groups (different states) is listed as equation (7):

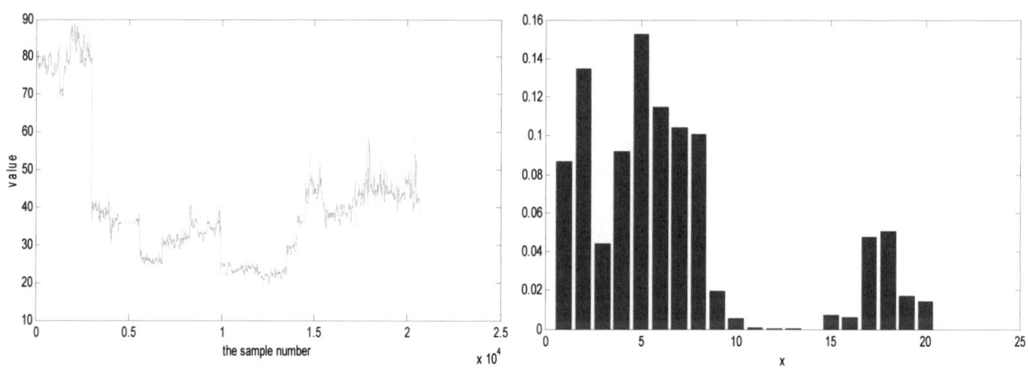

Figure 6. The curve and distribution bar graph of flow rate for a chemical process

$$S = \begin{cases} S_1 & x_{ik} \in [\min(X_i), X_{i.s1} + \gamma\sigma_{i.s1}] \\ S_2 & x_{ik} \in [X_{i.s2} - \gamma\sigma_{i.s2}, \max(X_i)] \end{cases} \quad k = 1, 2, ..., n \tag{7}$$

γ can be selected in the interval of (1, 3), and the total state region should cover the whole range of the sampled data. Ideally, the state regions should not have folded region or the folded region is narrow and with low probability. Or, the variable have no necessity for state differentiating, which is not the case discussed here.

By the rule of (7), the samples are separated into two groups. By the same way, the two groups can be further grouped with other variables. However, it's not suggested to divide the sample into many groups with too few elements in each group. Assume there are only two groups of samples are derived. They are marked as $X_a \in R^{n_1 \times m}$ and $X_b \in R^{n_2 \times m}$ ($n_1 + n_2 \le n$,some samples may be discarded because they don't belong to any state region).

ii. Determining the control limits of different state with PCA

Through grouping, the sample data is more convergence. First standardize X_a, denote:

$$X_a = (x_1, x_2, ..., x_m) \tag{8}$$

In it,, $x_1, x_2, ..., x_m \in R^{n_1 \times 1}$. Calculate each average and standard deviation:

$$M_a = (m_1, m_2, ..., m_m) \in R^{1 \times n_1}. \tag{9}$$

$$S_a = (s_1, s_2, ..., s_m) \in R^{1 \times n_1} \tag{10}$$

The measured data matrix is standardized as:

$$\tilde{X}_a = [X_a - (1 \quad 1 \quad ... \quad 1)^T M_a] diag(\frac{1}{s_1}, \frac{1}{s_2}, ... \frac{1}{s_m}) \tag{11}$$

Derive the eigenvalues of the covariance matrix $C = \frac{1}{n-1} \tilde{X}_a^T \tilde{X}_a$ and orthogonalize the matrix, then the model of PCA can be written as:

$$\tilde{X}_a = \hat{X}_a + E = T_a P^T + E \tag{12}$$

$$T_a = \tilde{X}_a P \tag{13}$$

In the equations, $P \in R^{m \times A}$ is the load matrix. If arrange the eigenvalues of C in descendent order, the former A elements are noted as $\lambda_a = (\lambda_1, \lambda_2, ..., \lambda_A)$. The corresponding unitized eigenvectors form the matrix of P, $P = (P_1, P_2, ..., P_A)$. A is the number of the selected components. $T_a \in R^{n_1 \times A}$ is the scored matrix and E is the error matrix..

The MSPC based on PCA defined two statistical variables: Hotelling's T2 and SPE (Q).

If the jth measurement vector is signed as $\tilde{X}_{aj}$ (row vector), the corresponding scored vector is signed as T_{aj} (ie. The jth row of matrix T_a), then T_{aj}^2 is defined as[7]:

$$T_{aj}^2 = T_{aj} \lambda_a^{-1} T_{aj}{}^T \tag{14}$$

When the testing level is α ,its control limit can be calculated according F distribution:

$$T_{A,n_1,\alpha}^2 = \frac{A(n_1 - 1)}{n_1 - A} F_{A,n_1-1,\alpha} \tag{15}$$

$F_{A,n_1-1,\alpha}$ is the critical value corresponding to testing level α, degree of freedom A,n_1-1.

SPE(also called statistical variable Q)is defined as:

$$Q_{aj} = e_{aj}e_{aj}^T = \tilde{X}_{aj}(I - PP^T)\tilde{X}_{aj}^{\ T} \tag{16}$$

The control limit can be calculated by:

$$\delta_\alpha^2 = \theta_1\left(\frac{c_\alpha\sqrt{2\theta_2 h_0^{\ 2}}}{\theta_1} + 1 + \frac{\theta_2 h_0(h_0 - 1)}{\theta_1^{\ 2}}\right)^{1/h_0} \tag{17}$$

In the equation $\theta_i = \sum_{j=A+1}^{m} \lambda_j^i (i = 1,2,3)$, $h_0 = 1 - \frac{2\theta_1\theta_3}{3\theta_2^2}$, c_α is the threshold value under the testing level α of normal distribution.When there's no fault data in the sample or the process normally functions, the two statistical variables satisfy the following inequalities.

$$T_{aj}^2 \le T_{A,n_1,\alpha}^2 \tag{18}$$

$$Q_{aj} \le \delta_\alpha^2 \tag{19}$$

the second data matrix X_b can also be processed to derive the control inequalities:

$$T_{bj}^2 \le T_{A,n_2,\alpha}^{\prime 2} \tag{20}$$

$$Q_{bj} \le \delta_\alpha^{\prime 2} \tag{21}$$

iii. The monitoring results for the experimental data

Three variables are considered and some of the data are modified on the original data. The no. 167-172 and no.350-355 samples show the transition procedure of two states. The no.360-365 samples are modified data which deviate from the normal constraint. Figure 7 shows the changing curves of the three variables and the scatter diagram in 3-dimention space.

Two experiments are designed to testify the advantage of differentiating states applied in this work. Figure 8 is the case no differentiating states, whilst figure 9 is the differentiating case. As the figures shown, the approach of differentiating states and determining the control limits respectively is more sensitive to the transition of states and fault data.

iv. Locate and isolate the fault data

In the case of MSPC, the fault data can be located and isolated by contribution diagram. The variable contribute to SPE and T_i^2 can be written as (22)(23). ξ_i is the ith column of unit matrix. The fault data is judged as the variable with larger contribution to SPE and T^2.

$$Cont_i^{SPE} = (\xi_i^T(I - PP^T)x)^2 \tag{22}$$

$$Cont_i^{T2} = x^T P^T \Lambda^{-1} P \xi_i \xi_i^T x \tag{23}$$

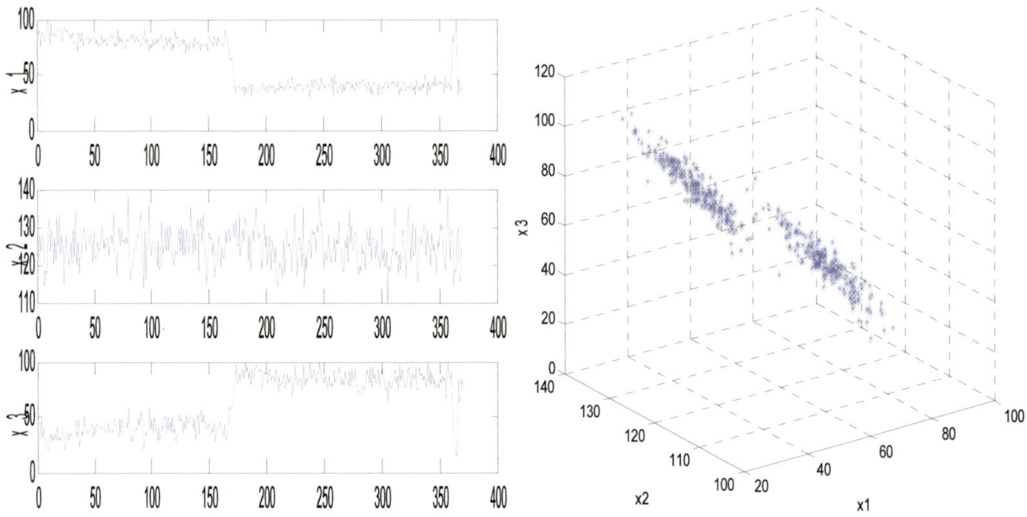

Figure 7. The changing curves and scatter diagram of three variables

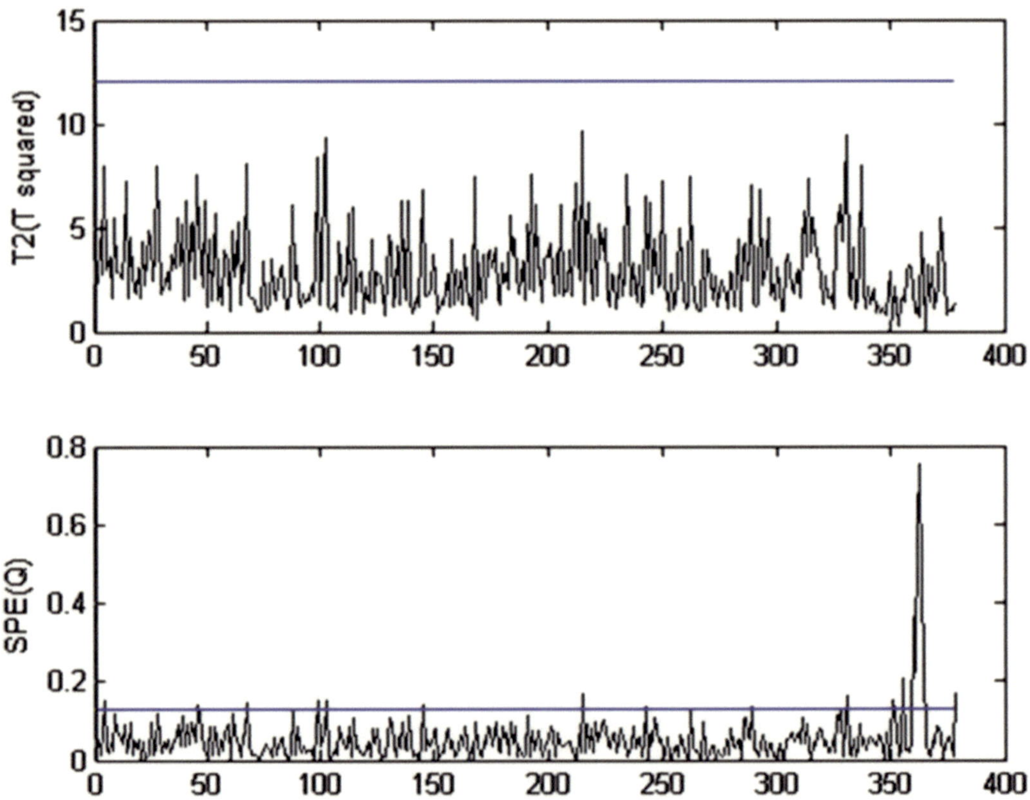

Figure 8. The result of monitoring no differentiating states

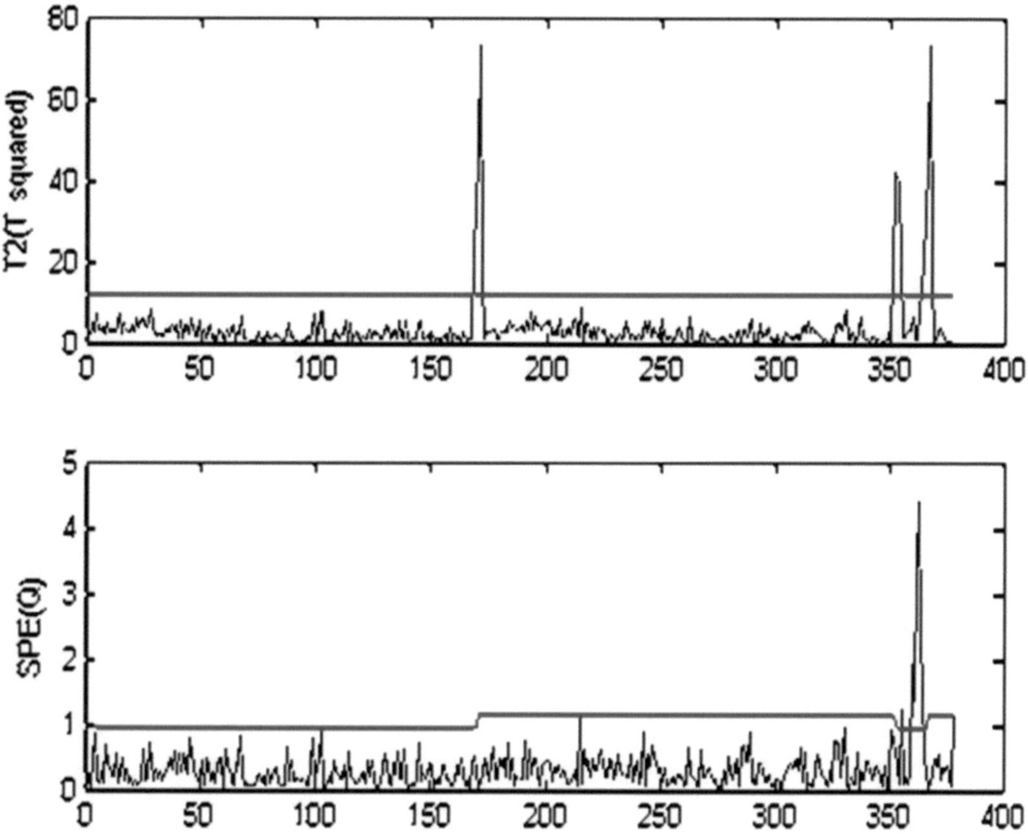

Figure 9. The result of monitoring no differentiating states

2.3. The steam pipe network modeling and calculating the fault data based on the model

After the fault data are detected, the next step is to estimate the true values of the variables. As the experience shown, the measured temperature and pressure values are usually accurate. So the object of the work is to calculate the flow rate value with the detected temperatures and pressures according to the steam network model.

2.3.1. The static model of steam pipe network

To deduce the model, the real conditions are simplified as: 1) The steam in the pipes is axial-direction one dimensional flow; 2) The network is composed of nodes and branches(pipes). 3) There is no condensate or secondary steam and the effect is neglected.

i. The hydraulic and thermal model of single pipe.
   - The hydraulic model of single pipe

     By the law of momentum conservation the hydraulic model can be written as [8]:

$$P_1^2 - P_2^2 = 1.25\times10^8 \frac{\lambda q^2 P_1 T_2 Z_2 L}{D^5 \rho_m T_1 Z_1} \tag{24}$$

P_1, P_2 —the input and output pressure of the steam in the pipe, Mpa;
T_1, T_2 —the input and output absolute temperature, K;
Z_1, Z_2 —compressibility factor of the input and output steam;
λ —frictional factor;
q —the mass flow rate, t/h;
L —the length of the pipe, m;
D —the inner diameter of pipe, mm;
ρ_m —the weighted mean density of steam, kg/m3.

Usually, there's little difference between the products of compressibility factors and temperatures in the same pipe. Considering the elbows, reducer extenders and other friction factors, the equivalent coefficient η is added to the equation. Equation (24) changes to (25).

$$P_1^2 - P_2^2 = 1.25\times10^8 \frac{\lambda q^2 P_1 (1+\eta) L}{D^5 \rho_m} \tag{25}$$

Thus:

$$q = \frac{D^5 \rho_m}{1.25\times10^8 \lambda q P_1 (1+\eta) L} P_1^2 - P_2^2 = C_P(P_1^2 - P_2^2) \tag{26}$$

$$C_P = \frac{D^5 \rho_m}{1.25\times10^8 \lambda q P_1 (1+\eta) L} \tag{27}$$

In accordance with Альтщуль equation, frictional factor λ:

$$\lambda = 0.11\times(\frac{\Delta}{D} + \frac{68}{\text{Re}})^{0.25} \tag{28}$$

Δ —the equivalent absolute roughness, mm; Re —the Reynolds number of the steam.

After transferring the units of D, q to mm and t/h (ton per hour), Reynolds number is :

$$\text{Re} = \frac{D u \rho_m}{\mu} = 354 \frac{q}{D\mu} \tag{29}$$

In the equation, u —the characteristic velocity of the steam in this pipe, m/s; μ — the mean dynamic viscosity coefficient. It's determined by equation (33)[9].

Denote ρ_1,ρ_2 as the input and output steam densities. ρ_1,ρ_2,μ,ρ_m can be calculated with following equations (31),(32),(33). The unclaimed symbols were defined in reference[9].

$$\rho_1 = \frac{P_1}{\pi(r_\pi^0 + r_\pi^r)RT_1} \tag{30}$$

$$\rho_2 = \frac{P_2}{\pi(r_\pi^0 + r_\pi^r)RT_2} \tag{31}$$

$$\rho_m = \frac{\rho_1}{3} + \frac{2\rho_2}{3} \tag{32}$$

$$\mu = \Psi(\delta,\tau)\eta^* \tag{33}$$

- The thermal model of single pipe

By the law of energy conversation, the static thermal model can be written as follows:

$$T_1 - T_2 = \frac{(1+\beta)q_l L}{1000c_p q \times \dfrac{1000}{3600}} = \frac{(1+\beta)q_l L}{278c_p q} \tag{34}$$

$$q = \frac{278c_p q^2}{(1+\beta)q_l L}(T_1 - T_2) = C_T(T_1 - T_2) \tag{35}$$

$$C_T = \frac{278c_p q^2}{(1+\beta)q_l L} \tag{36}$$

The definitions and units of T_1,T_2,L,q are the same as eqation (24).(note: changing equation (34) to (35) is to make thermal model has the same pattern with equation (26)).

β —the appending heat loss coefficient for the appendix of pipe, valve, and support, the value can be 0.15-0.25 depending on pipe laying methods;
c_p —specific heat capacity at constant pressure, kJ/kg.K;
q_l —the amount of heat loss along unit pipe length, W/m.

In term of equation IF-97, The unclaimed symbols are defined in the reference[9].

$$c_p = -R\tau^2(\gamma_{\gamma\gamma}^o + \gamma_{\gamma\gamma}^\gamma) \tag{37}$$

The amount of heat loss along unit pipe length can be calculated using equation (38)

$$q_l = \frac{T_0 - T_a}{\frac{1}{2\pi\varepsilon}\ln\frac{D_o}{D_i} + \frac{1}{\pi D_0 a_w}} \tag{38}$$

T_0, T_a —the out surface temperature of the pipe and the environment temperature, K;
D_0, D_i —the outer and inner diameters of the steam pipe, mm;
ε —the coefficient of thermal conductivity of the heat insulated material, W/m.K;

In equation (38), a_w is determined by:

$$a_w = 11.6 + 7\sqrt{v} \tag{39}$$

v —the velocity of environmental air, m/s;

ii. Steam pipe network synthetic model

- Incidence matrix of pipe network

  Draw out pipe network map, number the nodes in the order of steam sources, users, three-way nodes, and number the pipes in the order of leaf pipes, branch pipes. Suppose there are m_1 intermediate nodes and m_2 outer nodes, there are total $m = m_1 + m_2$ nodes and $p = m - 1$ pipes. Set the element of the ith row and the jth column in $A \in R^{m \times p}$ as a_{ij}, it's defined as:

$$a_{ij} = \begin{cases} 1 \ (flow \ \ in \ \ from \ \ node \ \ i) \\ -1 (flow \ \ out \ \ from \ \ node \ \ i) \\ 0 \ (unrelated) \end{cases} \tag{40}$$

- The flow rate balance equation of the pipe network

  Denote: $P = (P_1^2, P_2^2, ..., P_m^2)^T$, $P_i^2, (i = 1, ..., m)$ the square of the ith node's pressure, MPa²;

  $T = (T_1, T_2, ..., T_m)^T$, $T_i, (i = 1, ..., m)$ the ith node temperature, °C;
  $\vec{q} = (q_1, q_2, ..., q_p)^T$, $q_j, (j = 1, ..., p)$ the flow rate of the jth pipe, t/h;

  $Q = (Q_1, Q_2, ..., Q_m)^T$, $Q_i, (i = 1, ..., m)$ the flow rate of the ith node, unit t/h; For steam souce the value is minus q_j, and for the user the value is positive q_j, otherwise, the value is 0.

  $C_p^* = diag(C_{P1}, C_{P2}, ..., C_{Pp}), C_{Pj}, (j = 1, ..., p)$ parameters are determined by equation (27);

  $C_T^* = diag(C_{T1}, C_{T2}, ..., C_{Tp}), C_{Tj}, (j = 1, ..., p)$ the parameter are determined by equation(36).

  In term of mass conservation law, the total flow rate of each node should be 0, that is:

$$A\vec{q}+Q=0 \tag{41}$$

According to the hydraulic and thermal equations, for the pipe network equations:

$$\vec{q}=C_P^*A^TP=C_T^*A^TT \tag{42}$$

Substitute (42) into (41):

$$AC_P^*A^TP+Q=0 \tag{43}$$

$$AC_T^*A^TT+Q=0 \tag{44}$$

Equations (26), (27), (35), (36), (42), (43), (44) comprise the static model.

2.3.2. Hydraulic and thermal calculation based on searching

Industrial networks are equipped with temperature, pressure and flow meters except intermediate nodes. The proposed algorithm tries to calculate the flow rate of each pipe with the given condition. The flow rate meter readings are applied to evaluate of the algorithm.

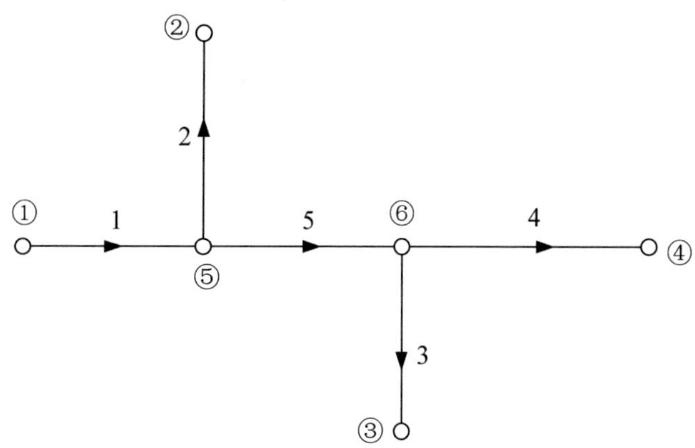

Figure 10. Layout of steam pipe network

Set figure 10 as an example. the index of the nodes and pipes are shown in the figure:

Step 1. Determine the elements of matrices A, Q as the preliminarily defined.

$$A=\begin{pmatrix} 1 & 0 & 0 & 0 & 0 \\ 0 & -1 & 0 & 0 & 0 \\ 0 & 0 & -1 & 0 & 0 \\ 0 & 0 & 0 & -1 & 0 \\ -1 & 1 & 0 & 0 & 1 \\ 0 & 0 & 1 & 1 & -1 \end{pmatrix} \tag{45}$$

$$Q = (-q_1, q_2, q_3, q_4, 0, 0)^T \tag{46}$$

As the previous definitions for equation (41)(42), $P_5, P_6, T_5, T_6, \vec{q}$ are unknown. From equation (27), it's easy to discover friction factor and density are relative with P_5, P_6, T_5, T_6. Only by presetting $\vec{q}$ can equations (42) (43) be applied for hydraulic iterative calculation. However, in thermal calculation only c_p relates with P_5, P_6, T_5, T_6. Changing equation (34) to (35), $\vec{q}$ can be calculated directly.

$$q_i = \frac{(1+\beta)q_{li}L_i}{278c_p\Delta T_i}, (i = 1, 2, ..., 5) \tag{47}$$

In equation (47), i —the pipe index, ΔT_i —the temperature difference of input and output.

Step 2. Thermal calculation

It can be inferred by the flow direction of steam that:

$$P_2 < P_5 < P_1, \max(P_3, P_4) < P_6 < P_5 \tag{48}$$

$$T_2 < T_5 < T_1, \max(T_3, T_4) < T_6 < T_5 \tag{49}$$

Presume:

$$P_5 = \frac{1}{2}(P_2 + P_1); P_6 = \frac{1}{2}(P_5 + \max(P_3, P_4)) \tag{50}$$

Set the searching start value for T_5, T_6 as T_1 and $\max(T_3, T_4)$ respectively, and set the searching step size as h then begin searching T_5, T_6 in the range (49). For each searching step size, calculate c_p (equation (37)) and flow rate of each pipe (equation (49),(50)), and test whether $\|A\vec{q} + Q\| \le \xi_1$ is satisfied. If it is not satisfied, update T_5, T_6 with one step to continue calculation, or retain the temperature value and the corresponding flow rates.

Step 3. Hydraulic calculation

Just as the thermal calculation, first presume the temperature of the intermediate nodes as the value searched by the former step, and set the start point and step size for the two intermediate pressures. Then begin searching in the region determined by inequality (48) with the step size to calculate the density (equation (32)), frictional factor ((28), (29), (33)), and flow rate ((26), (42)) of each pipe. When calculating the flow rate, presetting the initial values as the result of step 2 will reduce the iterative calculation time. Test whether $\|A\vec{q} + Q\| \le \xi_2$ is satisfied, If the inequality is not satisfied, update P_5, P_6 with one step size to continue calculation, or retain the pressure values and the corresponding flow rates.

Step 4. Thermal model Verification

Substitute the intermediate nodes' temperature (determined by step 2), pressure and flow rate values (determined by step3) into thermal model to verify the inequality

$\left\| AC_T^* A^T T + Q \right\| \leq \xi_3$. If it is satisfied, the calculation ends with the result of step 3. If not, presume the intermediate pressure values as the result of step 3, reduce step size and go to step 2 and start another calculation cycle.

In the algorithm, the value ξ_1, ξ_2, ξ_3 can be set as 0.05-0.3 depending on the demand for errors and calculation stability. The calculation flow chart is depicted with figure 11.

2.3.3. Comparisons of calculation results and measurements

The specifications of each pipe are shown in table 1. The parameters in the model and algorithm are initially set as: $\eta = 0.2$, $\Delta = 0.2$, $\beta = 0.15$, $\varepsilon = 0.035$, $\xi_{1,2,3} = 0.3$.

Substitute the specification data in Table 1 and temperature and pressure data in Table 2 to the algorithm calculating flow rate values.The comparision results are listed in Table 2. It shows the largest difference is less than 6%. Actually, the neglected factors in model, the parameter errors and measurement errors add to the difference.

Pipe No.	Length(m)	Inner Diameter (mm)	Outer Diameter (mm)	Heat Insulation Layer Thickness (mm)
1	600	700	720	110
2	2300	600	630	105
3	700	400	426	95
4	1300	500	529	100
5	1200	600	630	105

Table 1. The Specifications of the Pipes

Outer node No.	Pressure(MPa)	Temperature (°C)	Measured q(t/h)	Calculated q (t/h)	the relative difference
1	0.705	286	117	121	3.4%
2	0.679	195	19	20	5.2%
3	0.563	246	19	19	0
4	0.445	256	79	82	3.8%

Table 2. The Comparisons of the Measured Data and the Calculated Data.

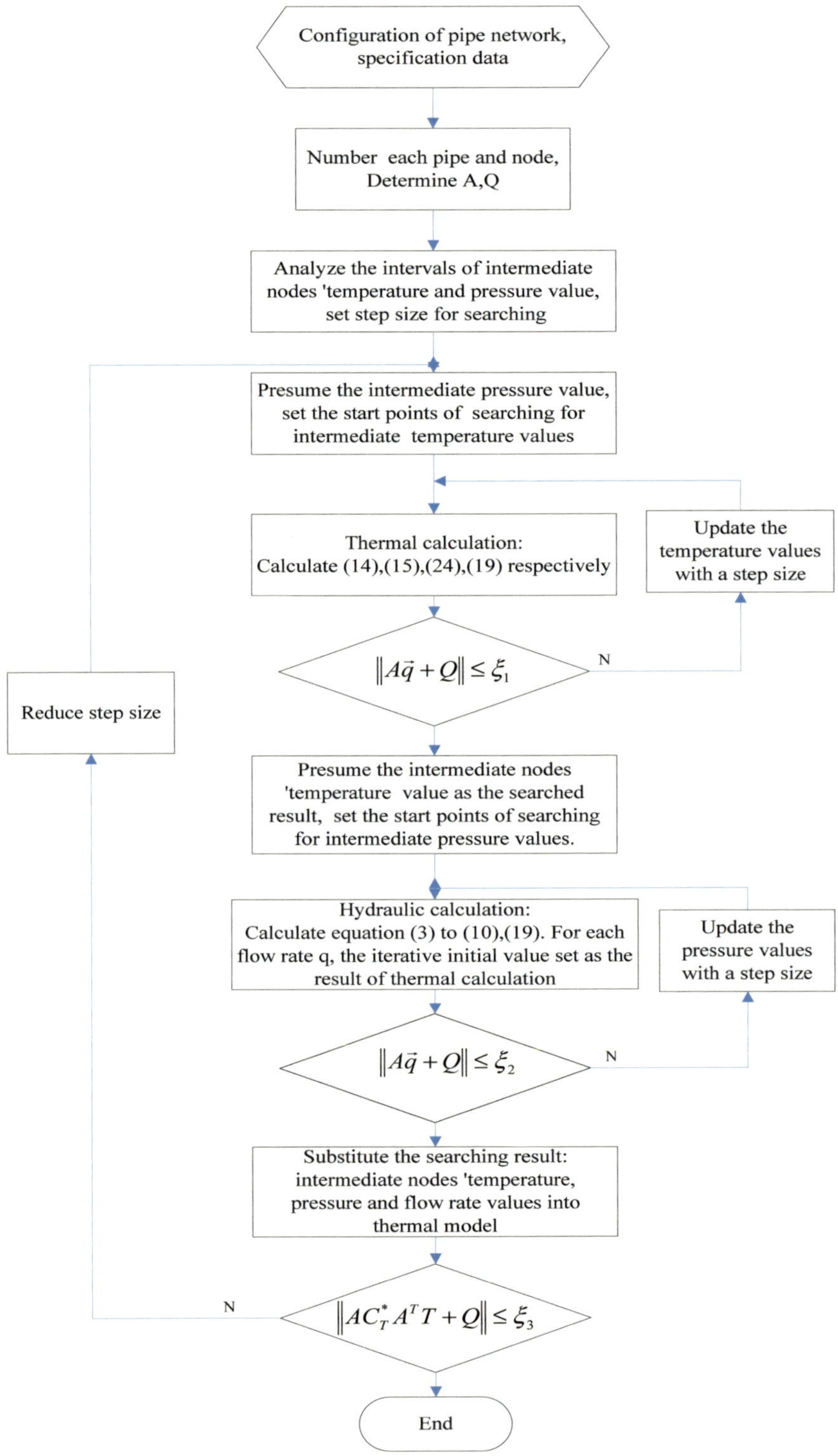

Figure 11. The flow chat of flow rate calculation

The results show the validation of the model and effectiveness of the algorithm, and the proposed model and algorithm can be applied to simulate the running of static steam pipe network, reconstruct the discovered fault data of the mass flow rates.

As for the larger scale steam network with more than 3 intermediate nodes, it's difficult to apply the algorithm directly. However, the pipe network can be divided into several smaller networks from the nodes with known temperature and pressure.

3. Gross errors detection

3.1. Problem definition

A section of the steam network named "S2" for an iron &steel plant in China is shown in figure 12. In the figure, N1-N7 represent the different production processes. The arrows point to the direction of steam flow, the variables Xi or Xij on behalf of the real steam mass flow. The electric valves are remotely controlled by the operators. Many industrial steam systems are similar in structure to this system, but the scale is much larger.

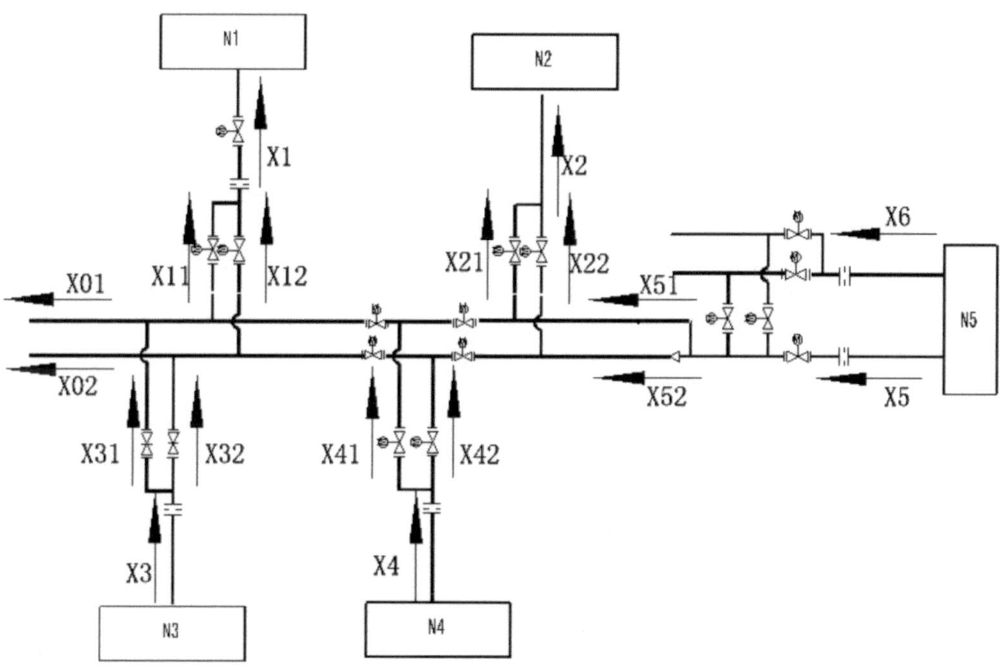

Figure 12. A section diagram of steam network

If all of the variables above have been measured, suppose that the electric valves are fixed at a certain position, and the pipeline leakage and the amount of condensate water can be neglected, the constraint equations can be written out on the basis of the mass balance.

$$\begin{cases} X1-X11-X12=0 \\ X2-X21-X22=0 \\ X3-X31-X32=0 \\ X4-X41-X42=0 \\ X5+X6-X51-X52=0 \end{cases} \tag{51}$$

The overall balance equation can also be written out as:

$$X5+X6+X4+X3-X1-X2-X01-X02=0 \tag{52}$$

If we remark

$$X=(X01,X02,X1,X11,X12,X2,X21,X22,X3,X31,X32,X4,X41,X42,X5,X6)^T=(X_1,X_2,...,X_{18})^T \tag{53}$$

as the vector of mass flow rates in the steam network.(51) (52) can be abbreviated as:

$$AX=0 \tag{54}$$

In (54), A is the incidence matrix. The matrix is composed by the elements of 1, -1, and 0.

If Y represents the vector of measured flow rates and X is the vector of true flow rates , then:

$$Y=X+W+\varepsilon \tag{55}$$

In the equation, W is the vector of the measurement gross errors (or systematic errors), and ε is the vector of random measurement errors with each element being normally distributed with zero means and known covariance matrix, Q. The approaches to detect and remove the gross error from the measurements are to be discussed in this section.

When there is no gross error in the measurement vector, finding a set of adjustments to the measured flow rates to satisfy equation (54) is the problem of data reconciliation. Denote the adjustment vector as a , and the adjusted flow rate vector as $\hat{X}$, we get:

$$\hat{X}=Y+a \tag{56}$$

Applying Least squares method, it can be stated as the constrained least-squares problems:

$$\min a^T Q^{-1} a \text{ subject to } A\hat{X}=0$$

Q is the weighed parameter matrix. Usually, Q is selected as:

$$Q=diag(\sigma_1^2,\sigma_2^2,...,\sigma_{18}^2) \tag{57}$$

$\sigma_i(1\le i\le 18)$ is the deviation of the measurement variable $X_i(1\le i\le 18)$.

The solution $\hat{X}^*$ to the problem can be obtained by Largrange multipliers[10]

$$\hat{X}^*=Y-QA^T(AQA^T)^{-1}AY \tag{58}$$

The vector of residuals e

$$e = Y - \hat{X}^* = QA^T (AQA^T)^{-1} AY \tag{59}$$

By gross error detection, the systematic errors in the measurements can be removed, and by data reconciliation the random errors can be reduced. However, if the gross errors are not removed, the gross errors in some variables will propagate to other accurately measured variables in the procedure of data reconciliation. So gross error detection had better be adopted in advance.There are two basic types of gross error detection methods, they are based on measurement test (MT) and nodal imbalance test (NT).

3.2. Gross error detection based on test of residuals

Detection based on test of residuals belongs to the statistical test and has been termed and evaluated [11]. Measurement test[12] is the basic algorithm.

Step 1. Apply the least-squares routine by using equation (58) (59) to compute $\hat{X}^*$ and e.
Step 2. Compute the variable for each pipe (or stream)

$$z_j = e_j / \sqrt{v_{jj}} \tag{60}$$

In the equation,

$$V = QA^T (AQA^T) AQ \tag{61}$$

On the hypothesis that the measured value in the jth stream doesn't contain gross error, z_j follows standard normal distribution.

Step 3. Compare z_j with a critical test value z_c. If $|z_j| > z_c$, denote stream j as a bad stream. z_c recommended[12] $z_c = z_{1-\beta/2}$, the $1-\beta/2$ point of the standard normal distribution.

$$\beta = 1 - (1-\alpha)^{1/n} \tag{62}$$

Where n is the number of measurements tested (currently, n=18), α (the recommended value is 0.05) is the overall probability of a type I error for all tests, and β is the probability of a type I error for each individual test. Denote by S the set of bad streams found by the above procedure. The measurement $y_j, j \in S$ is considered to contain gross error.

Step 4. If S is empty, proceed to step 7. Otherwise, remove the streams contained in S and aggregate the nodes connected with the stream. This process yields a system of lower dimension with compressed incidence matrix A', measurement vector Y', and weighed matrix Q'. Denote T as the set of streams with the measurement data in Y'.

Step 5. Replace A, Y, and Q with A', Y', and Q' respectively and Compute the least-squares estimates in T applying equation (58).

Step 6. Solve equation (54) to Compute the rectified values of the streams in S by substituting the estimated values computed in step 5 with the data of the steams in T. The original measured data are used for the streams in the set $R = U - (S \cup T)$, where U is the all stream set.

Step 7. The vector comprised of the results from steps 5, 6 and the original measured data for the streams in R is the rectified measurement vector. If S is empty, then $\hat{Y} = \hat{X}^*$, the rectification of the measured data is completed in step 1.

Notations to the algorithm:

1. In step 4, it's possible to remove the good streams into S in some cases.
2. The equation (62) and critical level value provides a conservative test since the residuals are generally not independent. It's not always applied [13].
3. The different significance levels will induce different results and effects.
4. For the shortage of MT that it tends to spread the gross errors over all the measurement and obtain unreasonable (negative or absurd) reconciled data, The Iterative Measurement Test (IMT) method and Modified IMT (MIMT) method were proposed. However, the main frames are the same with MT.

3.3. Gross error detection based on nodal imbalance test

The algorithms of gross error detection based on statistical test of nodal imbalance are mainly based on the work of literature [14]. Applying nodal imbalance test to each node and the aggregation node (ie. pseudonode) to locate and remove the gross errors. The basic algorithm named Method of pseudo nodes (MP) is listed as:

Step 1. Compute the nodal imbalances vector r and the statistical testing variable vector z. On the assumption that no systematic error exits, the defined variable z_i are standard normal distributed.

$$r = AY \tag{63}$$

$$z_i = r_i / \sqrt{g_{ii}} \tag{64}$$

Where

$$G = AQA^T \tag{65}$$

Step 2. Compare each z_i with a critical test value $z_c = z_{1-\alpha/2}$, which corresponding to the point of the significance testing level $(1-\alpha/2)$. For the instance a test at the 95% significance level, $\alpha = 0.05$ and $z_c = 1.96$. If $|z_i| \le z_c$ is satisfied, the ith node is regarded as a good node and denote all streams connected to the node i as good streams.

Step 3. If no bad nodes are detected in step 2, proceed to step5. Otherwise, repeat steps 1 ,2 (changing the matrices and vectors accordingly) for pseudonodes containing 2, 3... m nodes.

Step 4. Denote the set of all streams not denoted as good in the previous steps by S. The measurements $y_i, i \in S$ are considered containing gross errors.

Steps 5-8. The procedure is the same as the steps 4-7 of the MT algorithm.

Notations to the algorithm:

1. The principle assumption is that the errors in two or more measurements do not cancel.
2. In step 3, m is chosen by the effect of locating the gross errors. If the increasing of m doesn't gain any improvement, the step can be stopped.
3. By applying graph-theoretical rules[14] some streams can be determined as bad measured streams. The additional identification may be useful for the procedure of MP.
4. Equation (62) is not applied in the algorithm to control type I error. The probability of a type I error in the nodal imbalance test is not necessarily equal to the probability of rejecting a good measurement in MT [15].
5. If the set S obtained in step 4 is empty, but there is one or more nodes are truly bad. Usually the instances happen when there's leak in the network system or the measured errors canceling each other. To solve the problem, the Modified MP (MMP), MT-NT combined methods [16] are proposed.

As the above two methods shown, gross data detection and data reconciliation are inherently combined together. Gross errors detection is proceeded with the help of the least-square routine, which actually can obtained the optimal estimates.

4. Data reconciliation

4.1. The basic problems of data reconciliation

Just as mentioned in section 3, the data reconciliation problem is expressed as the solution of the constrained least-squares problems.However, the assumptions, the constraint equations, and the weighted parameter matrix have to be discussed.

4.1.1. On the assumptions of data reconciliation in the present application

For the application of the steam network, there are four latent assumptions. They are:

1. The process is at the steady state or approximately steady state.
   Suppose that the electric valves are fixed at a certain position, and the mass flow rate in each steam has been close to a constant for a period of time. So the constraint equation (54) can be written out on the basis of the balance for all the nodes (including real nodes and pseudonode) and the solution of the problem can be in accordance with the actual state.

2. The measurement data are serial uncorrelated. The assumption makes it easier to estimate the deviations of the measurements. Although this is usually not the real case, the serial data can be processed to reduce the confliction [17].

3. The Gross Errors have been detected and removed. If there's gross error in the measurement data, the procedure of data reconciliation will propagate the errors.

4. The constrained equations are linear. The style of equation (54) is linear; however the true constrained equation will be related with many other environmental variables and obviously not linear. So the assumption is to be approximately satisfied.
Only these four assumptions are nearly satisfied, can the solution of the problem (equation (58)) be closer to the true value.

4.1.2. On the constraint equation

Need to note equations (51) and (52) are based on the assumptions of steady state and linear constraints, no pipeline leakage or condensate water loss. But it cannot avoid the condition of the pipeline leakage and condensate water. So the equations (51)(52)should be written as:

$$\begin{cases} X1 - X11 - X12 - \delta 1 = 0 \\ X2 - X21 - X22 - \delta 2 = 0 \\ X3 - X31 - X32 - \delta 3 = 0 \\ X4 - X41 - X42 - \delta 4 = 0 \\ X5 + X6 - X51 - X52 - \delta 5 = 0 \end{cases} \tag{66}$$

$$X5 + X6 + X4 + X3 - X1 - X2 - X01 - X02 - \delta 0 = 0 \tag{67}$$

Here:

$$\delta i = f_i(\tau, D_i, l_i, T_i, P_i, X_i) \qquad (i = 0,1,\dots,5) \tag{68}$$

It represents the vector for the condensate water and leakage loss amount of each constraint equation. Each element has relations with the environmental temperature τ, pipe diameter D, pipe length l, steam temperature T, pressure P and the flow rates of main pipes.

It's difficult to set up the mathematical models of these loss amounts. However, by the first order of Taylor Expansion at the point of ($\tau_0, D_{i0}, l_{i0}, T_{i0}, P_{i0}, X_{i0}$):

$$\begin{aligned} \delta i &= f_i(\tau_0, D_{i0}, l_{i0}, T_{i0}, P_{i0}, X_{i0}) + \frac{\partial f_i}{\partial \tau_0}(\tau - \tau_0) + \frac{\partial f_i}{\partial T_0}(T - T_{i0}) + \frac{\partial f_i}{\partial P_{i0}}(P_i - P_{i0}) + \frac{\partial f_i}{\partial X_{i0}}(X_i - X_{i0}) \\ &= \delta_{i0} + k_{i\tau}(\tau - \tau_0) + k_{iT}(T - T_0) + k_{iP}(P_i - P_{i0}) + k_{ix}(X_i - X_{i0}) \end{aligned} \tag{69}$$

The constants in equation (69) can be determined by the method of multiple linear regression with a certain number of history data. As the constraint equation changed into:

$$AX = \delta \tag{70}$$

The solution to the least-squares problem of reconciliation is:

$$\hat{X}^* = Y - QA^T(AQA^T)^{-1}(AY - \delta) \tag{71}$$

4.1.3. On selection of the weighted parameter matrix Q

The selection of Q directly influences the result of the data reconciliation. In theory Q is recommended as the equation (57). However, the deviations are usually unknown or may vary as the time of instruments being used getting long. The deviation of each measurement can be estimated by the standard deviation of the sample data.

$$\sigma_i \approx s_i = \sqrt{\frac{1}{N-1}\sum_{j=1}^{N}(X_{ij} - \bar{X}_i)^2}, (1 \le i \le 8) \tag{72}$$

Where, X_{ij} is the j th sample of the variable X_i, and $\bar{X}_i$ is the average of X_i.

If a small number of high-precision instrument are applied,, the corresponding elements in the matrix of Q with smaller value, then the quality of data will be greatly improved. Though some literatures [18] [19] recommended the methods to determine or adjust the matrix Q, the methods or theories for general application is not available.

4.1.4. Simulation results

The simulation results of data reconciliation for the present application are shown in table 3. The measured data is presumed according to (73)

$$Y = X + 10\% \times X. \times Rand(18,1) \tag{73}$$

The standard deviations are set to be 0.5 percent of the true values. In the simulation, the pipe network loss is not considered. As shown in table 3, most of the rectified data are much closer to the true values. The simulation testifies the efficiency of data reconciliation.

	X01	X02	X1	X11	X12	X2	X21	X22	X3
true value	5	5	50	25	25	60	30	30	30
measured value	5.347	5.159	54.751	25.086	26.097	62.289	32.297	32.386	30.561
variance	0.0006	0.0006	0.0625	0.0156	0.0156	0.0900	0.0225	0.0225	0.0225
rectified value	5.347	5.150	52.228	25.090	25.514	62.178	31.946	31.730	31.174
	X31	X32	X4	X41	X42	X5	X51	X52	X6
true value	15	15	40	20	20	20	25	25	30
measured value	15.735	15.668	42.585	21.419	21.509	20.552	26.699	26.638	32.488

	X01	X02	X1	X11	X12	X2	X21	X22	X3
variance	0.0056	0.0056	0.0400	0.0100	0.0100	0.0100	0.0156	0.0156	0.0225
rectified value	15.582	15.592	42.306	21.340	21.566	20.946	26.086	26.236	31.375

Table 3. The Result of Data Rectification

5. Conclusion

In the chapter, three data processing approaches to improve data quality are demonstrated. For the importance of properly controlling the steam system performance normally, The data obtained from EMS should be accurate and reliable. However, the data may be influenced many outer factors. The approaches proposed in the chapter are to detect the fault data, locate and remove the gross errors and reduce the random errors.

Four main reasons induce the low accuracy of the mass flow rate measurement. Combining the principle of "3σ" and empirical distribution function to determine control limit is proposed for single variable monitoring, and applying PCA to determine the control limits for the multivariate process. With the limits, most of fault data can be identified easily. For the fault data of flow rates, the approach to setup the mathematical model of the steam network and calculate the flow rates is proposed. The simulation and experimental results show the effectiveness of the approaches.

Two approaches, MT and MP, to detect the gross errors are demonstrated. Both are preceded by selecting the statistical variables, which follow standard normal distribution, and applying hypothesis test. Some notations for the two algorithms are stated.

The constrained least-squares problem applied for present application is discussed. The four assumptions are approximately satisfied when the steam network is normally function and the state is nearly static. The pipe network loss can be considered to add to the constraint equations for more accurate results. The weighed parameter matrix has influence on the results of data reconciliation. To estimate the deviations of the instruments online and apply several instruments with high precision will improve the quality of reconciled data.

Author details

Luo Xianxi, Yuan Mingzhe and Wang Hong
Key Laboratory of Industrial Informatics, Shenyang Institute of Automation, Chinese Academy of Science, Shenyang, China

Luo Xianxi and Li Yuezhong
East China Institute of Technology, Nanchang, China

Luo Xianxi
Graduate School of the Chinese Academy of Sciences, Beijing, China

Yuan Mingzhe and Wang Hong
Shenyang Institute of Automation, Guangzhou, Chinese Academy of Science, China

Acknowledgement

This work is supported by the Knowledge Innovation Project of Chinese Academy of Science (No.KGCX2-EW-104-3), National Nature Science Foundation of China (No.61064013) and Natural Science Foundation of Jiangxi Province,China (No. 20114BAB201024).

6. References

[1] T. Tong, *The Analysis on the Flow Rate Measurement of Steam* Journal of Hebei Institute of Architectural Engineering, 2002. 6(20): p. 43-45.
[2] D. Zhenjian, On the Methods to Reduce the Steam Measurement Difference along the Transmission Line and the Existing Problems. China Metrology, 2003. 3: p. 67-68.
[3] Zhang J., Y.X.H., Multivariate Statistical Process Control. Peking: Chemical Industry Press (CIP). 2002.
[4] Chand, S.T.S., Estimating the limits for statistical process control charts: A direct method improving upon the bootstrap. European Journal of Operational Research, 2007. (178): p. 472-481.
[5] Kuo, H.C.W.-L., Comparisons of the Symmetric and Asymmetric Control Limits for X and R Charts. Computers & Industrial Engineering, 2010.（59）: p. 903-910.
[6] Chen, T., *On Reducing False Alarms in Multivariate Statistical Process Control.* Chemical Engineering Research and Design, 2010. (88): p. 430-436.
[7] Zhou Donghua, Li Gang, ed. Data Driven Fault Diagnosis Technology for Industrial Process--Based on Principal Component Analysis and Partial Least Squares. Peking: Scince Press. 2011.
[8] Zhang Zenggang, Research on Coupling Hydraulic & Thermal Calculation for Steam Pipe Network Theory & its Application(D). 2008: China University of Petroleum.
[9] W.Wagner, A.K., *Properties of Water and Steam.* 1998: Springer-Verlag Berlin Heiderberg.
[10] Kuhn, D.R., H.Davidson, *Computer Control II:Mathematics of Control.* Chem. Eng.Prog., 1961: p. 57.
[11] Iordache, C., R.S.H.Mah, Performance Studies of the Measurement Test for Detection of Gross Errors in Process Data. AIChE Journal, 1985. 31,1: p. 187.
[12] Mah, R.S.H., A.C.Tamhane, *Detection of Gross Errors in Process Data.* AIChE Journal, 1982. 28: p. 828.
[13] Crowe, C.M., Y.A.Garcia, *Reconciliation of Process Flow Rates By Matrix Projection.* AIChE Journal, 1983. 29: p. 881.
[14] Mah, R.S., G.M.Stanley and D.Downing, *Reconciliation and Rectification of Process Flow and Inventory Data.* IEC Proc. Des. Dev., 1976. 11: p. 197.
[15] Serth, R.W. and W.A. Heenan, Gross Error Detection and Data Reconciliation in Steam-metering Systems. AIChE Journal, 1986. 32(5): p. 733-742.

[16] Mei, C., H. Su, and J. Chu, *An NT-MT Combined Method for Gross Error Detection and Data Reconciliation.* Chinese Journal of Chemical Engineering, 2006. 14(5): p. 592-596.

[17] Kao.C.S, e., A General Prewhitening Procedure for Process and Measurement Noises. Chem. Eng.Prog., 1992. 118: p. 49.

[18] Wu Shengxi, Z.e., Estimation of Measurement Error Variances/Covariance in Data Reconciliation, in the 7th World Congress on Intelligent Control and Automation. 2008: Chongqing, China. p. 714-718.

[19] Narasimhan, S. and S.L. Shah, *Model identification and error covariance matrix estimation from noisy data using PCA.* Control Engineering Practice, 2008. 16(1): p. 146-155.

Chapter 13

Transport Intensity and Energy Efficiency: Analysis of Policy Implications of Coupling and Decoupling

Rafaa Mraihi

Additional information is available at the end of the chapter

http://dx.doi.org/10.5772/50808

1. Introduction

Economic growth based on use of non renewable energy constitutes a serious problem because of, especially, negative environmental externalities. Growth of energy consumption and gas emissions are the principal negative impacts of these modes of development. However, the sustainable development requires modes of development which demand few of energy and produce few of pollutant gas. Literature has interested of this problematic in an aggregate or disaggregate contexts. The first is concerning the relationship between economic growth, domestic energy consumption and gas emissions. The second is corresponding to the same relationship but per economic sector. Industry and transport sectors are more studied because of their important link with economic and environmental spheres. They have an important contribution in economic growth but they are responsible of several environmental externalities.

The transport is one of the major activities which consume more energy and produce gas pollutants. Majority of freight and passengers is transported by road mode which is considered an important source of fossil fuels consumption and CO_2 emissions. In order to make transport sector more sustainable, some strategies should be elaborated to reduce its energy consumption and gas emissions. In other terms, governments should apply a set of instruments, such as economic, fiscal, regulatory and technological instruments, to control driving factors of transport-related energy consumption and gas emissions.

Before any strategy, it's necessary to evaluate the sustainability degree of transport sector. Sustainable transport literature give us several indicators through them it's possible to measure energy demand and gas pollutants production associated to transport activity. Examples include transport intensity, transport energy intensity, transport energy emission

INTECH
open science | open minds

© 2012 The Author(s). Licensee InTech. This chapter is distributed under the terms of the Creative Commons Attribution License http://creativecommons.org/licenses/by/3.0), which permits unrestricted use, distribution, and reproduction in any medium, provided the original work is properly cited.

intensity, vehicle intensity and vehicle energy intensity. Other indicators, such as modal mix (road, rail, air and water shares), energy mix for every mode (gasoline, diesel, liquefied petroleum gas (LPG), electricity or other fuel types, such as bio-carburant), rate of motorization, transport energy consumption share and annual growth of Vehicle Park are also used to diagnostic sustainability from transport sector. Determine the driving factors of transport-related energy consumption and pollutants gas emission growth is a one of main phases. It helps to choice the optimal strategy which corresponds to every responsible factor. With all types of intensities which are evoked, economic, demographic, urban and technological factors are taken as the main factors which can influence energy consumption and so gas emissions of transport sector. Moreover, examples of influencing factors include also average travelling distance, vehicle types share (personal cars, bus, heavy and light trucks), average vehicle age, driving condition, urbanization, urbanized kilometers and national road network.

More existing methods has presented by sustainable transport literature to examine the relationship between economic growth, transport activity and environmental impacts and to determine the driving factors. We distinguish three principal approaches: first, causality approach that based on time series methods, Kuznets Environmental Curve model and third, decomposing analysis method.

The aim of this chapter is to analyze the sustainability of transport activity especially through the energy efficiency indicator and different associated influencing factors. To this end, the rest of the chapter is structured as followed: section 2 presents an overview of works which have focused on transport energy efficiency. Section 3 describes methodologies applied to study energy consumption from transport sector with more interesting to decomposing analysis. Section 4 presents the driving factors and policy options to ameliorate transport energy efficiency. Section 5 concludes.

2. Transport energy saving: literature review

Transport activity is strongly related to economic activities. Traffics realized permit to link markets of production and consumption through the satisfaction of persons and freight transport demand. However, transport services supply is often associated to many problems that affect economy, society and environment. Negative environmental externalities appear especially if economic activity and transport services are more coupled. In this context, *transport intensity* is often used to measure transport demand and to analyze negative consequences of the coupling relationship between economic growth and transport activity. It's defined as the ratio of gross mass movement to Gross Domestic Production (GDP). It can be measured separately for the passenger services (passenger kilometres, *pkm*) and freight (ton kilometres, *tkm*). Close relationship between the growth in transport and economic growth implied an increase of transport intensity and so an increase of transport-related energy consumption and gas emissions. Transport-related energy consumption is currently measured in the literature through *energy intensity* which calculated by the ratio of transport energy consumption to transport services supply. It indicates the demand of energy per unit of transport service (*tkm* or *pkm*). Amelioration of energy efficiency from transport sector implies reduction of energy intensity and so saving energy consumption for the sector.

Transport energy intensity has studied in several contexts. Examples include studies which interest to the relationship between economic growth and transport activity and transport energy consumption. This analyze is defined in two approaches. Firstly, the economic approach which aims to study the causality relationship between them and to analyse the expenditure in energy consumption. Secondly, the ecologic approach which aims to study the correlation between them in order to determine the coupling phenomena between them and to measure and analyse energy consumption and gas emissions. Moreover, examples include studies which decomposing the intensities in order to determine the influencing factors of the transport-related energy consumption growth.

More existing studies have focused on the causality and cointegration relationship between transport energy consumption, transport activity and some main factors. Their main objective is to determine the time tendencies of their trends, the sense of their causality, the degree of their cointegration and then the coupling problem (Meersman et Van De Voorde, 1999; Kulshreshtha et al., 1999; Stead, 2001; Banister and Stead, 2003; Léonardi et Baumgartner, 2004; Tanczos et Torok, 2007).

Recently, Akinboade, Ziramba and Kumo (2008) have used the co-integration techniques in order to analyze the long-run relationship among the variables which explicating the aggregate gasoline demand function over the period 1978-2005 in South Africa. The results confirm the existence of a co-integrating relationship. The estimation of the price and income elasticities of gasoline demand serves to develop appropriate energy policy. The estimated elasticities show that gasoline demand in South Africa is price and income inelastic. The important policy implication is the unreliability of the public transport system in South Africa. Yaobin (2009) explains cointegration relationship between transport energy consumption growth, population growth, economic growth and urbanization process for china over the period 1978 -2008. The results show a unidirectional Granger causality running from urbanization to energy consumption both in the long and short run. Lu, Lewis and Lin (2009) have estimated the development trends of the number of motor vehicles, vehicular energy consumption and CO_2 emissions in Taiwan during 2007-2025. They have adopted simulation of different economic growth scenarios in order to explore the influence of economic growth on energy consumption.

Pongthanaisawan and Sorapipatana (2010) have analyzed the relationship between motorcycle and car ownerships and level of economic development for the case of Thailand. They study the impacts of this relationship on fuel consumption and greenhouse gas emissions. Using semi-parametric techniques, the authors have shown that economic development affects the ownership of private vehicles which and so fuel consumption and gas emissions. The important conclusion of their study is that the amelioration of public transport system leads to reducing the traffic mobility of private vehicles, promotion of the vehicle efficiency and so reducing fuel consumption and gas emissions. Yan and Crookes (2010) have forecasted the future trends of energy demand of road vehicle and emissions under various strategies for reducing the impacts of china's road vehicles on energy resources and environment. These strategies have concerned on the fuel economy regulation, alternative fuels and vehicles, public and non-motorized transport and economic incentives. Rudra (2010) explores the causality relationship between transport infrastructure, energy consumption and economic growth in India over the period 1970-2007. He finds a unidirectional causality from transport infrastructure to energy consumption. This results mains that energy and transportation policies must be recognized. Marshall *et al.* (2011) explores the causal relationship between residential location and vehicle miles of travel, energy consumption and CO2 emissions in Chicago metropolitan area over the period 2007 - 2008. Reinhard *et al.* (2011) concluded that urban energy planning and urbanization management are strongly linked and must be coordinated to lead to sustainable energy development.

In order to study transport-related energy consumption, majority of works have used the decomposition method. It's one of the most effective applied tools used to investigate the factors influencing energy consumption and its environmental impacts. The intensity decomposition method dates back to studies undertaken in the 1980s. It has known an expansion with works interesting to evaluation of aggregate energy consumption caused especially by the preceding energy crisis. It had evoked especially in the industrial context (Howarth *et al.*, 1991, Parck 1992). However, in the 1990s and 2000s this technique has been generalized to be used and applied to other sectors such as transport sector. The main objective of this method is to identifying factors that influence directly or indirectly energy consumption. One of the important decomposition of energy efficiency is which had proposed by Kaya (1989) in the context of energy economy. Kolbs and Waker (1995) have used decomposition method to find determinants of energy consumption and greenhouse gas emissions[1].

For the case of transport, several studies have been interested to decomposing energy consumption in transport sector in order to show the contribution of traffic in pollution. Schipper, Scholl and Price (1997) have decomposed energy intensity on three factors; transport activity (tone kilometre), structure of transport (types of modals) and intensity (energy used per unit of transport). They have concluded that best energy efficiency cannot compensate increasing of transport and modal share of road transport. Kveiborg and Fosgerau (2004) have decomposed the energy intensity of road transport for Denmark along the period 1981-1997. They have found that energy efficiency has been ameliorated by reducing of industrial production share and travelling distance through implantation of logistic platforms. Steer Davies Gleave (2003) have concluded, for the case of Germany, France, Spain, Italy and England during the period 1970-2000, that reducing of the road transport share has ameliorated the energy efficiency for the countries group. Steenhof *et al.* (2006) have used decomposition of energy intensity in order to examine the determinants of GES caused by freight transport in Canada. They have concluded that technical progress is inadequate solution if share of freight road transport increase for USA. Tanczos and Torok

---

[1] Liu and Ang (2007) have exposed a large majority of studies which used the method of intensity decomposition.

(2007) have studied the relationship between road transport, energy consumption and CO_2 emissions for the case of Autriche. Sorrell *et al.* (2009) have found for the case of England during the period 1989-2004 that increasing of vehicle transportation capacity and reducing of vehicle average energy consumption have ameliorated the energy efficiency. Niovi *et al.* (2010) estimates the effect of the spatial structure of the economy and the degree of spatial concentration of activities on fuel demand by using decomposition analysis. The results show that urban density increases fuel consumption.

3. Methodologies used for study of transport energy efficiency

3.1. Causality and cointegration relationship between transport energy consumption, transport activity and economic growth

An open question for relationship between transport activity, economic growth and environmental effects is the correlation between their trends. Negative environmental effect will be increasing when correlation between economic growth and transport is largely important. Exam of this correlation is important in so far as it provides a several instruments to elaborate an efficient transport policy. In literature, a large majority of studies have interested to determine separately correlation between first, transport activity and economic growth (coupling problem) and secondly transport activity and energy consumption and gas emission. In this chapter we propose a demarche which treats simultaneously both the tow problems. We attempt to include in our analysis dimension of sustainable transport for which actual policies transport have been taken.

In this context, a large number of studies have interested to the problem of coupling. An important number of institutional reports had elaborated like REDEFINE (1999) and SPRITE (2000) in Europe. Otherwise, several scientific studies have been elaborated in two directions. First, studies which estimate relationships between transport of passengers or goods and economic growth using previous traffic models (Meersman and Van De Voodre (1999) for the case of Belgium and Klushershtha and al. (1999) for India). Second, studies which have interested to aggregated indicators in order to estimate the coupling (Baum (2000) and Baum and Kurte (2002) have used the intensity of road transport to measure coupling for the case of Germany and Stead (2001) for the case of Europe).

In order to study the relationship between transport energy consumption, transport activity and economic growth, we should test the stationary of the series. To this end tests which are common use are Augmented Dickey-Fuller (ADF) tests (Dickey and Fuller, 1979) and Phillips-Perron (PP) tests (Phillips and Perron, 1988). The objective is to know if the series are stationary in levels or in some order of differentiation. If the integration of the two series is of the same order, we should test whether the two series are cointegrated over the same period. Analyze of cointegration between the series is often realized through method of Johansen (1988).

The Trace and maximum eigenvalue test provide us the information concerning the presence or nor of the cointegration. Consequently, we can estimate a vector error correction model (VECM) that incorporates variables and variation levels for information on the speed of adjustment to equilibrium. Engle and Granger (1987) showed that if two series are cointegrated, the VECM for the two series can be written as follows:

$$\Delta y_t = \alpha + \sum_{i=1}^{k} \beta_i \Delta y_{t-i} + \sum_{i=1}^{k} \lambda_i \Delta x_{t-i} + \eta ECT_{t-1} + \mu_{1t} \tag{1}$$

$$\Delta x_t = \alpha + \sum_{i=1}^{k} \beta_i \Delta x_{t-1} + \sum_{i=1}^{k} \lambda_i \Delta y_{t-i} + \eta ECT_{t-1} + \mu_{2t} \tag{2}$$

In Eqs. (1) and (2), PCTS and PCGDP (or per capita transport energy consumption PCTEC) represent per capita transport services and per capita GDP, respectively, whereas ΔPCTS and ΔPCGDP are the differences in these variables that capture their short-run disturbances and k is the number of lags. μ_{1t} and μ_{2t} are the error terms. ECT is the error correction term that measures the magnitude of past disequilibrium. The coefficient η represents the deviation of the dependent variables from the long-run equilibrium. The significance of the explanatory variables coefficient (λ_i and β_i) confirms the presence of short-run causality.

The robustness of the VECM is evaluated by using the normality residual test of Jarque-Bera, the Portmanteau auto-correlation test, the autocorrelation LM test, and the White homoscedasticity test. All these tests help us to accept or not the null hypothesis of no serial correlation. The normality residual test statistics of Jarque-Bera indicate if we accept or not the null hypothesis of normality of the residuals. The joint test statistics of the White homoscedasticity test with the no cross terms indicates if we accept or not the null hypothesis of non-heteroscadasticity at a 5% confidence level. If the model passes all the tests successfully, the estimation of the VECM gives the cointegrating vector.

After tests of cointegration, Granger causality test should be applied in order to exam the causality relationship between series. The sources of causation can be identified from the significance test of independent variables coefficients in the VECM. Regarding the causality of the short- run, we can test the nullity of the parameters associated with independent variables in each equation of VECM using the χ^2 -Wald statistics. The Causality long-run can be tested by the significance of the speed of adjustment. We use the t-statistics on the coefficients of the ECT indicate the significance of the long-run causal effects. The test give us the values of the speed of adjustment coefficients in the two equations of the PCTS and PCGDP which indicate if any deviation of the balance of long run of the value of the growth rate of the income tends to accelerate to adjust themselves with the shock and to return on its level of balance in a way faster than the rate of growth of the transport services. The validation of the first equation makes it possible to affirm that it is better to explain the GDP by the transport services than the transport services by the income. After testing cointegration and causality between transport activity evolution and economic growth, we can conclude if the two series are coupled or uncoupled.

The same procedure can be applied between transport activity and transport energy in order to determine the relationship between them and so to conclude if the transport sector is sustainable or not.

3.2. Relationship between transport activity and transport energy consumption : estimation of EKC

The genesis of the EKC can be traced back to Kuznets (1955), who originally discussed the relationship between economic growth and income inequity and suggested when per capita increase, income inequity increases also at first stage but after a certain level, starts decreases. This relationship follows an inverted-U curve and has is known as the Kuznets curve. Since the early 1990s, this curve measurement has progressed to become more used in analysis of relationship between economic development and environment quality (Grossman and Krueger, 1991; Bandyopadhyay, 1992; Panayotou, 1997).

Sustainable transportation system literature has more focused on negative environmental impacts caused by transport activity. Increase of traffic leads to increase of energy consumption and so degradation of air quality. Several factors can explain the increase of traffic, such as the economic growth, growth of population, urbanization, change in the lifestyles, increase of road infrastructures, etc. All these factors can lead to increase of passenger and freight mobility (Carlsson-Kanyama and Lindén, 1999; Ramanathan, 2000; Storchmann, 2005; Van Dender, 2009).

In the EKC literature there is a few studies which focus on the relationship between transport-related energy consumption and gas emissions and economic growth. Among these studies, we can quote the study elaborated by Cole et al. (1997) which examine the between per capita income and local air pollutants, and between energy consumption from transport sector and traffic for European countries over the period 1970-1992. They conclude that EKC relationship exist for local air pollutants from transport. Hilton and Levinson (1998) test the existence of EKC for plumb emissions from transport sector for 48 countries during 20 years. They find that their relationship with economic growth supports an EKC and explain their evolution by the increase of the private cars use. Two types of factors are mentioned by the authors: first, the pollutant fuel intensity (pollutant content per fuel type) and second, vehicle fuel intensity (energy efficient vehicle). Khan (1998) shows the existing of an urban EKC (UEKC) for hydrocarbon emissions from urban traffic in California State. He explains the increase of these emissions through the growth of personal mobility per private cars and its related fuel consumption.

Recently, Rupasingha et al. (2004) examine the urban polluting emissions of 3029 American counties using the EKC model and urban size and daily mobility as important determinants. Liddle (2004) examines the EKC relationship between per capita road energy consumption and per capita income, using IEA statistics. They conclude that hypothesis of an inverted-U curve are not existed and then EKC can't be proved. Tanishita (2006) examines the existing of the EKC for energy intensity from passenger transportation and concerning a set of data during the period 1980-1995. He finds that the relationship between the energy intensity of private and public transportation and the per capita Gross Regional Product (GRP) corresponds to an inverted U-shape of the EKC.

In the EKC literature, many functional form of EKC model are presented. Some studies consider only a cubic equation of income per capita as those of Grossman and Krueger (1991,

1995) and Harbaugh et al. (2002), or a quadratic equation, such as studies of Selden and Song (1994), Holtz-Heakin and Selden (1995), and Stern et al., (1996). Other econometric studies estimate several empirical models (linear, quadratic, cubic). The significance of the cube per capita income is based on the assumption of an N relation. Moreover, two principal types of curves coexist in the EKC literature. The first is the "diachronic" one which used with times series and aims to study the evolution of transport sector energy consumption comparatively with the evolution of income. However, the second is the "synchronic" one which used with cross-section data. The quadratic functional form of EKC is assuming the traditional EKC functional form. Then, we present the following regression model to describe the interaction between economic growth and transport sector energy consumption:

$$\ln(\mathrm{E}) = \alpha_0 + \alpha_1 \ln(PCGDP) + \alpha_2 \left[\ln(PCGDP)^2 \right] \tag{3}$$

were E refers to energy consumption from transport activity and treated as dependent variable, $PCGDP$ refers to per capita gross domestic production (PPP) and treated as independent variable and ln indicates natural logarithmic transformation. If the regression coefficient α_2 is negative, functional form of regression model corresponds to the standard EKC model.

The inverted-U relationship implies that energy consumption is reduced and so environmental quality improves beyond a certain threshold of income per capita. Lind and Mehlum (2007) present the basic properties that must satisfy the U relation. Their main idea insists that the U- inverted should have a positive slope at the beginning of the turning point and negative thereafter. This condition ensures that the endpoint is in the range of data. This condition can be written as follows:

$$\alpha_1 + 2\alpha_2 \ln PCGDP_{Min} \geq 0 \tag{4}$$

$$\alpha_1 + 2\alpha_2 \ln PCGDP_{Max} \leq 0 \tag{5}$$

where $\ln PCGDP_{Min}$ and $\ln PCGDP_{Max}$ are, respectively, the minimal and maximum values of the variable $\ln PCGDP$.

3.3. Methodology of decomposition analysis

In order to determine coupling relationship between increase of transport activity and economic growth and also transport energy consumption between, the analysis based on time series models and EKC models are considered as an aggregated analysis which not provides an explanation of the sources of coupling problem and growth of energy consumption. To this end, more existing studies have explained this relationship by decomposing aggregated variables of transport demand (transport intensity) and transport energy demand (transport energy intensity) into coupling and decoupling factors. The main objective of this approach is to identify the main factors that influence energy consumption of the transport sector, especially of road mode, to evaluate their impacts and to propose some solutions some sustainable policy options to reduce energy intensity. Majority of studies have interested on consumption of fossil fuels namely, diesel and gasoline. Ang and Zhang (2000) have proposed a review of literature for this decomposition analysis. Banister and Stead (2002) have studied the energy efficiency and economic efficiency of transport sector for European countries. They have found that passengers-kilometer is not a driving factor for nine European countries; tonne-kilometer is a driving factor of energy transport efficiency growth for six countries and of energy transport efficiency deterioration for eight countries. Zhang et al. (2011) have decomposed the energy consumption in Chinese transportation sector and have found that the transportation activity effect is the important contributor to increase energy consumption in the transportation sector and the energy intensity effect drivers significantly the decrease of energy consumption.

Technique of energy decomposition is largely useful in sustainable transport studies. Several methods of decomposition have been proposed in the literature such as the refined Laspeyres techniques (Lin *et al.*, 2008), the Arithmetic Mean Divisia Index (AMDI) and the Logarithmic Mean Divisia Index (LMDI) techniques (Ang, 2005; Liu *et al.*, 2007, Hatzigeorgiou *et al.*, 2008; Timilsina and Shrestha, 2009).

In this section we derive the methodology to decompose transport sector energy consumption growth to the influencing factors, such as demographic, economic and urban characteristics. Examples include vehicle fuel intensity, vehicles intensity, urbanization, per capita GDP, motorization, road network length, modal mix, fuel mix and more other factors which will be discussed in the section 4.

Transport energy intensity in year $t\,(TEI_t)$, can be expressed as:

$$TEI_t = \sum_{it} TEI_{ijt} \tag{6}$$

where subscripts *i*, *j* and *t* refer to fuel type (e.g., diesel, gasoline and GPL), transport mode (e.g., road, rail, air and water) and year, respectively. In order to decomposing transport energy intensity into the contribution factors, several formulations can be proposed according the factors integrated in the decomposition and the transport services type (freight transport or passenger transport). It should be note that choice of potential factors is based on availability of time series data and the causal relationship test. For example Eq. 1 can be expressed as

$$TEI_{ijt} = \frac{TE_{ijt}}{GDP_t} = \frac{TE_{ijt}}{TE_{it}} \times \frac{TE_{it}}{TS_{it}} \times \frac{TS_{it}}{GDP_t} \times \frac{GDP_t}{RV_{it}} \times \frac{RV_{it}}{POP_t} \tag{7}$$

Eq.2 can also be rewritten as

$$TEI_{ijt} = \sum_{ijt} MM_{jt} \times EI_{it} \times TI_{it} \times VI_{it} \times M_{it} \tag{8}$$

were MM refers to modal mix which indicates the share of fuel consumption by a mode in total transport energy consumption, EI refers to transportation energy intensity per mode which indicated the demand of energy to produce unit of transport services per mode, TI represents transport intensity per mode which indicates the demand of modal transport services to produce unit of GDP, VI refers to vehicle intensity and measures the demand of vehicles by mode to produce unit of GDP and M refers to rate of motorization. The change in these factors summarizes their direct and indirect impacts on change in transport energy intensity. The indirect impacts pass through the influence of demographic, economic and urban characteristics on transport energy intensity.

Using LMDI method because their advantages comparatively to Laspeyres techniques as shown by (Ang, 2004)[2], change in transport energy intensity $\left(\Delta TEI_{ijt}\right)$ between two periods can be attributed to effects of:

- Change in modal mix $\left(\Delta MM_{ijt}\right)$ named effect coefficient MM_{eff},
- Change in vehicle intensity $\left(\Delta EI_{ijt}\right)$ named effect coefficient EI_{eff},
- Change in economic activity $\left(\Delta TI_{ijt}\right)$ named effect coefficient TI_{eff},
- Change in urbanized kilometers $\left(\Delta VI_{ijt}\right)$ named effect coefficient VI_{eff},
- Change in national road network $\left(\Delta M_{ijt}\right)$ named effect coefficient M_{eff},

Consequently,

$$\Delta TEI_{ijt} \equiv RTE(T) - RTE(0) \equiv MM_{eff} + EI_{eff} + TI_{eff} + VI_{eff} + M_{eff} \tag{9}$$

Then, effects can be calculated for example for MM_{eff} as:

$$\Delta TEI_{ijt} \equiv \left[TEI(T) - RTE(0)\right] \ln\left[MM(T) / MM(0)\right] / \left[TEI(T) - RTE(0)\right] \tag{10}$$

Growth of transport-related energy intensity can be analyzed among its sensibility to changes in the named direct factors.

4. Driving factors of transport energy efficiency and policy options

Before presentation of decomposition methodology, we present and discuss some potential factors which have impacts on transport energy intensity. This is following by policy options to reduce transport energy use.

4.1. Driving factors of transport energy intensity change

Several factors contribute to change in energy intensity of transport sector. Examples include:

---

[2] Contrarily to Laspersey method, the advantages of LMDI method are, for example, the residual-free decomposition and the accommodation of the occurrence of zero values in the data set to small positive constant.

Causality relationship analysis		
Study	Variables	Result
Léonardi and Baumgartner, 2004	Transport energy efficiency and CO_2 emission efficiency in road freight transport. transport energy efficiency and efficiency of vehicle usage and	r^2= 0.42 r^2= 0.39
Mraihi and Hourabi, 2011	Energy intensity in road freight transport and transport intensity	Existence of a long-term relationship between transport intensity and energy intensity in road freight transport. Elasticity = 0.71
Environmental Curve of Kuznets		
Study	Variables	The estimated turning point
Meuni and Pouyann, 2009	Polluting emission due to the urban daily mobility (quadratic form).	Existence of an inverted-U form is significant for CO and NOx: $23.739 at 2000 constant price. $27.433 at 2000 constant price.
Ubaidillah, 2011	Trend of CO emission from road transport.	Existence of an inverted-U shape function of income for road transport: $21,402 at 2000 constant price.
Decomposing analysis method		
Study	Variables	Results
Timilsina and Shrestha, 2009	Change in CO_2 emission from transport for Asian countries. Driving factors: CO_2 intensity of a fuel, share of a fuel in a transportation mode, share of fuel consumption by a mode in total transport sector energy consumption, transportation energy intensity, economic activity as captured by per capita GDP and population.	Transport energy efficiency is considering an important driving factor of emission growth for 7 Asian countries. Economic growth and population growth are the critical factors in the growth of transportation sector CO2 emissions in all countries except Mongolia.
Banister and Stead, 2002	Change in transport energy intensity for European country. Driving factors: passengers-kilometer and tones-kilometer.	Pass-km is not driving factor for nine European countries (with an impact of 10%). Tones-km is a driving factor of energy transport efficiency growth for six European countries (with an impact more than 10%) and of energy transport efficiency deterioration for eight countries (with an impact lower than 10%.

Table 1. Examples of empirical results of transport energy efficiency analysis

- *Transport intensity per mode*: transport intensity is measured by the ratio of transport services to the GDP. Transport services can refer to the number of passengers/kilometer for passenger transport and tones/ kilometer for freight transport. The transport intensity measures the transport demand necessary to create a unit of GDP. Coupling between economic growth and transport activity development leads to the increase of transport intensity. The transport energy consumption, as an externality of coupling relationship, depends largely on the growth of this demand. However, modal transport energy consumptions are different and the road mode may the important consumer of energy. Then, transport energy intensity depends on modal transport intensity. The modal mix of transport services indicates the mode that is more used and if the transport policy approved by public authorities helps to shifting over to the mode which consumes less of energy.
- *Modal energy intensity*: energy intensity is measured by the ratio of energy consumption per mode to transport services per mode. It measures the amount of energy necessary to achieve a transport services output. It can be calculated from each mode and so illustrates the evolution of their fuel consumption shares. The analysis of this indicator needs also the analysis of the modal mix for energy consumption. In this context fuel prices, fuel taxes, clean fuel subsidizes, clean vehicles subsidizes and many other economic, fiscal and regulatory factors contribute to energy intensity changes. Motor gasoline and diesel are more used especially in road mode and the switching to renewable fuels can reduce the gas emissions, especially, because of the more use of.
- *Economic structure*: economic structure growth could be one of important potential driving factors of transport energy consumption growth. Shares contribution of sectors influence the use of transport services. Then, change in economic structure which place services sector in first place could be one of the main factors which could reduce transport energy intensity. However, economic growth based on industrial and agriculture sectors improve the use of transport services and lead to growth of energy intensity.
- *Vehicle fuel intensity*: it measures the average demand of fuel per unit of vehicle. There is a great deal of interaction between road transport-related energy consumption and efficient energy use of vehicles. Amelioration of fuel consumption efficiency of vehicle can be realised through vehicle design and technology. Vehicle fuel intensity depend on fuel prices, taxes imposed to fossil fuels use, subsidizes given to clean fuel use, growth of clean vehicles use and comportment of driving behaviour.
- *Vehicles intensity*: it measures the demand of vehicles necessary to produce one unit of GDP or of transport services output. Transport energy demand is closely link to national vehicle park. It indicated if the economic growth needs more trucks for freight transport and personal cars for passenger transport. Travelling distance is also depending on vehicle demand. Several factors can be contributing to change of vehicle intensity, such as particular credits for ownership cars, leasing credits for road freight vehicles, and other fiscal and financial facilities which contribute to the annual increase of injected supplementary vehicles.
- *Economic and motorization growths*: the evolution of per capita GDP (based on purchasing power parity (PPP)) influence the motorization level through the buying power.

Increase of buying power leads to the growth of private vehicles number and so of personal mobility. Then, transport energy intensity is linking to the level of personal motorization. Rapid growth of both motorization and travelling distance improve energy consumption and gas emission of transport sector. Moreover, increase of motorization in urban areas leads to more urban congestion with which the energy intensity will be reinforced for urban transportation.

- *Demographic and urbanisation growths*: growth of population, especially in urban areas, has an influence on transport energy consumption. If the demographic change is going through a growth of urbanisation, urban density tends to growing at a higher average annual growth rate. The percentage of urbanized kilometres will increases[3] because of growth of urban density and spatial distribution of households and activities behind the urban roads. It should be note, that in urban areas, urban planning and transport planning favourite the accessibility of urban activities by roads more than other modes. Then with the less supply of rail mode, the evolution of urbanized kilometres and road transport-related energy consumption may be high correlated (Newman and Kenworthy, 1989).
- *National road network*: if the cities and activities are more accessible by road infrastructures, this encourages using of road mode. The length of national road network (local, regional, national and express highway) can explain the growth of road transport-related energy demand. Public choices in term of transport investments repartition contribute significantly to the length of road network and so to use of road mode.

4.2. Policy options

In this section we derive some policy options which can reduce energy intensity from transport sector and so ameliorate the energy efficiency of transport activity. These policy options aim to reduce coupling factors effects and improve decoupling factors effects. As we have showed, from decomposing analysis, it become possible to determinate the principal contributors of transport energy consumption growth, named driving factors. The analysis of critical factors trends helps to identify the instruments to reduce their effects and so to save energy. From these instruments, it will be possible to elaborate some energy sustainable policies for transport sector. Examples of instruments include technological, technical, regulatory, fiscal and economic instruments.

Decoupling the transport energy consumption from economic growth is one of the important solutions if the economic growth is identified as a driving factor. Several practices are suggested in this context and depend on the type of decoupling (relative or absolute decoupling)[4]. Then many policy options are available to ameliorate the transport sector energy efficiency:

---

[3] Urbanized kilometers measure the concentration of population behind the urban road network.

[4] Decoupling is said absolute if negative externalities are stable or decreasing simultaneously with an economic growth. However, it said relative if the growth of negative externalities is less than of economic activity.

Logistic solutions, such as relocation of production units, multimodality, intermodality, optimization of the entire transportation chain from origins to final delivery and rescheduling of transport operations for companies are some examples of instruments which aim to reduce the travelling distances but not transported tons.

Modal shifting is also a decoupling instrument which aims to improve the demand of less energy-intensive modes of transport. In the large majority of countries, road transport is the predominant mode. It has often the important share in transport sector fuel consumption. To reduce its energy intensity, authority should shift over to rail mode. To this end some tools should be applied, for examples, reinforcing of rail infrastructures, the amelioration of services quality (availability, speed, regularity, tariffs, security, accessibility) and subsidizes for users of rail services.

Economic structure change which increases the share of service sector in the GDP can reduce transport intensity and so ameliorate the energy efficiency of transport sector. Volume of physical production and its movements from production market to consumption market have a determining influence on travel distances. When economic growth is driving especially by tertiary sector and trucks with high transportation capacities are used for freight transport, travels can be reduced and energy necessary to satisfying the demand of economic growth in term of transportation can be also reduced.

Transport planning can be reduce personal mobility and reduce distances especially in urban areas. It should be adopted by local authorities in order to make sustainable their transport system. Shifting over public transport is one of the important solutions of urban transport planning. Ameliorating supply and quality of public passengers transport could reduce personal motorization. Developing of public transport network by extension of public transport lines and through integration of private investment could be improving the public transportation. Moreover, revising the spatial distribution of households and activities in order to reduce travels can be deteriorating the urban density, the urbanized kilometers number and so road transport-related energy consumption in urban areas. New equity in term of spatial repartition of activities between all cities could reduce the concentration of populations and economic activities in megacities and so urbanized kilometers.

Fiscal and economic instruments could ameliorate the transport energy efficiency. They can be use to encourage the shifting over to energy efficient mode and the switching to clean fuels. In order to ameliorating the vehicle fuel efficiency and so reducing the vehicle fuel intensity, governments can impose some fiscal instruments. For examples, taxation of fossils fuels for personal cars use can increase the personal fuel expenditures and so encouraging the substitution of collective transportation to individual transportation. In addition, fossil fuel taxes for road freight transport could encourage the companies to promote the use of vehicle with high capacities transportation and less energy consumption. These taxes could be applied along with subsidies given for users of renewable energy. Government can also apply economic instruments, such as the increase of diesel and gasoline prices in order to substitute of the clean fuels to fossil fuels.

Regulatory instruments can be also help to improve the transport sector energy efficiency. Restrictions of older road vehicles importation and encouragements of vehicle fuel economy standard can be use as successful tools to reduce energy intensity. Additionally, control of leasing credits given to road freight companies and particular credits for personal cars could reduce road vehicles demand and improve public transport demand.

Technological instruments have shown their positive impacts on vehicle energy efficiency. Research and development have given several technologies which aim to reduce energy consumption of vehicles. Fuel economy can be realized by ameliorating of energy efficiency of drive train. So, new technologies which applied on the power of vehicle's engine could ameliorate the engine efficiency. Moreover, fuel economy increase also with reducing of the amount of energy necessary to move the vehicle. New technologies used to reduce vehicle weight and rolling resistance permit to increase energy efficiency. We should be note that these technologies are more accessible for developed countries and more expansive for developing countries.

5. Conclusion

As stated at the beginning of this chapter, the relationship between economic growth, transport activity and energy consumption is analyzed by more existing studies. In this chapter, majority of methodologies used to study energy efficiency for transport sector has been examined. From the sustainable transport point of view, there are large set of factors which influence transport sector energy consumption. All these factors are defined in the contexts of coupling and decoupling relationship between transport development and economic growth.

Ameliorating of transport sector energy efficiency depends on economic, urban, technological and fiscal factors. Decisions in land use planning and transport planning, prices and quality of fuels, taxes and subsidizes and investments on new technologies have a significant impacts on control of energy efficiency from transport sector and their environmental effects. However, in developing countries, improving the energy efficiency in transport sector requires more attention by government authorities, investors and civil society. Policymakers in these countries give more importance to transport accessibility to population and search to satisfying the economic growth in term of freight transport without any policy of energy economy.

Author details

Rafaa Mraihi
Higher Institute of Transport and Logistics of Sousse, University of Sousse, Sousse, Tunisia

6. References

Agras J., and Chapman, D., 1999. A dynamic approach to the environmental Kuznets curve hypothesis. *Ecological Economics*, 28, 267-277.

Akinboade, O., Ziramba, E., and Kumo, W., 2008. The demand for gasoline in South Africa: An empirical analysis using co-integration techniques. *Energy Economics*, 30, 3222-3229.

Ang, B.W., 2004. Decomposition analysis for policymaking in energy: which is the preferred method?. *Energy Policy*, 32, 1131-39.

Ang, B.W., 2005. The LMDI approach to decomposition analysis: a practical guide. *Energy Policy*, 33, 867-871.

Ang, B.W., and Zhang, F.Q., 1999. Inter-regional comparison of energy-related CO2 emissions using the decomposition technique. *Energy*, 24, 297-304.

Banister, D., and Stead, D., 2002. Reducing transport intensity. *European journal of transport and infrastructure research*, N° 3/4, 161-178.

Baum, H., 2000. Decoupling transport intensity from economic growth. In: ECMT (éd.), Key issues for transport beyond 2000, 15th International symposium on theory and practice in transport economics. Paris. OECD.

Baum, H., and Kurte, J., 2002. Transport and economic development. ECMT, Transport and economic development, OECD, 5-49.

Bentzen J., 2004. Estimating the rebound effect in US manufacturing energy consumption,

Dickey D., and Fuller W., 1979. Distribution of the estimators for autoregressive time series with a unit root. *Journal of the American Statistical Association*, 74, 427-431.

Energy Economics, 26, 123-134.

Engle, R.F., and Granger, C.W.J., 1987. Cointegration and error correction: representation, estimation and testing. Econometrica 55: 251–276.

Fosgerau, M., and Kveiborg, O., 2004. A review of some critical assumptions in the relationship between economic activity and freight transport. *International Journal of Transports Economics*, 31, 247-261.

Grossman, G., and Krueger A., 1991. Environmental impacts of a North American free trade agreement. Working paper, N° 3194. Cambridge: National Bureau of Economics Research.

Grossman, G., and Krueger, M., 1995. Economic growth and the environment. *Quarterly Journal of Economics*, 110, 353-377.

Harbaugh W. Levinson A., and Wilson D., 2002, Re-examining the empirical evidence for an environmental Kuznets curve. *The Review of Economics and Statistics*, 84, 541-551.

Hatzigeorgiou, E., Polatidis, H., and Haralambopoulos, D., 2008. CO2 emissions in Greece for 1990-2002: a decomposition analysis and comparison of results using the arithmetic mean divisia index and logarithmic mean divisia index techniques. *Energy*, 33, 492-499.

Holtz-Eakin, and Selden T., 1995. Stoking the fires? CO2 emissions and economic growth. *Journal of Public Economics*, 57, 85-101.

Howarth, R.B., Schipper, L., Duerr, P.A., and Strom, S., 1991. Manufacturing energy use in eight OECD countries: decomposing the impacts of changes in output, industry structure and energy intensity. *Energy Economics*, 13 (2), 135-142.

Johansen, S., 1988. Statistical analysis of cointegration vectors. *Journal of Economic Dynamics and Control*, 12, 231-254.

Kaya, Y., 1989. Impact of carbon dioxide emission control on GNP growth: interpretation of proposed scenarios. Paper presented to the Energy and Industry Subgroup, Response Strategies Working Group, Intergovernmental Panel on Climate Change, Paris, France.

Kolb, A., and Wacker, M. 1995. Calculation of energy consumption and pollutant emissions on freight transport routes. *The Science of the Total Environment*, 169, 283-288.

Kulshreshtha, M., Nag, B., and Kulshreshtha, B., 1999. A multivariate cointegrating vector auto regressive model of freight transport demand: evidence from Indian railways. *Transportation Research Part A*, 35, 29-45.

Kuznets, S., 1955. Economic growth and income inequality. American Economic Review 45 (1), 1-28.

Léonardi, J., and Baumgartner, M. 2004. CO2 efficiency in road freight transportation: status quo, measures and potential. *Transportation Research Part D*, 9, 451-464.

Lind, J. T. and Mehlum, H., 2007. With or Without U? - The appropriate test for a U shaped relationship. MPRA Paper 4823, University Library of Munich, Germany.

Liu N., and Ang B., 2007. Factors shaping aggregate energy intensity trend for industry: Energy intensity versus product mix. *Energy Economics*, 29, 609-635.

Liu, L.C., Fan, Y., Wu, G., and Wei, Y. M., 2007. Using LMDI method to analyze the change of China's industrial CO2 emissions from final fuel use: an empirical analysis. *Energy Policy*, 35, 5892–5900.

Lu, I.J., Lewis, C., and Lin, S.J., 2009. The forecast for motor vehicle, energy demand and emission from Taiwan's road transportation sector. *Energy Policy*, 37, 2952-2961.

M. Zhang, H, Li, M. Zhou, H. Mu, 2011. Decomposition analysis of energy consumption in Chinese transportation sector. Applied Energy, Vol. 88, 2279-2285.

Marshall, L., Joseph, L.S., Pablo, D., and Kimberly, A.G., 2011. The effect of residential location on vehicle miles of travel, energy consumption and greenhouse gas emissions: Chicago case study. *Transportation Research Part D*, 16, 1-9.

Meersman, H., and Van De Voorde, E., 1999. La croissance des transports de marchandises est-elle évitable? In : CEMT, (éd) Quels changements pour les transports au siècle prochain ? 14è symposium international sur la théorie et la pratique dans l'économie des transports. Paris: OECD, 23-51.

Mraihi, R., and Hourabi, I. Transport Intensity and energy efficiency: analysis of coupling and decoupling polycies implications. Logistiqua'2011, Hammamet, Tunisia, 2011.

Newman, P., and Kenworthy, J.R., 1988. The transport energy trade off: fuel-efficient traffic versus fuel-efficient cities. *Transportation Research part A*, 3, 163-174.

Niovi, K., Daniel, J.G., and Robert, B.N., 2010. Estimating the effect of urban density on fuel demand. *Energy Economics*, 32, 86-92.

Nur Zaimah Ubaidillah,N., Z. The Relationship between Income and Environment in UK's Road Transport Sector. Is There an EKC?. 2011 International Conference on Economics and Finance Research IPEDR.

Park, S.H., 1992. Decomposition of Industrial Energy Consumption: An Alternative Method. *Energy Economics*, 14, 265-270.

Phillips, P.C.B., and Perron, P., 1988. Testing for unit root in time series regression. *Biometrica*, 75, 335-346.

Pongthanaisawan, J., and Sorapipatana, C., 2010. Relationship between level of economic development and motorcycle and car ownerships and their impacts and greenhouse gaz emission in Thailand. *Renewable and Sustainable Energy Reviews*, 14, 2966-2975.

REDEFINE, 1999. Relationship between Demand for Freight-transport and Industrial Effects. Summary Report, NEI.

Reinhard, M., and Yasin, S., 2011. Impacts of urbanization on urban structures and energy demand: What can we learn for urban energy planning and urbanization management?. *Sustainable Cities and Society*, 1, 45-53.

Rudra, P.P., 2010. Transport infrastructure, energy consumption and economic growth Triangle in India: Cointegration and Causality Analysis, *Journal of Sustainable Development*, 3, 167-173.

Schipper, L., Scholl, L., and Price, L., 1997. Energy use and carbon emissions from freight in 10 industrialized countries: an analysis of trends from 1973 to 1992. *Transportation Research part 2 D*, 1, 57-76.

Seldon, T., and Song, D., 1994, Environmental quality and development: is there a Kuznets curve for air pollution emissions. *Journal of Environmental Economics and Management*, 27, 147-162.

Sorrell, S., Lehtonen, M., Stapleton., L., Pujol. J., and Champion. T., 2009. Decomposing road freight energy use in the United Kingdom. *Energy Policy*, 37, 3115-3129.

SPRITE, 2000, Separating the Intensity of Transport from Economic Growth. Final Publishing Report, ITS.

Stead, D., 2001. Transport intensity in Europe: indicators and trends. *Transport policy*, 8, 29-46.

Steenhof, P., Woudsma, C., and Sparling, E., 2006. Greenhouse gas emissions and the surface transport of freight in Canada. *Transportation research*, 11 D (5), 369- 376.

Steer Davies Gleave, 2003. Freight Transport Intensity of Production and Consumption. Report no EUR 20864EN, Institute for Prospective Technological Studies. Joint Research Centre, Seville, Spain.

Stern, D., Common, M.S., and Barbier, E.B., 1996. Economic growth and environmental degradation: the environmental kuznets curve and sustainable development. *World Development*, 24, 1151–1160.

Tanczos, K., and Torok, A., 2007. The linkage between climate change and energy consumption of Hungary in the road transportation sector. *Transport*, 22, 134-138.

Timilsina G. R., and Shrestha A., 2009b. Transport sector CO2 emissions growth in Asia: Underlying factors and policy options. *Energy Policy*, 37, 4523-4539.

Yan, X., and Crookes, R.J., 2010. Energy demand and emissions from road transportation vehicles in China. *Progress in Energy and Combustion Science*, 36, 651-676.

Yaobin, L., 2009. Exploring the relationship between urbanization and energy consumption in China using ARDL (autoregressive distributed lag) and FDM (factor decomposition model). *Energy*, 34, 1846-1854.

Chapter 14

Tools for Categorizing Industrial Energy Use and GHG Emissions

Teuvo Aro

Additional information is available at the end of the chapter

http://dx.doi.org/10.5772/48766

1. Introduction

The political target to cut energy use and greenhouse gas (GHG) emissions has various expressions. For example, the European Union some years ago set the target that energy efficiency must be improved by 20% by 2020. In good policy making, regional strategies must be parallel with national strategies. When the approaches in the national strategies are top-down, the regional approaches ought to be bottom-up. Therefore, the issue is to have policies that work in practice or "in the real world" (Johansson, 2006).

The policies whether they are carried out in a company, or at a regional, a national or even continent-wide level need tools. Industry is diversifying all the time. Does this development path mean that industrial energy use is diversifying as well? At first glance when going very deeply into energy use this seems to be true. The energy use may be diversified when looking at the details, but to conduct an energy-efficiency policy or GHG emission reduction policy with a wide scope requires generalisations. This is certainly the case when we withdraw from the detailed level. The energy use must be categorised. This article will mainly discuss these tools and how to generalise and categorize industrial energy use.

2. Energy policy levels and decision-making

One can set many policy levels when looking at the policies that aim to mitigate climate change and cut energy use. On a national basis, the levels for industry can be as follows:

Company level

This can be either one company or an enterprise or a group of enterprises that have the same owner. At present, only companies belonging to the EU Emissions Trading System (the EU ETS) have a direct responsibility to control their CO_2 emissions. Companies that do not belong to the ETS have no direct responsibility other than the country of their location.

INTECH
open science | open minds

© 2012 The Author(s). Licensee InTech. This chapter is distributed under the terms of the Creative Commons Attribution License http://creativecommons.org/licenses/by/3.0), which permits unrestricted use, distribution, and reproduction in any medium, provided the original work is properly cited.

Regional level

Regional may mean different things in different contexts. For example, in Finland one province has one million inhabitants, another one has a few hundred thousand, whereas in some bigger countries one city may have many millions of inhabitants.

National level

Countries make agreements on GHG mitigation and they are responsible for fulfilling the agreements under the Kyoto Protocol or under some more limited agreements such as among the EU countries.

International level

A level where conclusions are made on the GHG emissions mitigation goals and how the mitigation targets are divided among the countries or groups of countries.

The service and public sectors are taking an ever-bigger role as an employer. In 2007, industry directly employed only 18% of the workforce in Finland. This is still a moderate number for an industrialised country among the EU-15, where the average is 17%. The decline in industrial employment has been very fast. In 2000, industrial employment in Finland was 20% (Eurostat, 2008). Whenever fewer people have a direct relation to industry, it is more and more difficult for industry to have a fair and effective communication with citizens and the authorities.

These thresholds for fair communication due to limited information will always exist. To lower the thresholds, information must be clear and jargon must be avoided. One precondition is that the subsidiary principle is applied. The decision-makers must understand the key points of industrial energy use in the target area of the policy as well as other conditions affecting industry. For those conducting the actions of the selected policy, choosing the correct policy level is even more important. The policy level must be selected so that the area of decision-making is understandable "in one man's head". The level is a very personal question. It depends on the persons and their personal experiences and skills. In Finland, one reasonable level is the province so as to cover "the scope of in one man's head".

3. Opportunities to improve industrial energy efficiency by 2020

The main goal of this review of the past is to forecast for the year 2020 with today's knowledge. The review is based on an article (Aro, 2009).

3.1. Pumps and fans

The electric motors of pumps and fans consume electrical energy. When looking at only the improvements opportunities in the motors, pumps and fans, no large consumption reductions can be expected by 2020. However, the situation is not necessarily all that bad. The system level can give opportunities. This means how the components are used as a part of technical systems. The lower the air and liquid flows with fewer pressure lifts, the smaller the electricity consumption. At the systems level, frequency converters and more energy efficient motors give a good option to achieve the best possible level of efficiency with fans and pumps. At the systems level, there are always endless opportunities for energy savings and efficiency improvements.

3.2. Compressed air

It is well known that compressed air systems require good maintenance to run efficiently. If not, air leakages may be more than 70% of the total compressed air production. Therefore, there always exists potential savings with compressed air systems. In the past, compressors were piston or screw types, as they are today. However, there have been efficiency improvements due to novel control systems and the use of frequency converters. Some 10-30% improvements have been seen compared to the past. In total, over 10% of improvements can be expected in the near future.

3.3. Heat recovery from exhausted waste gas flows

Heat recovery from exhausted waste gas flows is used to heat fresh air for air-conditioning or process air use. During the last 30 years, no remarkable success has been achieved in the efficiency of the various types of heat recovery equipment applicable to air-conditioning.

Recovery of process exhausts is more demanding because of corrosive substances and particles. For these applications, new types of heat recovery have been developed as well as new materials tested. Although in the past success on the equipment level has been rather limited and it does not seem that it is going to be much better by 2020, there have been improvements due to advanced control systems that help to better run the systems than in the past.

Heat recovery is meant to decrease heat consumption. Heat recovery equipment causes pressure losses in gas flows, which means that the fans and pumps must produce higher pressure and, therefore, more electricity is consumed. However, if they are well designed, the increase in electricity consumption is clearly less than the decrease in heat consumption.

3.4. Cooling and heat pumps

Compressors for cooling and heat pumps were piston-operated ones in the past, whereas today they are mainly scroll or screw types. The latter ones are easy to control. The new compressors and improved control technology have given opportunities for an improvement in energy efficiency of 10-20% compared with the past. In Finland, free cooling by outdoor air or lake and river water has become more popular compared with the past. Free cooling still has many opportunities, especially in industrial process cooling and also in offices, where computers require cooling also during the heating season. Through free cooling, the electricity consumption of cooling can be reduced by dozens of percents. Although there have been improvements in cooling applications, the demand for electricity for cooling can be expected to increase due to the need to improve working environments and due to new process requirements.

Heat pumps have in principle still many opportunities in industrial processes but heat sources i.e. liquids and gas flows for heat pumps are difficult to exploit. There are blocking and corrosion problems with heat exchangers. No leap forward can be expected by 2020.

3.5. Heat production

Heat production and heat use have been areas where energy efficiency improvements have been remarkable since the 1970s. There are various reasons for this positive development in Finland:

1. Transfer from steam to hot water, hot oil, and electricity.
2. Transfer from a factory's own boilers to district heating. This type of outsourcing has in most cases caused improvements in efficiency or at least savings in operating costs.
3. Increased use of natural gas. Natural gas is easy and clean to burn compared with other fuels.
4. Outsourcing of boiler plants is comparable to district heating. There are opportunities for efficiency improvements when heat production is outsourced to a company specialised in heat production.

Boiler efficiencies with gas and oil were at a good level even 30 years ago. When one thinks about the future, no remarkable improvements can be expected. With solid fuel boilers, there are opportunities for improvements. With all types of boilers and heat distributing systems, there is always some potential as a result of good maintenance and operation.

3.6. Lighting systems

Incandescent lamps have disappeared in general lighting, but fluorescent and mercury lamps are still on the market with more efficient applications. High -pressure sodium lamps are taking more and more of the market due to their good energy efficiency. Many people think that the future is in light emitting diodes (LEDs). High expectations have been set for the good energy efficiency and long service life of the LED lamps. It is not certain what their market penetration will be by 2020. For the moment, barriers to market penetration include limited LED lamp applications for general industrial lighting as some of the existing lamps (such as sodium lamps) already have rather good energy efficiency and that the LED lamps need their own light fixtures. However, general lighting may give dozens of percents in saving opportunities through novel lamps, lighting fixtures, control of lighting, and good maintenance.

3.7. Conclusions

For a variety of reasons, the changing of individual technologies to more energy efficient ones is not an easy way to achieve high reduction cuts by 2020. The service life of individual technologies is 10-30 years, which means that most of the motors, pumps, fans, etc that are now in use will still be in use in 2020. With many individual technologies, there are not very remarkable efficiency improvement expectations. If a deep cut in energy consumption and CO_2 emissions is pursued, it will be found through novel system thinking. This in turn means a lot of work for skilful people. Based on the past, the general energy efficiency in industry has improved 1% per year (Blok, 2007) and in future the expectations on the growth of production are clearly more than 1%. An equation to be solved means something else than what we have seen in the past. In future, energy efficiency policy must be more target-oriented, and not more or less a by-product of normal industrial development, as it has been until now. A study by Blok (2004) discusses the preconditions by which new equipment will achieve an energy efficiency improvement rate higher than 5% per year. According to the study, it may be possible but it will require substantial efforts from all parts of society.

4. Sectoral and cross-sectoral approach

In a study (Stigson et al, 2008), the concept of a sectoral approach was seen to depend on the person who defines the concept. In the study, the following scope was categorized:

Sector-wide transnational approaches, e.g. transnational industry-led approaches that aim to engage a sector on a broad international basis or global sectoral industry approach; bottom-up *country commitment*, possibly combined with no-lose targets; and top-down sectoral crediting as an incentive mechanism, e.g. sectoral Clean Development Mechanism

The same study found three common features typical of sectoral approaches: 1) collection of data and information about the sector to establish performance indicators or benchmarks; 2) sharing and distributing best practices within companies to enhance monitoring, reporting and verification of emissions and operational efficiency; and 3) engaging with major companies in emerging economies, where the greatest emissions growth and reduction potential lie.

Of these three, the first one lays an information foundation for the other two, where the target is to achieve practical improvements in GHG mitigation, in energy efficiency, or in other fields of energy policy.

Sectoral approaches are most useful especially in the fields of industry where rather homogenous products are handled, such as in the steel and other metal industries, the cement industry, and in the pulp and paper industry. Sectoral approaches provide useful background information on industry but they are time consuming, need a lot of work and a constant updating of the information. The main defect is the collection of reliable data, especially in global comparisons. Collected data is particularly poor even from "the easy sectors" such as the iron and steel, chemical and petrochemical, and pulp and paper sectors (Tanaka, 2008).

Sectoral analysis based on economical figures, e.g. value added (€) or turnover (€) per tonne of steel produced, is easier to collect but not so useful in emission-reduction target setting compared to physical data such as steel produced per consumed form of energy or per CO_2 emissions, see for example (Worrel et al., 1997).

Thus far, much of the discussion on GHG mitigation has been targeted at international or national levels where sectoral approaches illuminate the origins of CO_2 emissions and are useful for general industrial GHG policymaking. To achieve real results in the mitigation policy, more and more activities must be set at local or regional levels. That is where the real results in the tackling of climate change will take place.

In the United States, where commitment to international agreements is weak, the sub-national GHG policies have developed strongly. It has been estimated that if those states, which have set their own GHG emission reduction targets, achieve those targets, nationwide US GHG emissions would be stabilized at 2010 levels by 2020. And this, without any serious mitigation action taken by over half of the states (Lutsey and Sperling, 2008).

At a local or regional level, successful policy means co-operation among different industries and not only among specific industrial sectors. This is because at the local level there are many industrial sectors and one sector may have only one or very few separate companies. Furthermore, co-operation is needed between industry and other sectors of society. A cross-sectoral approach is a must.

5. Industrial cross-sectoral approach

How should one develop a target-oriented and bottom-up approach to reduce the CO_2 emissions and energy consumption of industry? First of all, what kind of tools is needed to conduct a target oriented policy? Because the problem is energy use, we should have a view on industrial energy use. What is common in general? One way is to classify and categorise the sectors and companies according to their energy use. In Aro (2009), this is done in the following way: building energy users (HVAC and lighting), process heat users, process electricity users, and direct combustion users, see table 1. This classification based on the form of energy use is useful when designing regional energy efficiency policies, since energy efficiency improvement/CO_2 reduction strategies can be built specifically for each of these four categories. These policies can also be, however, applicable to several industrial sectors as long as they belong to the same category of energy use.

At local or regional levels this categorization can be used in many ways such as at company, energy utility, and zoning levels. Of course, there are no limits to use it also at national or international levels whenever it is seen to be useful.

Building energy users are good for district heating and zoning must be targeted to collecting these kinds of industries in areas where district heating is possible. Companies using heat in production are good as a part of district heating or (CHP) where they can guarantee constant heat load throughout the year and/or they are considered to be a good target for biofuel power plants. It is beneficial to locate direct combustion users near a natural gas network.

Form of energy use	Description
Building energy users	Small amounts of electricity and heat are used in the production. HVAC and lighting are clearly the main energy consumers. Assembly lines, the production of equipment and machines are typical industrial sector representatives for this category. In general, many industries that are often described as non-energy intensive can be considered building energy users.
Major users of electricity for process/production	Electricity use in process/production is clearly bigger than the building electricity consumption. Typical branches of industry falling into this category are pulp and paper, metal production, the production of plastic products, and glass making.
Major users of heat for process/production	Heat use in the process/production is clearly bigger than the building heat consumption. Heat means energy forms that are transmitted by pipes such as water, steam, and hot oils. Typical branches of industry belonging to this category are pulp and paper, dairies, part of the textile industry, chemical industry, production of rubber products.
Direct combustion users	In some applications, the product can be heated directly or indirectly by fire and/or flue gases. Especially natural gas is good in many applications. Typical representatives are cement and lime production, glass and brick production, bakeries, and the production of metals.

Table 1. Ways to use energy in industry (Aro, 2009).

Table 2 shows an example of this categorization as applied to various industries. Although the companies may belong to different industrial sectors, they may have common aspects in the ways they use energy. For this purpose, the categorisation is useful. It can be used for benchmarking and exchanging information between industries and industrial sectors. At local and regional levels, it is good to have co-operation among industries. For this, categorisation gives opportunities to build up workshops and common development projects under the same theme of energy use in spite of being from different industrial sectors.

If a cross-sectoral approach is needed among industries, the regional energy policy also needs a cross-sectoral approach between industry and other sectors of society. This approach means district heating, biofuel use, and other society-wide energy projects, where advantages are achieved if industry is involved in the projects.

5.1. A company level

Categorization is a tool to reduce energy use and CO_2 emissions. The reductions are always realised at the company or plant levels because energy is used there.

If one sets a general target to reduce energy consumption to a certain level such as the EU target of 20%, improvements in energy efficiency, and reductions in CO2 emissions, what do the targets look like at the company level? The driving forces for a company are external and internal ones. The external ones are, for example, EU targets and the internal ones are the company's own policies to reduce CO2 emissions. Therefore, the question is how to react to the external ones. In principle, company-level energy related CO2 emissions are formed by a multiplication of the form of energy and specific CO2 emissions of the energy form. The development path of company-level CO2 emissions is a phased process where in every step the quantity of the energy form or specific emissions of the energy form or both are changed (Fig. 1).

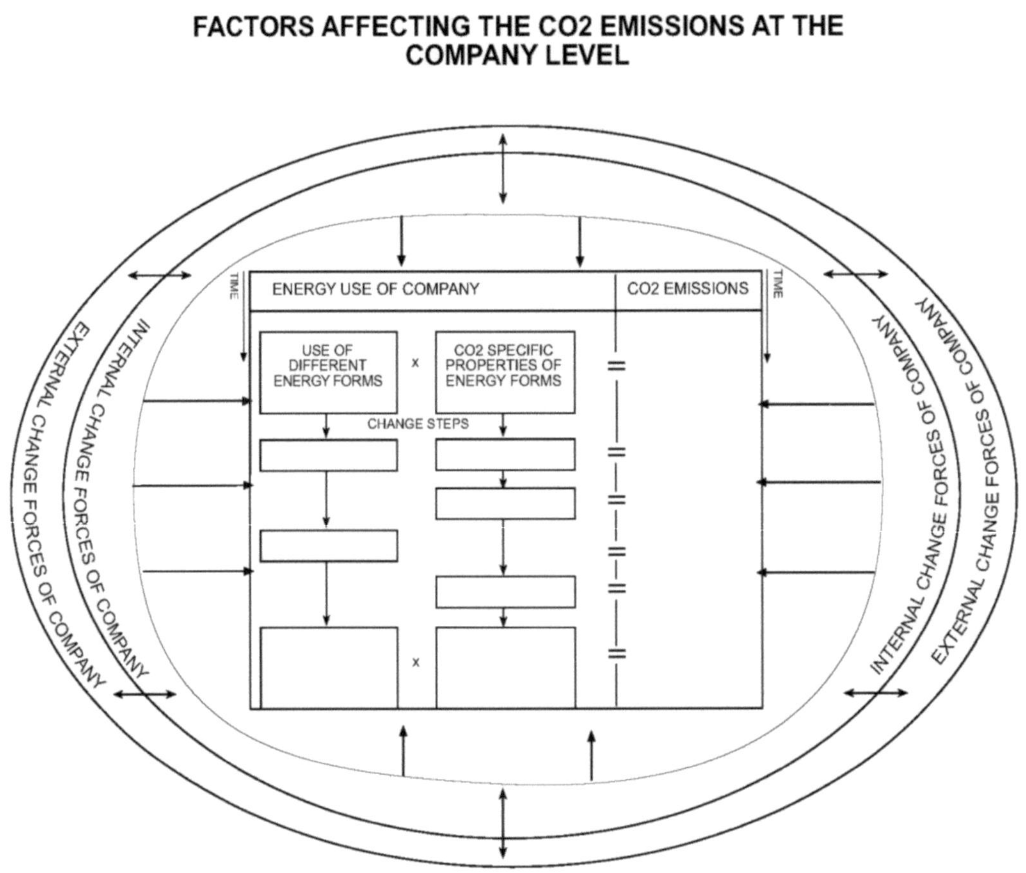

Figure 1. Development of a company's CO2 emissions path.

Standard Industrial Classification 2002	Industry	Significant building energy user, typical user of the district heating	Significant process electricity consumption	Significant process heat consumption (water, steam, hot oil)	Significant process direct combustion user (oil, natural gas and solid fuels)
15	**Manufacture of food products and beverages**	If the processes can be managed with electrical heating, the usage of district heating is reasonable. If a boiler is needed for the process, it is often used to heat up the buildings as well.	Electrical ovens Cooling Grinding machines Mixing machines Concentration plants Pumping	Cooking Washing Sterilization, acid and alkali washes Pasteurization Dewatering Manufacturing of cheeses	Baking and rising
17	**Manufacture of textiles**	Typical building energy users	Drying Production machines	Dye works Drying Manufacture of special textiles	Drying The number of factories using direct combustion is diminishing.
18	**Manufacture of wearing apparel; dressing and dyeing of fur**	Typical building energy users	Machines and devices	Small steam generators are able to provide enough steam for pressing	
19	**Tanning and dressing of leather; manufacture of luggage, handbags, saddlery, harness and footwear**	Typical building energy users	Drying	Low temperature level water heating and leather drying	
20	**Manufacture of wood and of products of wood and cork, except furniture; manufacture of articles of straw and plaiting matter**	Building energy users. Only some of the buildings are heated.	Wood processing Drying Sawdust removal	Drying	Drying
21	**Manufacture of pulp, paper and paper products**	Some paper products processors are mainly building energy user.	Processes use a significant amount of electricity: Wood	Drying Water heating Pulp production	Drying

Standard Industrial Classification 2002	Industry	Significant building energy user, typical user of the district heating	Significant process electricity consumption	Significant process heat consumption (water, steam, hot oil)	Significant process direct combustion user (oil, natural gas and solid fuels)
			processing Grindery Pulp and paper machines Drying Pumping		
22	**Publishing, printing and reproduction of recorded media**	Small plants are typical building energy users.	Printing presses Drying	Drying	Drying
24	**Manufacture of chemicals and chemical products**	Small plants are typical building energy users	Pumping Fans Negative and positive pressures Heating Cooling Drying	Process heating Drying	Drying
25	**Manufacture of rubber and plastic products**	Small plants are typical building energy users	Extruders and other melting procedures Process cooling especially in the summer time	Process heating	
26	**Manufacture of other non-metallic mineral products**	Small plants are typical building energy users	Refiners Grinders Pumping Thermal treatments Melting processes	Water heating Thermal treatments	Incineration Melting Thermal treatments
27	**Manufacture of basic metals**		Processes use a significant amount of electricity: Melting Thermal treatments		Melting Thermal treatments
28	**Manufacture of fabricated metal products, except machinery and equipment**	Some plants are building energy users	Surface finishing Thermal treatments Processing and shaping Welding	Drying	Thermal treatments

Standard Industrial Classification 2002	Industry	Significant building energy user, typical user of the district heating	Significant process electricity consumption	Significant process heat consumption (water, steam, hot oil)	Significant process direct combustion user (oil, natural gas and solid fuels)
29	**Manufacture of machinery and equipment n.e.c.**	Some plants are building energy users	Shaping Thermal treatments Surface finishing Welding	Drying	Thermal treatments
30 and 31	**Manufacture of office machinery and computers, electrical machinery and apparatus n.e.c.**	Typical building energy users			
32	**Manufacture of radio, television and communication equipment and apparatus**	Typical building energy users			
33	**Manufacture of medical, precision and optical instruments, watches and clocks**	Typical building energy users			
34 and 35	**Manufacture of motor vehicles, trailers and semi-trailers, other transport equipment**	Some plants are typical building energy users	Surface finishing Welding	Surface finishing	
36	**Manufacture of furniture; manufacturing n.e.c.**	Some plants are typical building energy users	Drying Surface finishing Machines	Drying Surface finishing	

Table 2. Example of energy use categorising in different industries (Aro, 2009).

In the article (Aro, 2009), the energy consumption and CO_2 emissions of 6 industrial plants located in the Pirkanmaa region, Finland is reported, see table 3. Plant number 1 is a typical building energy user and plant number 6 belongs to the category of heat in process user, see table 2. The others were in the middle. Their energy use was analysed and, on the basis of the analysis for each plant, a Sankey diagram was drawn of the origin of energy related CO_2 emissions. The diagrams are shown for plants 1 and 6 in fig. 2. The potential for reduction of CO_2 emissions was estimated on the basis of the energy/CO_2 emission analysis. For plant 1, the economic reduction potential (payback period less than 5 years), based on energy prices only was some 15% and for plant 6 2-3 %. For the other four plants, it was between 4% and 25%.

	Energy consumption [MWh / a]	CO_2 emissions [t CO_2 / a]	Energy consumed in production	Own energy production	Sector
Plant no. 1	2,500	510	3%		Machinery and equipment
Plant no. 2	10,000	2,250	65%		Other non-metallic mineral product
Plant no. 3	20,000	4,300	73%	x	Food products and beverages
Plant no. 4	20,000	5,100	70%	x	Other non-metallic mineral product
Plant no. 5	25,000	5,200	40%		Rubber and plastic products
Plant no. 6	500,000	150,000	over 95%	x	Chemicals and chemical products

Table 3. Key figures of six industrial plants in the Pirkanmaa Region, Finland (Aro, 2009).

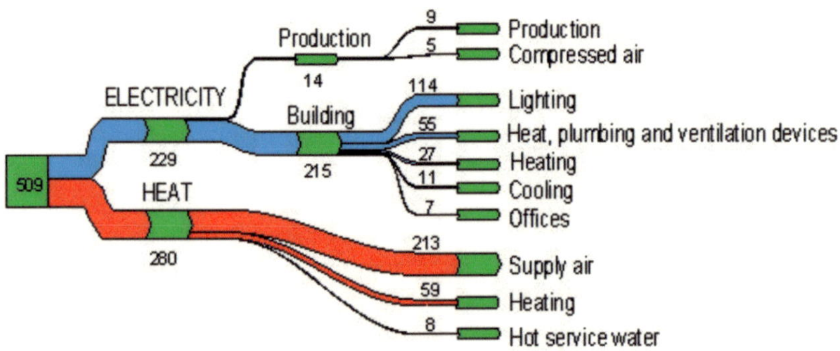

Plant no. 1

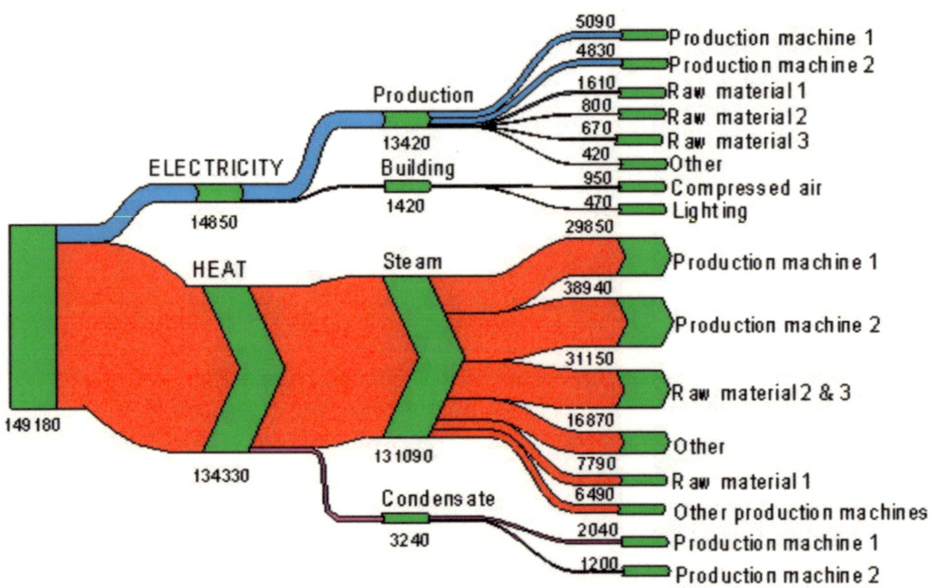

Plant no. 6

Figure 2. Sankey diagrams for plants 1 and 6. Origin of CO_2 emissions, t CO_2(Aro,2009).

What does the general target of a 20% reduction in CO_2 emissions mean for the 6 plants? Only plant 6 belongs to the EU ETS. For this plant, it was possible to calculate - other than the indirect CO_2 emissions - what the 20% reduction will mean for the plant economy in relation to the EU ETS allowance price (€/CO_2 tonne). For the other five, the relation is only indirect (through acquired electricity and heat and fuel prices) and, therefore, hypothetical. The economic effects of the allowance price (5-40 €/CO_2 tonne) for different reduction targets (5-30%) were calculated for the six plants' economy as a share of turnover. Plants 4 and 5 can even cover the 20% reduction with their own energy saving measures, but the others cannot. Except for plant 6, the burden of the reduction is not very demanding. Plant 6 has to expect costs of more than 1% of turnover whereas for the others it is less than 1%.

In interviews with the employees of the 6 plants, most of them saw that there is an opportunity to manage with a 20% reduction by 2020 at the existing production levels. However, successful business requires an increase in production. They see that this growth demand will be the main threat for achieving the reduction target.

Other barriers for the target are excessive outsourcing and the reduction of staff and lack of knowledge. Another barrier to rapid change is the rather slow rate in the construction of new industrial buildings comprising some 2% per year of the existing industrial building stock.

5.2. A regional level

5.2.1. Finland and Pirkanmaa region

Finland's population is rather small, only 5.3 million. With an area of 340 000 sq km, Finland is the 6th largest country in Europe. Finnish industry is versatile. We have light industry like the telecommunication industry, but we also have very heavy industry like the pulp and paper and steel industry. Most of the heavy industries belong to the EU Emission Trade System (EU ETS), while the light industries mainly do not.

In Finland, we have two schemes that promote the rational use of energy that are partly funded by the government: energy audits and energy investments regimes to improve energy efficiency and to increase the production of renewable energy. Furthermore, we have a voluntary agreement to improve energy efficiency. All these activities are applicable also for industry.

Finland consists of 15 provinces. Finnish regional energy policies have primarily focused on the promotion of biofuels. There have, however, been practically no regional activities for industry, especially as regards energy efficiency. In the past, the efforts to improve energy efficiency were mainly motivated by corporate economy and to some extent by the nation's fuel reserve supply stock. Today, because of climate change, the government of Finland is more interested in what is happening in industry. Improving energy efficiency means tackling climate change.

In a sparsely populated country with a lot of energy intensive industry such as Finland, it is a challenge to formulate a regional energy policy with a focus on industry. If we want to tackle climate change in the long run – as many estimate the GHG emissions must be cut by some 60% – the policy must be very comprehensive and well-organised, and one must keep one's finger on the pulse of what is going on in the industry.

What does the regional level give in fight against climate change? The aim is to build foundations for starting a regional carbon dioxide reduction programme and to discuss what opportunities the provincial aspect offers to the reduction of the industrial CO_2 emissions related to energy use. The approach is limited to D-sectors of the industrial statistics: the manufacturing industry. The fishing, farming, forestry, mining, construction industries, and the electricity, gas, and water supply industries are excluded from the study. Fuels used by industrial vehicles are also left out. The Pirkanmaa region, some 150 kilometres to the north from Helsinki, served as a target province. The centre of the region is the city of Tampere. The area of The Pirkanmaa region is some 4% and the population some 9% of the whole country. The work partly serves as one of the contributors to the establishment of the Pirkanmaa region energy programme.

Carbon dioxide emissions arising from the use of energy are divided in emissions originating from the combustion of fossil fuels and in emissions from the production of heat and electricity acquired by industry. Acquired heat and electricity are the sources of indirect emissions allocated to the industries. The classification of industrial statistics made by Statistics Finland was used in the calculations. The goal was to look at the energy related carbon dioxide emissions generated by different industrial sectors in the Pirkanmaa region. The emissions are also compared between the Pirkanmaa region and the whole country. Of all the greenhouse gases, the energy related emissions have the biggest impact and they represent 80% of Finland's greenhouse gas emissions (Finland's FNC, 2006).

5.2.2. CO_2 emissions by industry and source

Carbon dioxide emissions from the use of energy were calculated from the industrial statistics (Statistics Finland, 2006) and they were complemented with data from other public sources. When estimating the emissions of industrial sectors, both the use of fossil fuels and emissions related to the acquired heat and electricity used by the companies were taken into account. Following these principles, carbon dioxide emissions related to the energy use of industry in the Pirkanmaa region were estimated at 1.5 million tons in 2004. This is about 2% of Finland's total greenhouse gas emissions. In Finland as a whole, the energy related CO_2 emissions of industry are some 35% of the total GHG emissions.

Acquired electricity and heat and natural gas were identified as the main sources of CO_2 emissions. In total, some 70% of the emissions are indirect, i.e., originating from acquired electricity and heat (for example district heat). In Finland, the average CO_2 value for acquired electricity is 200 CO_2 g /kWh and for acquired district heat 220 CO_2 g /kWh (Motiva Oy, 2008).

The distribution of carbon dioxide emissions in both the Pirkanmaa region and the whole country was calculated on the basis of industrial statistics from 2004. The distributions between industrial sectors and emission sources are visualised in figures 3 and 4. Clearly, the biggest CO_2 emitting sector is the manufacturing of pulp, paper, and paper products. Labour intensive sectors, such as the manufacturing of machinery and equipment and the manufacturing of fabricated metal products are the sources with the least emissions. The use of electricity is the biggest source of CO_2 emissions.

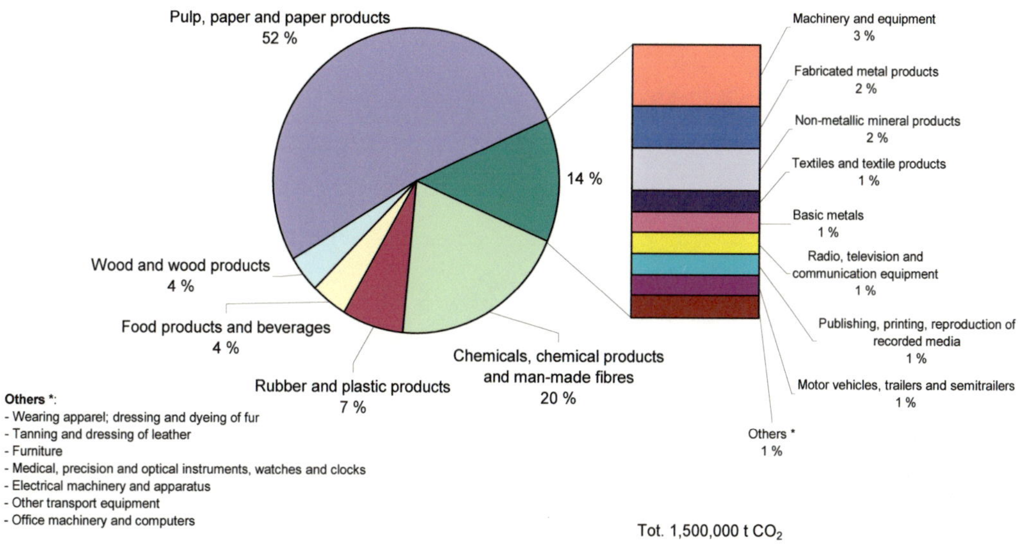

Figure 3. CO_2 emissions distribution by industrial sector in the Pirkanmaa region (Aro, 2009).

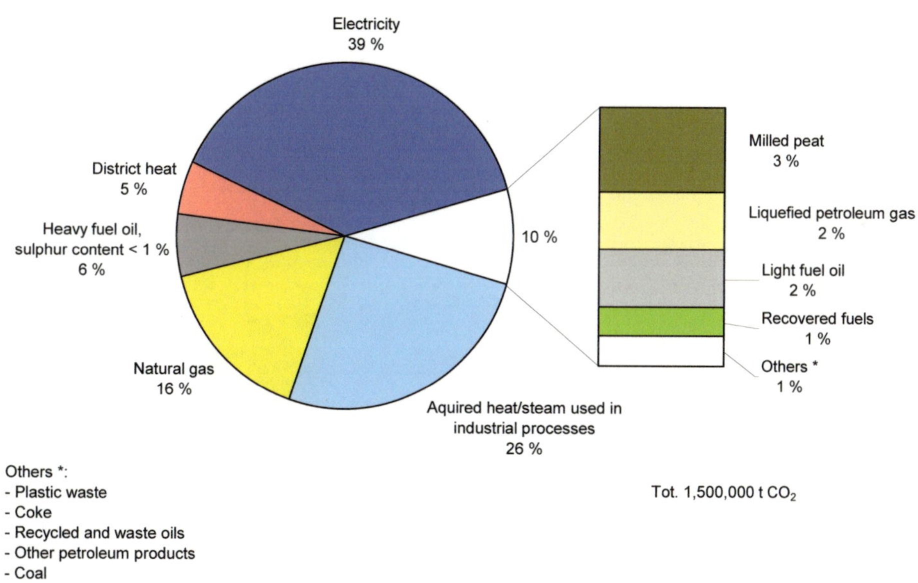

Figure 4. CO_2 emissions distribution by source originating from the industrial energy use in the Pirkanmaa region (Aro, 2009).

The emission sources for different branches of industry were visualised both for the Pirkanmaa region and for the whole country. Such pie charts are vital for understanding the differences between the regions and the whole country. Figure 5 is an example of the pie chart for the food products and beverage industry.

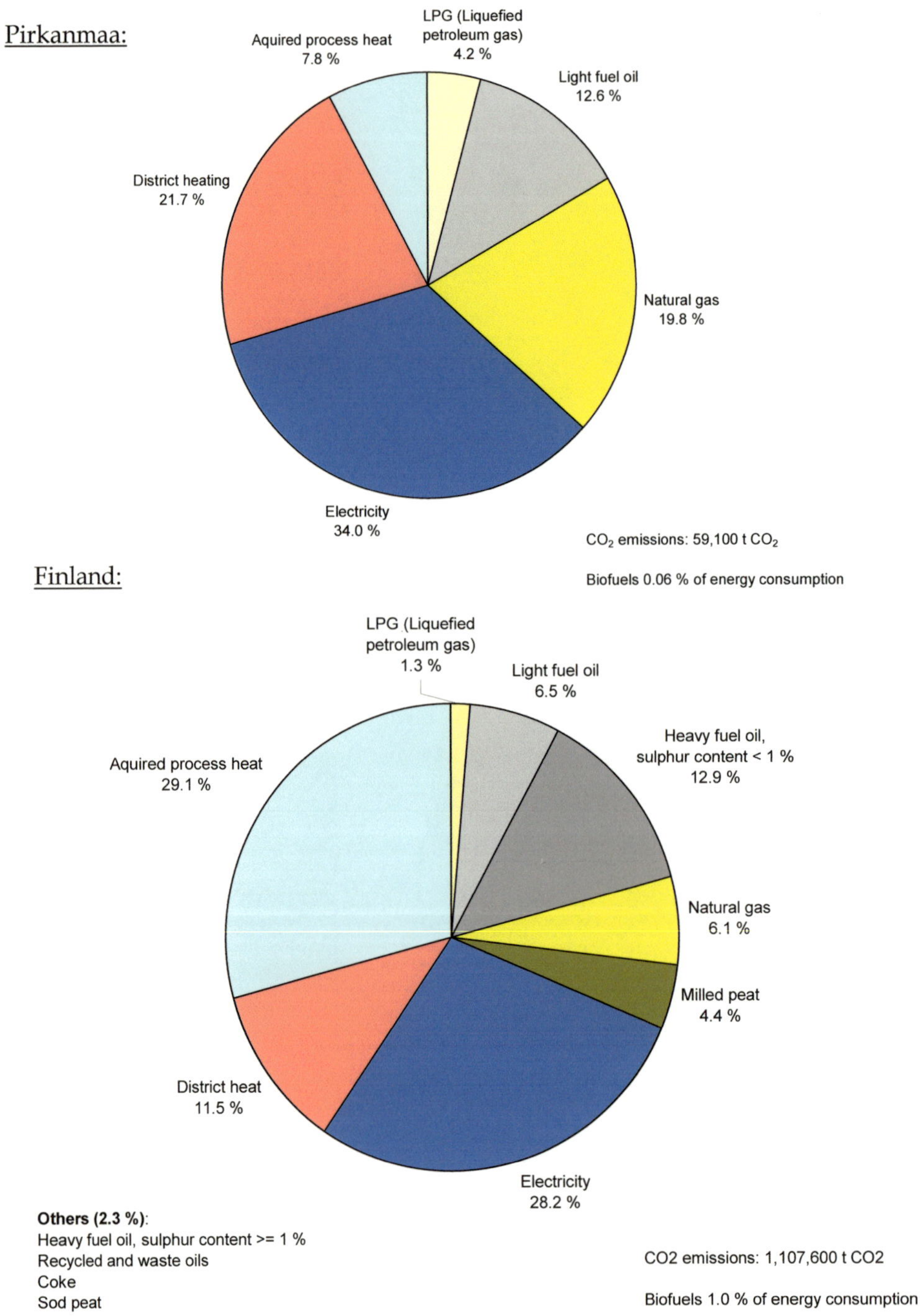

Figure 5. CO_2 emissions sources for food products and beverage industry in the Pirkanmaa region and in the whole country (Aro, 2009).

5.2.3. *Parameters as a tool in energy efficiency and GHG control policies*

Some key parameters were calculated and studied for the industrial sectors as a tool for the target-oriented approach, for example:

- District heat usage / total energy usage, %
- Value added / total energy usage, 1000 €/GWh
- Value added / CO_2 emissions, 1000 €/t CO_2
- Value added / electricity usage, 1000 €/MWh or GWh
- Total energy usage / value added, kWh/€
- Employment / CO_2 emissions, number of employees/1000t CO_2
- Employment / electricity usage, number of employees /GWh

The key parameters were also found suitable for comparing other characteristics of sectors and companies in relation to their CO_2 emissions. These comparisons showed, for example, that the Pirkanmaa region is more industrialised than provinces on average but industry in the Pirkanmaa region is less energy intensive than Finnish industry on average: the added value of the industry of the Pirkanmaa region is 12% of the whole country whereas its population is 9%. As regards the industry energy intensity, in the Pirkanmaa region the energy intensity is lower than in the whole country, see table 3. Table 3 also indicates that Finnish industry is in general very energy intensive because the average number is 9 kWh/€, when one definition of an energy intensive industry is 6 kWh/€ or higher (Blok, 2007).

Industry	**Whole Finland kWh/€**	**Pirkanmaa region kWh/€**
Food products and beverage	2	2
Textiles and textile products	2	1
Wood and wood products	6	7
Pulp, paper and paper products	29	8
Chemical and chemical products	6	15
Rubber and plastic products	1	1
Non-metallic mineral products	5	1
Production of basic metals	16	2
Fabricated metal products	1	1
Machinery and equipment	1	0.4
Average	9	3.5

Table 4. Energy consumption of the main industries per value added in the Pirkanmaa region and in the whole of Finland (Aro, 2009).

The structure of the industry has an influence on how much CO_2 emissions are "needed" to create a certain amount of employment opportunities or added value. For example, the manufacturing of machinery and equipment produces up to 20 times more jobs and wealth than the manufacturing of pulp, paper, and paper products with the same amount of CO_2 emissions, see figure 6. Deducing from this, the structure of industry should be steered towards sectors such as machinery and equipment. In this way, industry could reduce a remarkable portion of the current emissions with the present number of jobs and amount of welfare. In fact, industrialized countries have already taken this route. For example, parts of the steel industry, which produces significant amounts of CO_2 emissions, have been shut down or moved abroad. There has been a lot of debate about this intended or unintended carbon leak since the start of the EU ETS. In any case, global warming is indeed a global problem. Moving emitting sources from one place to another is the wrong way of dealing with global warming in both Finnish regional politics as well as in global politics. The Pirkanmaa region looks greener than the other provinces of Finland on average but it will not manage without the products of the less green Finland or the less green world.

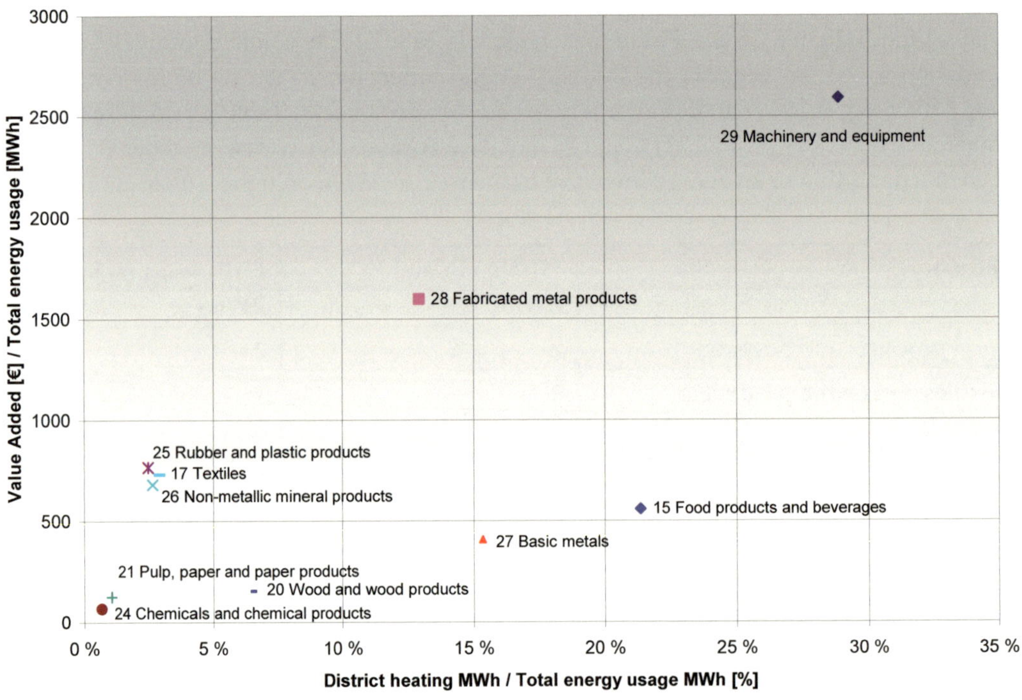

Figure 6. Value added per total energy consumption and the share of district heating from total energy consumption in the main industries of the Pirkanmaa region (Aro, 2009).

District heating produced in CHP plants is commonly known as an efficient way to generate power. In the Pirkanmaa region, district heating is common in industries that fall into the building energy users category, see figure 6. In these industries, no heat is needed in the process and, therefore, they do not necessarily need their own heat production. On average, the share of district heating may represent 30% of the total energy consumption among "building energy users" but on average in the energy intensive industries only a few percent. Due to Finland's long tradition of CHP, the opportunities for using this technology for the production of district heating are today limited, but there is still unused potential for small scale CHP in enterprises that need heat in their production.

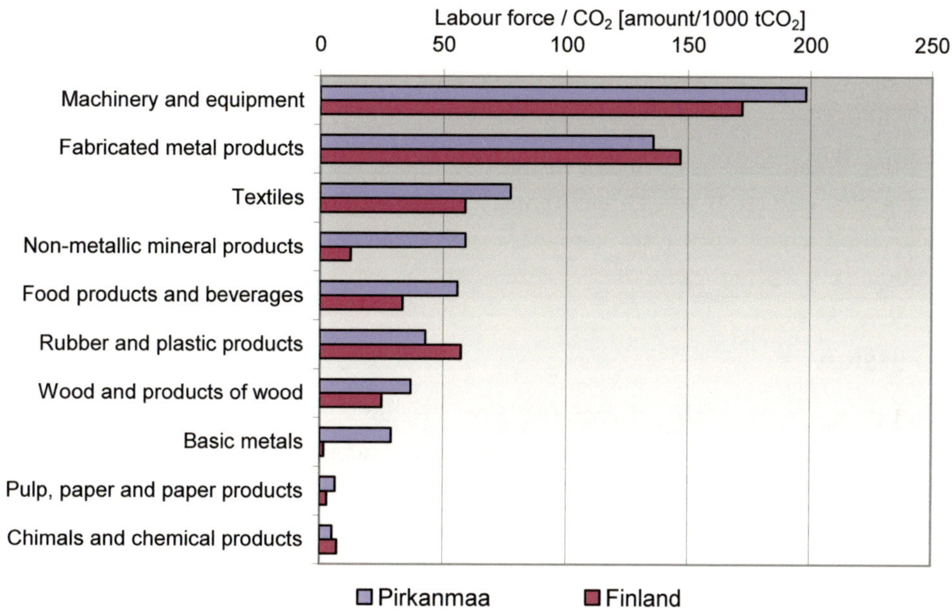

Figure 7. Number of jobs in relation to CO_2 emissions in the Pirkanmaa region and in the whole country (Aro, 2009).

5.2.4. Drafting a scenario for Pirkamaa region

In the article (Aro, 2009), the opportunities for the Pirkanmaa region to achieve a 20% reduction are estimated. In the region about 70% of industry's energy related CO_2 emissions are indirect. This means that these emissions come with acquired electricity and heat. Companies have only limited possibilities to influence the specific emissions of indirect emission sources. In practice, the best way to reduce indirect emissions is to improve energy efficiency. Biofuels are rarely used. Only the manufacturing of pulp, paper, and paper products and the manufacturing of wood and wood products use a significant amount of biofuels compared with their total energy use.

With the existing level of production in the Pirkanmaa region, a 20% cut in CO_2 emissions means some 300 000 tonnes annually. How can we get rid of these tonnes?

Heavy fuel oils are a source of some 100 000 tonnes of CO_2 emissions annually, which can be reduced at least by half by increasing biofuel-based energy production.

During the project, a case study of six companies was carried out. This study and other experiences show that there are energy saving opportunities – with a payback period of less than 5 years - between 4 – 25% depending on the company. On average, it can be estimated that 10% energy savings are possible.

Of the emissions, 70% are indirect. In Finland, there are many plans targeted at reducing the specific CO_2 emissions of electricity. By 2020, Finland will have at least one new nuclear power plant and the share of wind power and biofuels will have increased. It is very likely that the specific CO_2 emissions of electricity will decrease by around 10 – 15% which translates into some 100 000 tons of CO_2 emissions.

This drafting "road map" shows that in the Pirkanmaa region it is possible to achieve the target of a 20% reduction in CO_2 emissions by 2020 with constant production, but if the industrial production grows by some 2% per year as expected, this target will be challenging.

6. Discussion

Whatever policy is conducted it requires generalisations and categorizations. Details are starting point and a base for the categorization and the categorization is a tool to change the details. This is also true with the target oriented energy policy whether you are conducting it at company, regional, national. or even international levels.

In this article, an approach that starts from the ways in which industry uses energy has been discussed. The energy use is categorised for four ways: building energy users (HVAC and lighting), process heat users, process electricity users, and direct combustion users. This categorisation can be applied at company, regional, or national levels. Of course, the details must be in balance with a selected level. This approach and other tools are only to conduct some policy, the tools are not the policy. The approach described in this article is used for the drafting of the energy management of industrial parks in Flanders. As a part of other tools, it has been seen as valuable (Maes, T., et al 2011).

For benchmarking and navigating their position in energy efficiency and GHG emissions, companies need parameters and specific numbers. For these purposes, there is still much to do. One source of information is industrial statistics. Based on these statistics, CO_2 emissions and energy use parameters for national and regional comparisons can be developed. An approach, which combines company level CO_2 emissions and energy analyses, energy use categorization and parameters, gives opportunities to develop a comprehensive industrial energy and GHG policy at the regional level.

Industrial cross-sectoral and energy use categorization approaches support an industrial symbiosis that has been well documented at Kalunborg (Ehrenfeld and Certle, 1997). In Finland, a common example of this symbiosis is the use of saw mill residuals for heat production and as raw material for other areas of the forest industry. These symbioses have thus far developed more or less for economic reasons. The same is also true of the use of district heat by the industries belonging to the building energy users category.

To conduct an energy policy, means that we have to build sandboxes where to play and to use the described tools. Unfortunately, in today's world this is more and more difficult. The life cycles of industrial plants have become shorter and production may be transferred from one place to another very quickly without forewarning. For example, the financial crisis has been with us since 2008 and there are no good predications when it will end. Anyway, it is quite sure that there will be changes in how the industry is located in the future.

Author details

Teuvo Aro
AX Consulting, Axovaatio Ltd., Finland

7. References

Aro, T., 2009. Preconditions and tools for cross-sectoral regional industrial GHG and energy efficiency polic – A Finnish standpoint. Energy Policy , 37,2772-2733.

Blok, K., 2004. Improving Energy Efficiency by Five Percent and More per Year? Journal of Industrial Ecology, 8:4, 87-99.

Blok, K., 2007. Introduction to energy analysis. Techne Press, Amsterdam.

Ehrenfeld, J. and Gertler, N.,1997. Industrial Ecology in Practice: The Evolution of Interdependence at Kalunborg. Journal of Industrial Ecology, 1:1,67-79.

Eurostat, 2008. Employment by sex, age groups and economic activity. http://epp.eurostat.ec.europa.eu. September 22, 2008.

Finland's FNC, 2006. Finland's Fourth National Communication under the United Nations Framework Convention on Climate Change, 2006. Hämeen Kirjapaino Oy, Tampere.

Johansson, B., 2006. Climate policy instruments and industry - effects and potential responses in the Swedish context. Energy Policy, 32, 2344-2360.

Lutsey, N. and Sperling, D., 2008. America's bottom-up climate change mitigation policy. Energy Policy, 36, 673-685.

Maes,T., et al. 2011. Energy management on industrial parks in Flanders Renewable and Sustainable Energy Reviews. 15,1988-2005.

Motiva Oy, 2008. A state-owned limited company to promote energy efficiency and use of renewable energy.
http://www.motiva.fi. September 22, 2008.

Stigson, B., et al., 2008. Global sectoral industry approach to climate change: the way forward, CEPS task force report. Centre for European policy studies, Brussels.

Tanaka, N., 2008. Worldwide trends in energy use and efficiency, Key insights from IEA indicators analysis. OECD/IEA, Paris.

Worrell, E., et al., 1997. Energy intensity in the iron and steel industry, a comparison of physical and economic indicators. Energy Policy, 26, 727-744.

Section 4

Energy Efficiency – Software and Sensors

Chapter 15

Hierarchical Adaptive Balanced Routing Protocol for Energy Efficiency in Heterogeneous Wireless Sensor Networks

Said Ben Alla, Abdellah Ezzati and Ahmed Mohsen

Additional information is available at the end of the chapter

http://dx.doi.org/10.5772/47789

1. Introduction

Wireless sensor networks (WSNs) are composed of many homogeneous or heterogeneous sensor nodes with limited resources. Routing techniques are the most important issue for networks where resources are limited. WSNs technology's growth in the computation capacity requires these sensor nodes to be increasingly equipped to handle more complex functions. Each sensor is mostly limited in their energy level, processing power and sensing ability. Thus, a network of these sensors gives rise to a more robust, reliable and accurate network. Lots of studies on WSNs have been carried out showing that this technology is continuously finding new application in various areas, like remote and hostile regions as seen in the military for battle field surveillance, monitoring the enemy territory, detection of attacks and security etiquette. Other applications of these sensors are in the health sectors where patients can wear small sensors for physiological data and in deployment in disaster prone areas for environmental monitoring. It is noted that, to maintain a reliable information delivery, data aggregation and information fusion that is necessary for efficient and effective communication between these sensor nodes. Only processed and concise information should be delivered to the sinks to reduce communications energy, prolonging the effective network lifetime with optimal data delivery.

An inefficient use of the available energy leads to poor performance and short life cycle of the network. To this end, energy in these sensors is a scarce resource and must be managed in an efficient manner. In this chapter we propose Hierarchical Adaptive Balanced energy efficient Routing Protocol (HABRP) to decrease probability of failure nodes and to prolong the time interval before the death of the first node (stability period) and increasing the lifetime in heterogeneous WSNs, which is crucial for many applications. We study the

INTECH
open science | open minds

© 2012 The Author(s). Licensee InTech. This chapter is distributed under the terms of the Creative Commons Attribution License http://creativecommons.org/licenses/by/3.0), which permits unrestricted use, distribution, and reproduction in any medium, provided the original work is properly cited.

impact of heterogeneity of nodes, in terms of their energy, in wireless sensor networks that are hierarchically clustered. In these networks some high-energy nodes called NCG nodes (Normal node|Cluster Head| Gateway) are elected "cluster heads" to aggregate the data of their cluster members and transmit it to the chosen "Gateways" that requires the minimum communication energy to reduce the energy consumption of cluster head and decrease probability of failure nodes and properly balance energy dissipation. Simulation result shows an improvement in effective network lifetime and increased robustness of performance in the presence of energy heterogeneity.

The organization of this chapter is as followings: We briefly review related work in section 2. Section 3 describes heterogeneous sensor network. Sensor network models is analysed in section 4. In section 5, we present our HABRP protocol. Simulation results of the proposed protocol are discussed in terms of energy consumption, Length of stable region for different values of heterogeneity, number of alive nodes per round, variation of the Base Station location, sensitivity to degree of heterogeneity in large scale networks, improvement of stability period in section 6. Finally, in section 7, we conclude the chapter.

2. Related works

The main aim of hierarchical routing is to efficiently maintain the energy consumption of sensor nodes by involving them in multi-hop communication within a particular cluster and by performing data aggregation and fusion in order to decrease the number of transmitted messages to the sink.

LEACH (W. Heinzelman et al., 2000) is one of the first hierarchical routing approaches for sensors networks. The idea proposed in LEACH has been an inspiration for many hierarchical routing protocols. We explore recent works in this section.

2.1. Low-energy adaptive clustering hierarchy (LEACH)

Low-energy adaptive clustering hierarchy (LEACH) (W. Heinzelman et al., 2000) is one of the most popular hierarchical routing algorithms for sensor networks. The idea is to form clusters of the sensor nodes based on the received signal strength and use local cluster heads as routers to the sink. This will save energy since the transmissions will only be done by such cluster heads rather than all sensor nodes.

All the data processing such as data fusion and aggregation are local to the cluster. Cluster heads change randomly over time in order to balance the energy dissipation of nodes. This decision is made by the node S choosing a random number x between 0 and 1. The node becomes a cluster head for the current round if the number x is less than the following threshold:

$$T(S) = \begin{cases} \dfrac{P}{1 - P * (r \bmod \frac{1}{P})} & \text{if } S \in G \\ 0 & \text{otherwise} \end{cases} \tag{1}$$

Where P is desired percentage of cluster head nodes in the sensor network, r is current round number, and G is the set of nodes that have not been cluster heads in the last $1/p$ rounds.

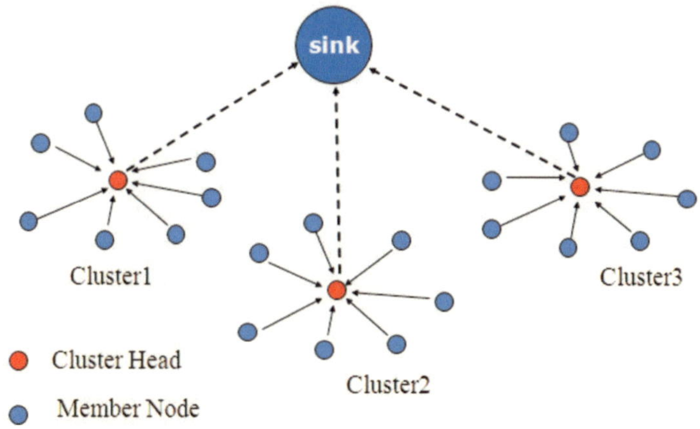

Figure 1. Network model with clustering

LEACH achieves over a factor of 7 reduction in energy dissipation compared to direct communication and a factor of 4–8 compared to the minimum transmission energy routing protocol (Akkaya & Younis, 2005). The nodes die randomly and dynamic clustering increases lifetime of the system. LEACH is completely distributed and requires no global knowledge of network. However, LEACH uses single-hop routing where each node can transmit directly to the cluster-head and the sink. Therefore, it is not applicable to networks deployed in large regions. Furthermore, the idea of dynamic clustering brings extra overhead, e.g. Head changes, advertisements etc., which may diminish the gain in energy consumption.

2.2. Power-Efficient Gathering in Sensor Information Systems (PEGASIS)

This is improved version from LEACH. The main idea of PEGASIS (Lindsey & Raghavendra, 2002) is that nodes are formed into a chain where each node receive from and transmit to closest neighbor only. The distance between sender and receiver is reduced as well as decreasing the amount of transmission energy. To construct a chain, PEGASIS uses a greedy algorithm that starts from the farthest node from the base station.

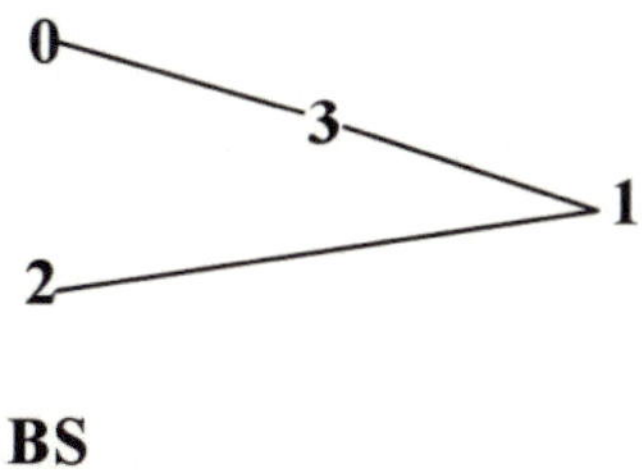

Figure 2. Chain is constructed using the greedy algorithm

In Fig.2, the algorithm starts with node 0 that connects to node 3. Then, node3 connects to node 1 and node 1 connects to node 2, which is the closest one to the base station *BS*. Because nodes already in the chain cannot be revisited, the neighbor distance will increase gradually. When a node dies (out of battery), the chain will be reconstructed by repeating the same procedure to bypass the dead node.

In one round of transmission, a randomized node is appointed to be the leader to transmit data to *BS*. If the *BS* locates outside the range of this node, multi-hop transmission will be employed. The leader will be changed randomly in every round, so that overall energy dissipation is balanced out. For transmitting a packet in each round, a token is used that passing from the one end of the chain to the other end of the chain. Only node that has a token can transmit a data packet to its intermediate node in the chain. When intermediate node receives data from one neighbor along with a token, it fuses the data packet with its own data and transmits a new data packet to the next node in the chain.

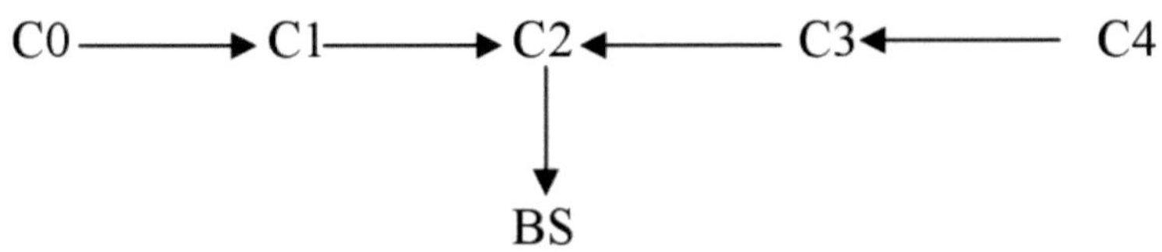

Figure 3. Token passing approach in PEGASIS

In fig.3, C0 will pass its data and token to C1. C1 fuses a data packet with its own data and pass a new data packet to the leader C2. C2 does not transmit a data packet to BS yet, but rather it passes a token to C4. When C2 receives a data from C4 and C3, it fuses and transmits the sensed data to BS.

2.3. A Stable Election Protocol (SEP)

A Stable Election Protocol (SEP) (G. Smaragdakis et al, 2004) is improved version of LEACH protocol. Main aim of it was used heterogeneous sensor in wireless sensor networks. This protocol have operation like LEACH but with this difference that, in SEP protocol sensors have two different level of energy. SEP based on weighted election probabilities of each node to become cluster head according to their respective energy. This approach ensures that the cluster head election is randomly selected and distributed based on the fraction of energy of each node assuring a uniform use of the nodes energy.

In SEP, two types of nodes (normal and advanced) are considered. It is based on weighted election probabilities of each node to become cluster head according to the remaining energy in each node. This prolongs the stability period i.e. the time interval before the death of the first node.

2.4. Distributed energy efficient clustering algorithm for heterogeneous wireless sensor networks

(Qing et al, 2006) proposed a distributed energy efficient clustering scheme for heterogeneous wireless sensor networks, which is called DEEC. In DEEC, the cluster heads are elected by a probability based on the ratio between residual energy of each node and the average energy of the network. The epochs of being cluster heads for nodes are different according to their initial and residual energy.

The authors have assumed that all the nodes of the sensor network are equipped with different amount of energy, which is a source of heterogeneity. DEEC is also based on LEACH protocol, it rotates the cluster head role among all nodes to expend energy uniformity.

Two levels of heterogeneous nodes are considered in the algorithm and after that a general solution for multi-level heterogeneity is obtained. To avoid that each node needs to know the global knowledge of the networks, DEEC estimates the ideal value of network lifetime, which is used to compute the reference energy that each node should expend during a round. Simulation results show that DEEC achieves longer lifetime and more effective messages than LEACH, SEP and LEACH -E.

2.5. Improved and balanced LEACH for heterogeneous wireless sensor networks

(S. Ben alla et al. 2010) proposed Improved and Balanced LEACH (IB-LEACH), in which some high energy nodes elect themselves to be gateway at any given time with a certain probability. Base station confirms that whether those nodes suit to be gateway. These nodes broadcast their status to the other sensors in the network using advertisement message (ADV). The non -gateway nodes elect themselves to be cluster heads with a certain probability. These cluster head nodes broadcast their status to the other sensors in the network using advertisement message (ADV). The non-cluster head nodes wait the cluster head announcement from other nodes. Each sensor node determines to which cluster it wants to belong by choosing the cluster head that requires the minimum communication energy, and send the join -request (Join- REQ) message to the chosen cluster head, and the cluster head nodes wait for join-request message from other nodes.

Each cluster head collect and aggregate the data of their cluster members and transmit it to the chosen gateways that requires the minimum communication energy to reduce the energy consumption of cluster head and decrease probability of failure nodes. Simulation results show that this protocol performs better than LEACH and SEP in terms of network lifetime.

2.6. Cluster head relay routing protocol for heterogeneous sensor networks

(Du & Lin, 2005) proposed a cluster head relay (CHR) routing protocol for heterogeneous sensor networks. This protocol uses two types of sensors to form a heterogeneous network with a single sink: a large number of low-end sensors, denoted by L-sensors, and a small number of powerful high-end sensors, denoted by H-sensors. Both types of sensors are static and aware of their locations using some location service. Moreover, both L-sensor and H-sensors are uniformly and randomly distributed in the sensor field.

The CHR protocol partitions the heterogeneous network into clusters, each being composed of L-sensors and led by an H-sensor. Within a cluster, the L-sensors are in charge of sensing the underlying environment and forwarding data packets originated by other L-sensors toward their cluster head in a multi-hop transmission. The H-sensors, on the other hand, are responsible for data fusion within their own clusters and forwarding aggregated data packets originated from other cluster heads toward the sink in a multi-hop transmission using only cluster heads. While L-sensors use short-range data transmission to their neighboring H -sensors within the same cluster, H-sensors perform long-range data communication to other neighboring H-sensors and the sink. Simulation results demonstrate that CHR performs better than directed diffusion and SWR.

2.7. Energy efficient cluster head election protocol for heterogeneous wireless sensor networks

(LI Han, 2010) proposed an energy efficient cluster head election protocol for heterogeneous wireless sensor networks and using the improved Prim's algorithm to construct an inter cluster routing. He has considered three types of sensor nodes. Some fraction of the sensor nodes are equipped with the additional energy resources than the other nodes. He has assumed that all the sensor nodes are uniformly distributed.

In this protocol, the cluster head node sets up a TDMA schedule and transmits this schedule to the nodes in the cluster. This ensures that there are no collisions among data messages and also allows the radio components of each non-cluster head node to be turned off at all times except during their transmit time, thus minimizing the energy dissipated by the individual sensors.

In order to reduce the energy consumption of the cluster heads which are far away from the base station and balance the energy consumption of the cluster heads which are close to the base station, a multi-hop routing algorithm of cluster head has been presented, which introduces into the restriction factor of remainder energy when selects the interim nodes between cluster heads and base station, and also the minimum spanning tree algorithm has been included. The protocol can not only reduce the consumption of transmit energy of cluster head, but also the consumption of communication energy between non-cluster head and cluster head nodes. Simulation results show that this protocol performs better than LEACH and EECHE in terms of network lifetime.

3. Heterogeneous wireless sensor networks

This section presents a paradigm of heterogeneous wireless sensor network and discusses the impact of heterogeneous resources (Yarvis ,2005)(V. Katiyar, 2011)

3.1. Types of heterogeneous resources

There are three common types of resource heterogeneity in sensor nodes: computational heterogeneity, link heterogeneity and energy heterogeneity.

Computational heterogeneity means that the heterogeneous node has a more powerful microprocessor and more memory than the normal node. With the powerful computational resources, the heterogeneous nodes can provide complex data processing and longer-term storage.

Link heterogeneity means that the heterogeneous node has high bandwidth and long-distance network transceiver than the normal node. Link heterogeneity can provide a more reliable data transmission.

Energy heterogeneity means that the heterogeneous node is line powered or its battery is replaceable.

Among above three types of resource heterogeneity, the most important resource heterogeneity is the energy heterogeneity because both computational heterogeneity and link heterogeneity will consume more energy resource.

If there is no energy heterogeneity, computational heterogeneity and link heterogeneity will bring negative impact to the whole sensor network, i.e ., decreasing the network lifetime.

3.2. Impact of heterogeneity on wireless sensor networks

Placing few heterogeneous nodes in the sensor network can bring following benefits:

Decreasing latency of data transportation: Computational heterogeneity can decrease the processing latency in immediate nodes and link heterogeneity can decrease the waiting time in the transmitting queue. Fewer hops between sensor nodes and sink node also mean fewer forwarding latency.

Prolonging network lifetime: The average energy consumption for forwarding a packet from the normal nodes to the sink in heterogeneous sensor networks will be much less than the energy consumed in homogeneous sensor networks.

Improving reliability of data transmission: It is well known that sensor network links tend to have low reliability and each hop significantly lowers the end-to-end delivery rate.

With heterogeneous nodes, there will be fewer hops between normal sensor nodes and the sink. So the heterogeneous sensor network can get much higher end-to-end delivery rate than the homogeneous sensor network.

3.3. Performance measures

Some performance measures that are used to evaluate the performance of clustering protocols are listed below(R. Sheikhpour et al., 2011).

Network lifetime (stability period): It is the time interval from the start of operation (of the sensor network) until the death of the first alive node.

Number of cluster heads per round: Instantaneous measure reflects the number of nodes which would send directly to the base station, information aggregated from their cluster members.

Number of alive nodes per round: This instantaneous measure reflects the total number of nodes and that of each type that has not yet expended all of their energy.

Throughput: This includes the total rate of data sent over the network, the rate of data sent from cluster heads to the base station as well as the rate of data sent from the nodes to their cluster heads.

Figure 4 shows the heterogeneous model for wireless sensor ne twork.

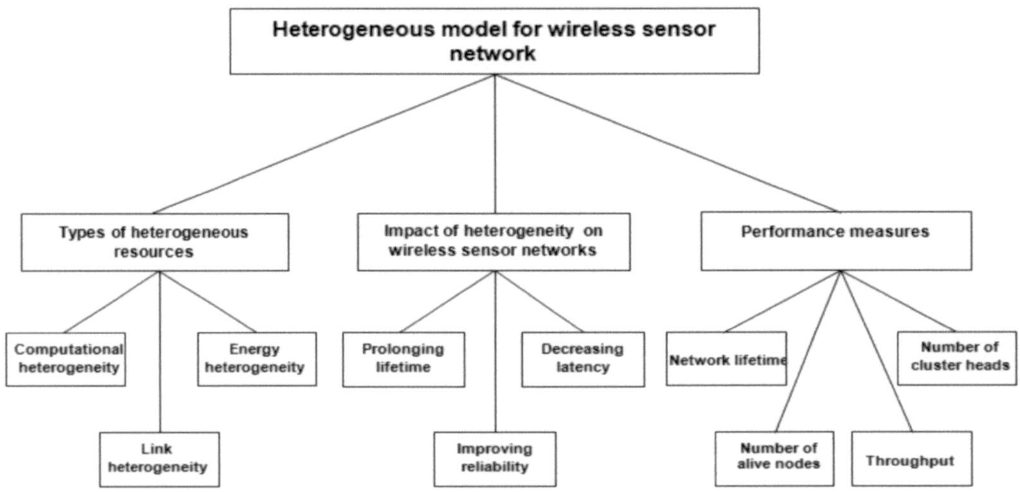

Figure 4. Heterogeneous model for wireless sensor network.

4. Wireless sensor network models

4.1. Network model

In this chapter we assume a sensor network model with following properties:

- The sink locates at the centre of sensor nodes and has enough memory and computing capability.
- The WSNs consist of the heterogeneous sensor nodes. Percentage of sensor nodes are equipped with more energy resources than the rest of the nodes. Let m be the fraction of the total number of nodes N which are equipped with a times more energy than the others.
- The distance can be measured based on the wireless radio signal power.
- All sensor nodes are immobile and have a limited energy.
- All nodes are equipped with power control capabilities to vary their transmitting power.

4.2. Radio energy dissipation model

For this project, three routing protocols, namely LEACH and SEP and our protocol Hierarchical Adaptive Balanced energy efficient Routing Protocol (HABRP), We use the same radio model shown in Fig.5 for the radio hardware energy dissipation in order to achieve an acceptable Signal-to-Noise Ratio (SNR) in transmitting a k bit message over a distance d .

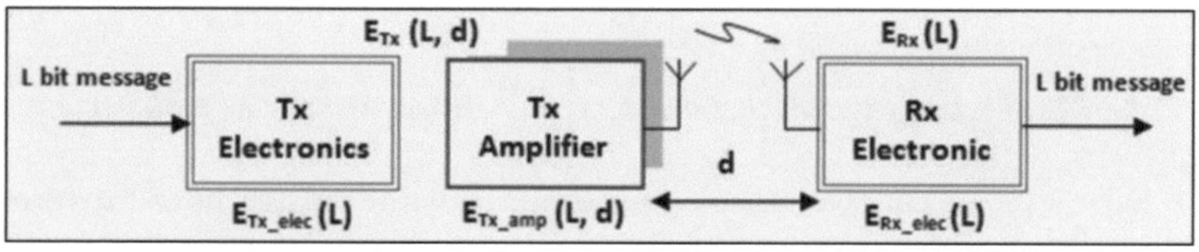

Figure 5. Radio Energy Dissipation Model.

In Fig.5 k is the number of bits per packet transmission and d is distance between the sender and the receiver. Electronics energy consumption is same for transmitting and receiving the data, is given by,

$$E_{Tx}(k) = E_{Rx}(k) = E_{elec} * k \tag{2}$$

E_{elec} is the energy dissipated per bit to run the transmitter or the receiver circuit.

Transmission cost to transmit K-bit message between any two nodes over distance d is given by the following equation:

$$E_{Tx}(k,d) = E_{elec} * k + E_{amp}(k,d) \tag{3}$$

$E_{amp}(k,d)$ is the amplifier energy consumption and it can be further expressed in terms of ε_{fs} or ε_{mp}, depending on the transmitter amplifier mode that applied. They are power loss factors for free space (d^2 loss) when $d < d_0$; and multipath fading (d^4 loss) when $d \geq d0$, respectively.

To transmit k-bit message within d distance, a node expends:

$$E_{Tx}\,(k,d) = \begin{cases} (E_{elec} * k) + \left(\varepsilon_{fs} * k * d^2\right) & if\ d \leq d_0 \\ (E_{elec} * k) + \left(\varepsilon_{mp} * k * d^4\right) & if\ d > d_0 \end{cases} \tag{4}$$

By equating the two expressions at $d=d_0$, we have:

$$d_0 = \sqrt{\frac{\varepsilon_{fs}}{\varepsilon_{mp}}} \tag{5}$$

To receive a k-bit message, a node expends:

$$E_{Rx}(k) = E_{elec} * k \tag{6}$$

Where E_{elec} is the energy used by the receiver electronics.

4.3. Optimal number of cluster

We assume there are N nodes distributed uniformly in MxM region. If there are C clusters, there are on average N/C nodes per cluster. Each cluster-head dissipates energy receiving signals from the nodes, beamforming the signals, and transmitting the aggregate signal to the base station. Therefore, the energy dissipated in the cluster-head node during a single frame is:

$$E_{CH} = \left[k * E_{elec} * \frac{N}{C}\right] + \left[k * E_{DA} * \frac{N}{C}\right] + [k * E_{elec} + k * \varepsilon_{mp} * d_{toBS}^4] \tag{7}$$

Where $k * \text{Eelec} * (\frac{N}{C} - 1)$ is the energy consumed by the cluster head to receive k bits information from $\left(\frac{N}{C} - 1\right)$ non cluster heads and E_{DA} represents the processing of data aggregation cost of a bit per signal. The expression for the energy spends by a non cluster head is given by:

$$E_{non-CH} = k * E_{elec} + k * \varepsilon_{fs} * d_{toCH}^2 \tag{8}$$

Where d_{toCH} is the distance from the node to the cluster head. The area occupied by each cluster is approximately M^2/C . In general, this is an arbitrary-shaped region with a node distribution $\rho(x, y)$.

The expected squared distance from the nodes to the cluster head (assumed to be at the center of mass of the cluster) is given by:

$$d_{toCH}^2 = \iint \left((x^2 + y^2) * \rho(x, y)\right) dx\, dy = \iint r^2 \rho(r, \theta) r\, dr\, d\theta \tag{9}$$

If we assume this area is a circle with radius $R = (M/\sqrt{\pi C})$ and $\rho(r, \theta)$ is constant for r and θ, (9) simplifies to:

$$d_{toCH}^2 = \rho \int_{\theta=0}^{2\pi} \int_{r=0}^{(M/\sqrt{\pi C})} r^3\, dr\, d\theta = \frac{\rho}{2\pi} \frac{M^4}{C^2} \tag{10}$$

If the density of nodes is uniform throughout the cluster area, then $\rho = (1/(M^2/C))$and

$$d_{toCH}^2 = \frac{1}{2\pi} \frac{M^2}{C} \tag{11}$$

Therefore, in this case the expression for the energy spends by a non cluster head is:

$$E_{non-CH} = k * E_{elec} + k * \varepsilon_{fs} * \frac{1}{2\pi} \frac{M^2}{C} \tag{12}$$

The energy dissipated in a cluster per round, $E_{cluster}$, is expressed by:

$$E_{cluster} = E_{CH} + \left(\frac{N}{C} - 1\right) * E_{non-CH} \approx E_{CH} + \frac{N}{C} * E_{non-CH} \tag{13}$$

Therefore, the total energy dissipated in the network per round, E_{total}, is expressed by:

$$E_{total} = C * E_{cluster} \tag{14}$$

We can find the optimum number of clusters by setting the derivative of E_{total} with respect to C to zero (W.Heinzelman et al., 2002):

$$\frac{\partial E_{total}}{\partial C} = 0 \tag{15}$$

$$C_{opt} = \sqrt{\frac{N}{2\pi}} \sqrt{\frac{\varepsilon_{fs}}{\varepsilon_{mp}}} \frac{M}{d_{toBS}^2} \tag{16}$$

The optimal probability of a node to become a cluster head, P_{opt}, can be computed as follows:

$$P_{opt} = \frac{C_{opt}}{N} \tag{17}$$

The optimal probability for a node to become a cluster head is very important. The authors (W.Heinzelman et al., 2000) showed that if the clusters are not constructed in an optimal way, the total consumed energy of the sensor network per round is increased exponentially either when the number of clusters that are created is greater or especially when the number of the constructed clusters is less than the optimal number of clusters.

5. Hierarchical Adaptive Balanced energy efficient Routing Protocol (HABRP)

In this section we describe HABRP which is an extension of the LEACH, which improves the stability period of the clustering hierarchy and decrease probability of failure nodes using the characteristic parameters of heterogeneity.

Routing in HABRP works in rounds and each round is divided into two phases, the Setup phase and the Steady State phase, each sensor knows when each round starts using a synchronized clock. Let us assume the case where a percentage of sensor nodes are equipped with more energy resources(advanced nodes) than the rest of the nodes (normal nodes). We refer to these powerful nodes as NCG nodes (nodes selected as normal nodes or cluster heads or gateways). Let m be the fraction of the total number of nodes N which are equipped with a times more energy than the others and the rest $(1-m)*N$ as normal nodes and E_0 the initial energy. We assume that all nodes are distributed uniformly over the sensor field. The total number of nodes and total energy in network is expressed by:

$$\begin{aligned} N &= N * m[NCG\,nodes] + N * (1-m)[normal\,nodes] \\ E[total] &= N * m * E_0 * (1+a) + N * (1-m) * E_0 \end{aligned} \tag{18}$$

5.1. HABRP network model

The basic system model of the protocol HABRP is depicted in Fig.6. Each sensor node sends the sensed data to its cluster head. The cluster head aggregates the collected data and transmits the aggregated information to closest gateway that will send data to the base station.

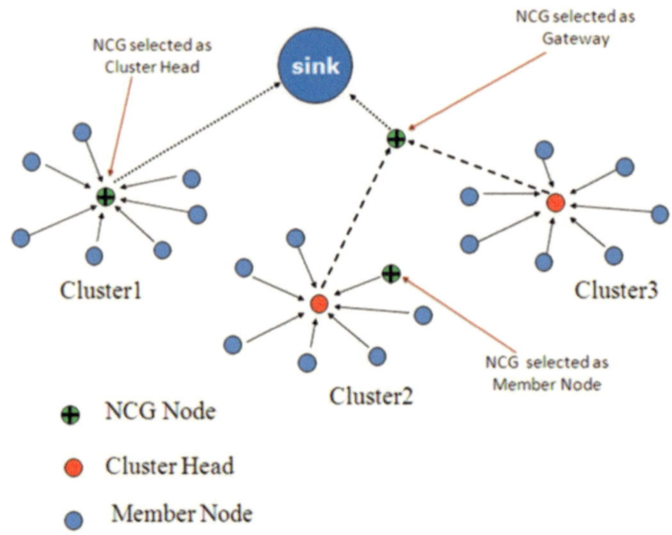

Figure 6. The HABRP Network model

The operation of HABRP is divided into rounds. Each round begins with a set-up phase followed by a steady-state phase, as shown in Fig.7.

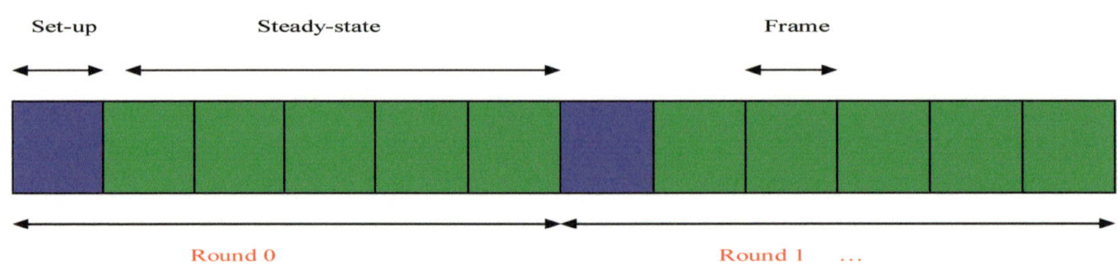

Figure 7. Time line showing HABRP operation

During the set-up phase the gateways are elected and the clusters are organized. It is constituted by gateway selection algorithm, cluster selection algorithm and cluster formation algorithm.

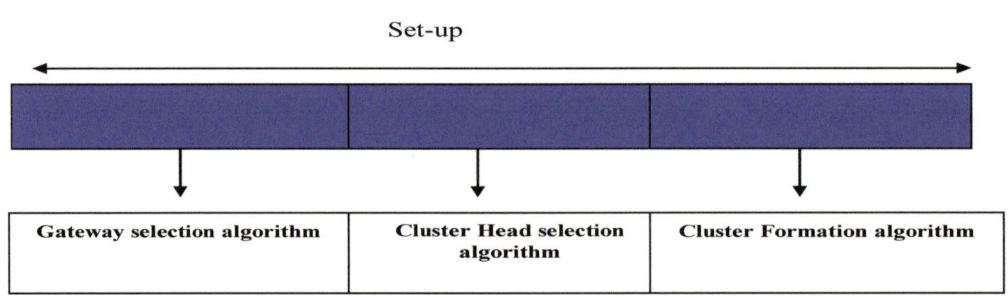

Figure 8. Time line showing set-up phase

After the set-up phase is the steady-state phase when data are transmitted from the nodes to the cluster head to agregate data and transmit it to the base station through the gateway that requires the minimum communication energy. The duration of the steady phase is longer than the duration of the setup phase in order to minimize overhead.

5.2. Gateway selection algorithm

Each sensor S elects it self to be a gateway at the beginning of each round, round. This decision is based on the suggested percentage of gateways for the network (determined a priori) and the number of rounds the node has been a gateway so far. The desired percentage of gateways is chosen such that the expected number of gateway nodes for each round is Ng. Thus, if there are $N * m$ NCG nodes (advanced nodes) in the network, the desired percentage of gateways is:

$$P_g = \frac{N_g}{N * m} \tag{19}$$

Decision to become gateway is made by the node S choosing a random number x between 0 and 1. The node becomes a gateway for the current round if the number x is less than the following threshold:

$$T(S_{gat}) = \begin{cases} \dfrac{P_g}{1 - P_g * (r \bmod \frac{1}{P_g})} * \dfrac{E_{s\_current}}{E_{s\_initial}} & \text{if } S \in G_{gat} \\ 0 & \text{otherwise} \end{cases} \tag{20}$$

We define as $T(S_{gat})$ the threshold for gateway node S, r is the currenet round, G_{gat} is set of nodes which have not been gateways in $1/P_g$ rounds , $E_{s\_current}$ is the current energy of the node and $E_{s\_initial}$ is the initial energy of the node.

5.3. Cluster head selection algorithm

The main idea is for the sensor nodes to elect themselves with respect to their energy levels autonomously. The goal is to minimize communication cost and maximizing network resources in other to ensure concise information is sent to the sink. Each node transmits data to the closest cluster head and the cluster heads performs data aggregation. Assume an optimal number of clusters C_{opt} in each round. It is expected that as a cluster head, more energy will be expended than being a cluster member. Each node can become cluster head with a probability P_{opt} and every node must become cluster head once every $1/P_{opt}$ rounds.

The optimal probability of a node to become a cluster head, P_{opt}, can be computed as follows:

$$P_{opt} = \frac{C_{opt}}{N - Ng} \tag{21}$$

Ng is a number of gateway nodes, C_{opt} is the optimum number of clusters that is expressed by:

$$C_{opt} = \sqrt{\frac{N - Ng}{2\pi}} \sqrt{\frac{\varepsilon_{fs}}{\varepsilon_{mp}}} \frac{M}{d_{toBS}^2} \quad (22)$$

Our approach is to assign a weight to the optimal probability P_{opt}. This weight must be equal to the initial energy of each node divided by the initial energy of the normal node. Let us define as P_{nrm} the weighted election probability for normal nodes, and P_{adv} the weighted election probability for the advanced nodes.

Virtually there are $N * (1 + a * m)$ nodes with energy equal to the initial energy of a normal node. In order to maintain the minimum energy consumption in each round within an epoch, the average number of cluster heads per round per epoch must be constant and equal to C_{opt}. The weighed probabilities for normal and advanced nodes are, respectively:

$$P_{nrm} = \frac{P_{opt}}{1 + a * m} \quad (23)$$

$$P_{adv} = \frac{P_{opt}}{1 + a * m} * (1 + a) \quad (24)$$

In Equation (1), we replace P_{opt} by the weighted probabilities to obtain the threshold that is used to elect the cluster head in each round.

We define as $\mathrm{T(S_{nrm})}$ the threshold for normal nodes, and $\mathrm{T(S_{adv})}$ the threshold for advanced nodes. Thus, for normal nodes, we have:

$$\mathrm{T(S_{nrm})} = \begin{cases} \dfrac{\mathrm{P_{nrm}}}{1 - \mathrm{P_{nrm}} * (\mathrm{r} \bmod \frac{1}{\mathrm{P_{nrm}}})} * \dfrac{\mathrm{E_{s\_current}}}{\mathrm{E_{s_{initial}}}} & \text{if } \mathrm{S_{nrm}} \in \mathrm{G_{nrm}} \\ 0 & \text{otherwise} \end{cases} \quad (25)$$

Where r is the current round, $\mathrm{G_{nrm}}$ is the set of normal nodes that have not become cluster heads within the last $1/\mathrm{P_{nrm}}$ rounds of the epoch and $\mathrm{T(S_{nrm})}$ is the threshold applied to normal nodes.

Similarly, for advanced nodes, we have:

$$\mathrm{T(S_{adv})} = \begin{cases} \dfrac{\mathrm{P_{adv}}}{1 - \mathrm{P_{adv}} * (\mathrm{r} \bmod \frac{1}{\mathrm{P_{adv}}})} * \dfrac{\mathrm{E_{s\_current}}}{\mathrm{E_{s_{initial}}}} & \text{if } \mathrm{S_{adv}} \in \mathrm{G_{adv}} \\ 0 & \text{otherwise} \end{cases} \quad (26)$$

Where r is the current round, $\mathrm{G_{adv}}$ is the set of advanced nodes that have not become cluster heads within the last $1/\mathrm{P_{adv}}$ rounds of the epoch, and $\mathrm{T(S_{adv})}$ is the threshold applied to advanced nodes.

5.4. Summary of HABRP

HABRP is a self-organizing, adaptive clustering protocol that uses randomization to distribute the energy load evenly among the sensors in the network. Each sensor elects it self to be a gateway at the beginning of each round with a certain probability.

These gateway nodes broadcast their status to the other sensors in the network using advertisement message (ADV). The non-gateway nodes elect themselves to be cluster-heads with a certain probability.

These cluster-head nodes broadcast their status to the other sensors in the network using advertisement message (ADV). The non-cluster-head nodes wait the cluster-head announcement from other nodes. Each sensor node determines to which cluster it wants to belong by choosing the cluster-head that requires the minimum communication energy, and send the join-request (Join-REQ) message to the chosen cluster head, and the cluster-head nodes wait for join-request message from other nodes.

Once all the nodes are organized into clusters, each cluster-head creates a schedule for the nodes in its cluster. This allows the radio components of each non-cluster-head node to be turned off at all times except for its transmit time, thus minimizing the energy dissipated in the individual sensors. Once the cluster-head has all the data from the nodes in its cluster, the cluster-head node aggregates the data and then transmits the compressed data:

- To the chosen gateway that requires the minimum communication energy to reduce the energy consumption of cluster head and decrease probability of failure nodes if :

$$E_{CH\_to\_BS} > E_{CH\_to\_Gat} + E_{Gat\_to\_BS} \quad (27)$$

These collected data are transmitted to the base station using cluster head-gateway-base station routing.

- To the Base station directly if :

$$E_{CH\_to\_BS} \leq E_{CH\_to\_Gat} + E_{Gat\_to\_BS} \quad (28)$$

Where $E_{CH\_to\_BS}$ is total energy dissipated for send data from cluster head to the base station and $E_{CH\_to\_Gat}$ is total energy dissipated for send data from cluster head to the Gateway and $E_{Gat\_to\_BS}$ is total energy dissipated for send data from Gateway to the base station as shown in Fig.9.

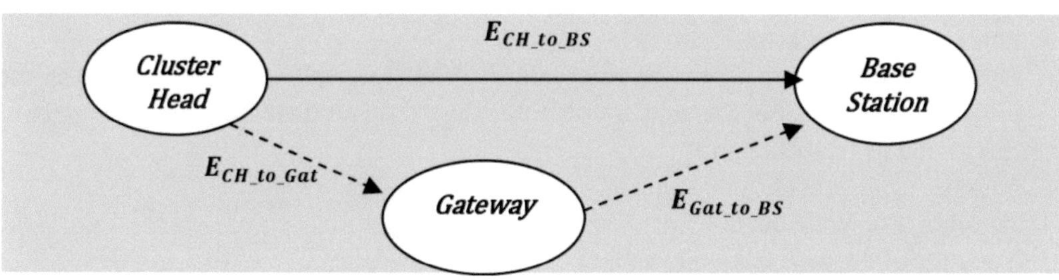

Figure 9. . Cluster head would transmit to Base Station through Gateway if $E_{CH\_to\_Gat} + E_{Gat\_to\_BS} < E_{CH\_to\_BS}$

6. Simulation

6.1. Parameter settings

We use a 100m×100m region of 100 sensor nodes scattered randomly. MATLAB is used to implement the simulation. To make a fair comparison, we introduce advanced energy levels to LEACH and SEP nodes with same settings as in our HABRP protocol, so as to assess the performance of these protocols in the presence of heterogeneity.

Specifically, we have the parameter settings:

Notation	Description	Value
M x M	Area	100x100, 300x300
N	Number of the sensors	100, 900
sinkX, sinkY	Sink node location	50x50, 50x200, 50x300, 300x300
E_0	Initial energy	0.5 J
Eelec	Electronics energy	50nJ/bit
E_{DA}	Energy of data aggregation	5nJ/bit
d_0	The threshold distance	87m
ε_{fs}	Amplified transmitting energy using free space	10pJ/bit/ m 2
ε_{mp}	Amplified transmitting energy using multipath	0.0013pJ/bit/ m4
k	Data packet size	500bytes
k_{broad}	Broadcast packet size	25 bytes
P_{opt}	Probability	0.05
C_{opt}	optimum number of clusters	5
N_g	number of gateway nodes	4

Table 1. Simulation parameter

6.2. Simulation metrics

Performance metrics used in the simulation study are:

- Energy consumption analysis
- Length of stable region for different values of heterogeneity. Stability period is the period from the start of the network operation and the first dead node. We also refer to this period as "stable region"
- Number of alive nodes per round.
- Percentage of Node death
- Variation of the Base Station Location
- Sensitivity to degree of heterogeneity in large-scale networks.
- Improvement of Stability period:

$$Improvement = \frac{Stable\ period\ of\ HABRP - Stable\ period\ of\ LEACH}{Stable\ period\ of\ LEACH} \quad (29)$$

6.3. Simulation results

6.3.1. Energy consumption analysis

The performance of HABRP is compared with that of the original LEACH and SEP in terms of energy and is shown in Fig.10 and Fig.11.

With the use of gateway nodes for data transmission from cluster heads to the sink, the energy consumption of the network is decreased. This is due to the gain of the energy dissipated by cluster heads to the base station. From the graph it is clear that HABRP can achieve twice the energy savings than LEACH and SEP protocol.

Fig.10 illustrates the energy performance of HABRP in homogeneous WSNs.

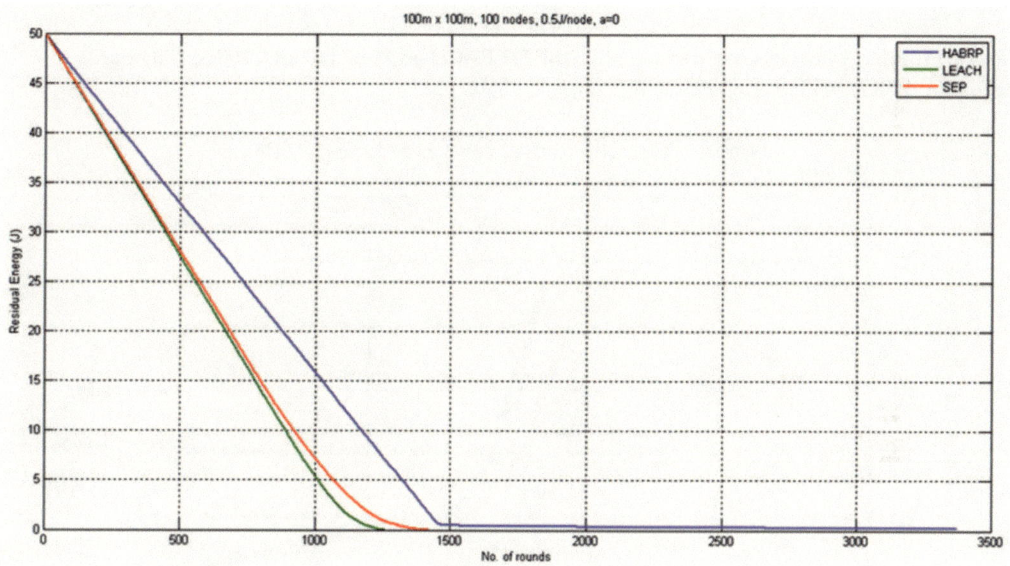

Figure 10. Energy analysis comparison of HABRP, LEACH and SEP in Homogeneous WSN (100m x 100m, 100 nodes, 0.5J/node, a=0(Homogeneous WSNs))

Fig.11 illustrates the energy performance of HABRP in heterogeneous WSNs.

6.3.2. Network lifetime

The number of nodes alive for each round of data transmission is observed for the HABRP protocol to evaluate the lifetime of the network. Fig.12 shows the performance of HABRP compared to LEACH and SEP. It is observed that the HABRP outperforms LEACH and SEP due to balanced energy dissipation of individual node through out the network.

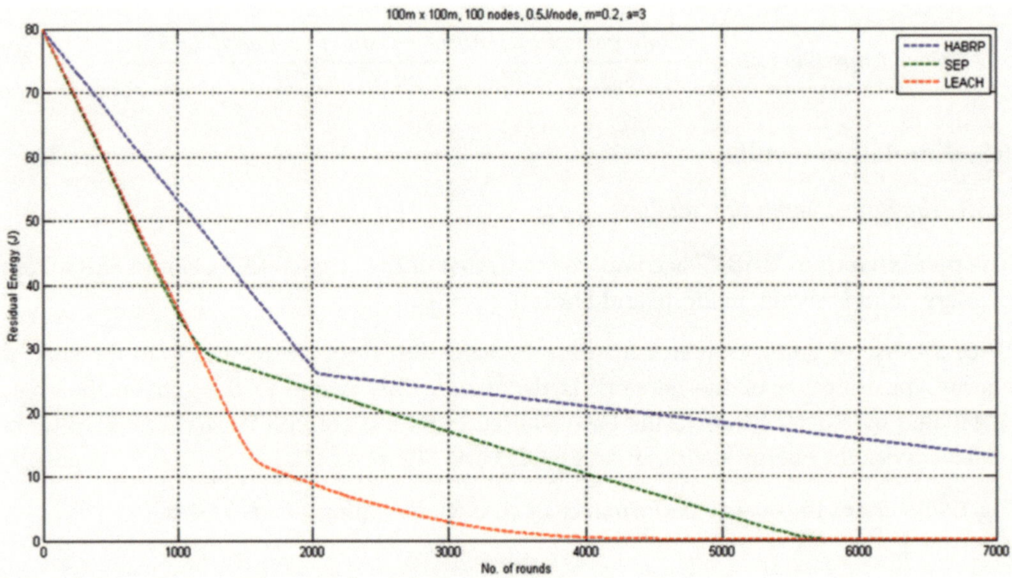

Figure 11. Energy analysis comparison of HABRP, LEACH and SEP (100m x 100m, 100 nodes, 0.5J/node, m=0.2, a=3(Heterogeneous WSNs))

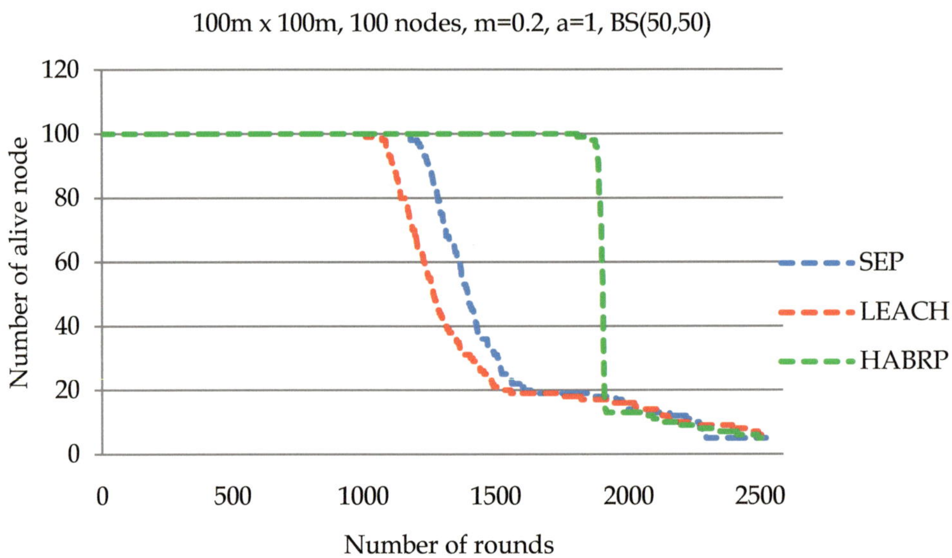

Figure 12. Number of alive nodes per round with 100m x 100m, 100 nodes, 0.5J/Normal node, 1J/Advanced node m=0.2, a=1, BS(50,50)

6.3.3. Variation of the Base Station Location

The results presented in the previous section show that HABRP is more energy-efficient than LEACH and SEP routing. Is this just a function of the simulation parameters? What happens if the base station is actually located within the network or very far away from the network? To answer these questions, we ran simulations where we varied the location of the base station from (x= 50, y= 50) to (x =50, y=300).

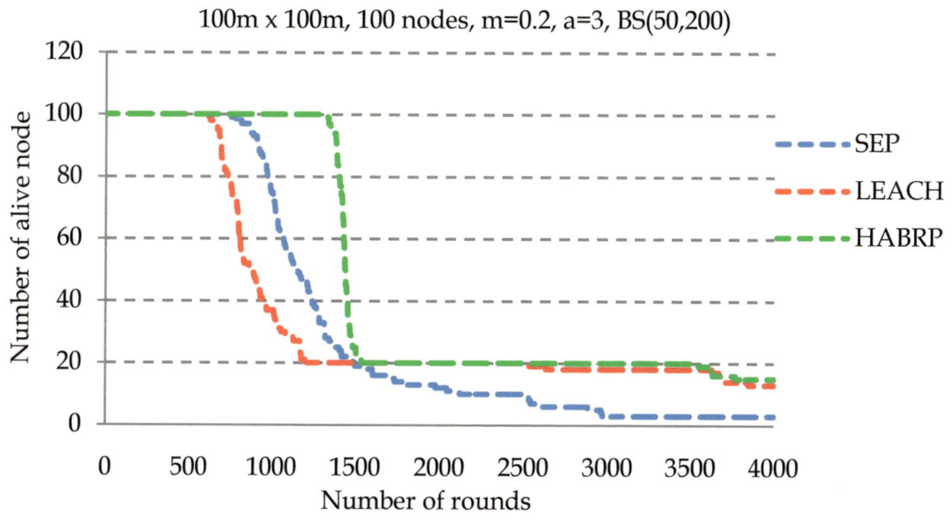

Figure 13. Number of alive nodes per round with 100m x 100m, 100 nodes, 0.5J/Normal node, 2J/Advanced node, m=0.2, a=3, BS(50,200)

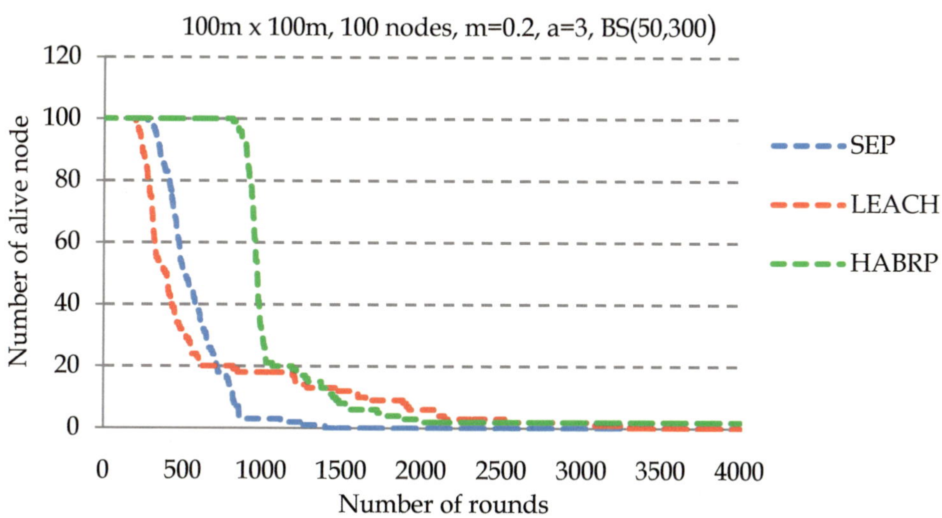

Figure 14. Number of alive nodes per round with 100m x 100m, 100 nodes, 0.5J/Normal node, 2J/Advanced node, m=0.2, a=3, BS(50,300)

For all base station locations we simulated, as the base station moves further away from the network, the performance of HABRP improves compared to LEACH and SEP.

6.3.4. Percentage of Node death

The number of rounds for 1%, 20%, 50%, 80% of node death is observed for HABRP, LEACH and SEP in Fig.13 and Fig.14. From the results of Fig.13 the stability period of LEACH and SEP protocols is limited to 892 rounds and the HABRP protocol extents up to 1335 rounds in homogeneous WSNs. In heterogeneous WSNs HABRP provides an extended lifetime of approximately twice LEACH protocol. HABRP has longer lifetime than LEACH and SEP.

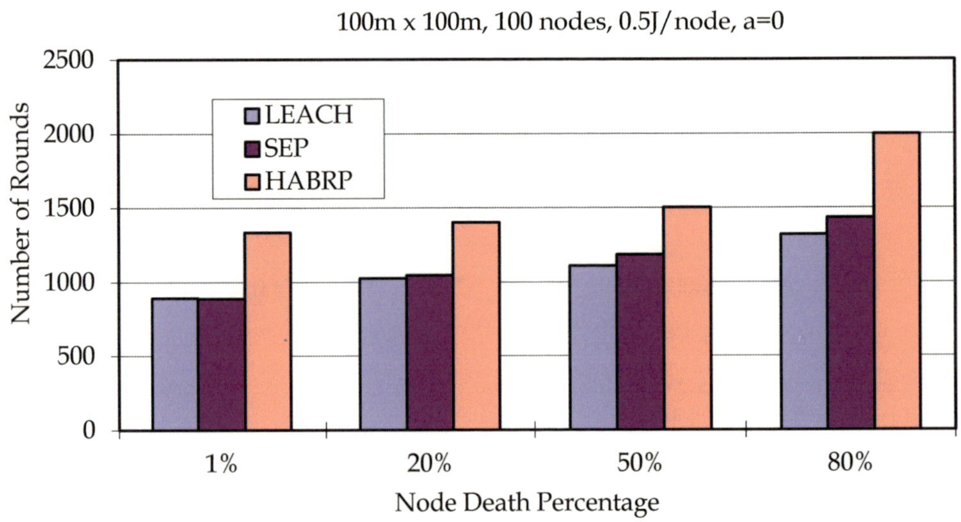

Figure 15. Node death percentage per number of Rounds with 100m x 100m, 100 nodes, 0.5J/node, a=0

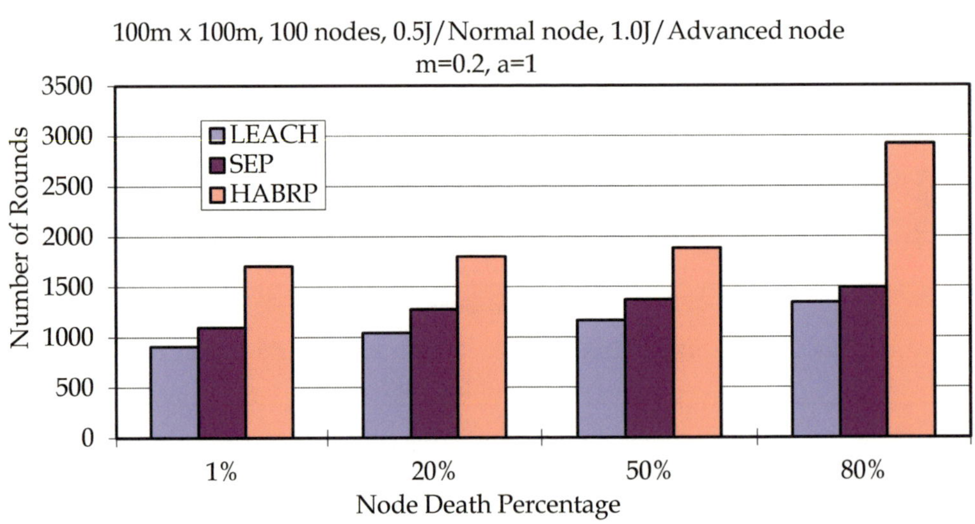

Figure 16. Node death percentage per number of Rounds with 100m x 100m, 100 nodes, 0.5J/Normal node, 1.0J/Advanced node m=0.2, a=1

6.3.5. Stable region in heterogeneous WSNs

In fig.15 the length of stable region for differnet values of energy heterogeneity is simulated, we observed that if we increase the number of NCG nodes with $a = 1$, the stability period is extended of approximately twice as LEACH protocol. In heterogeneous WSNs, HABRP has longer stable region than LEACH and SEP for differnet values of energy heterogeneity.

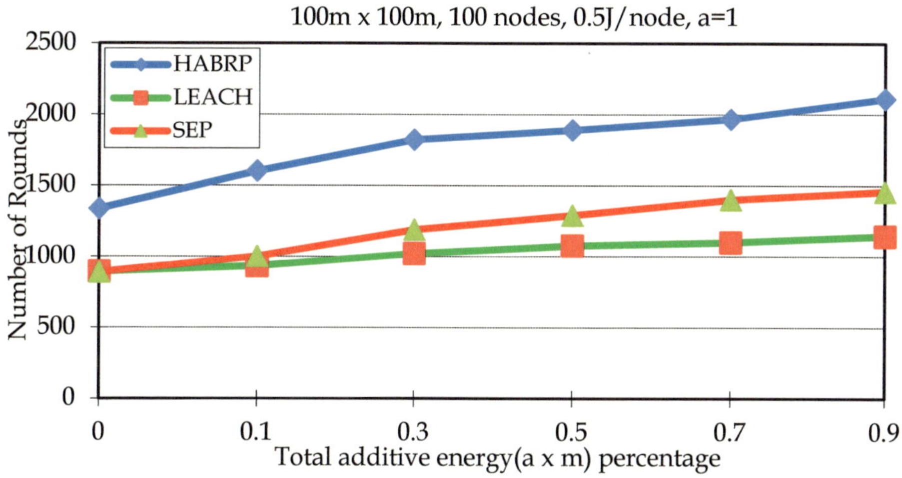

Figure 17. Length of stable region for differnet values of energy heterogeneity with 100m x 100m, 100 nodes, 0.5J/node, a=1

6.3.6. Heterogeneity in large-scale networks

We simulated the performance changes in large network with 900 nodes in area 300mx300m.

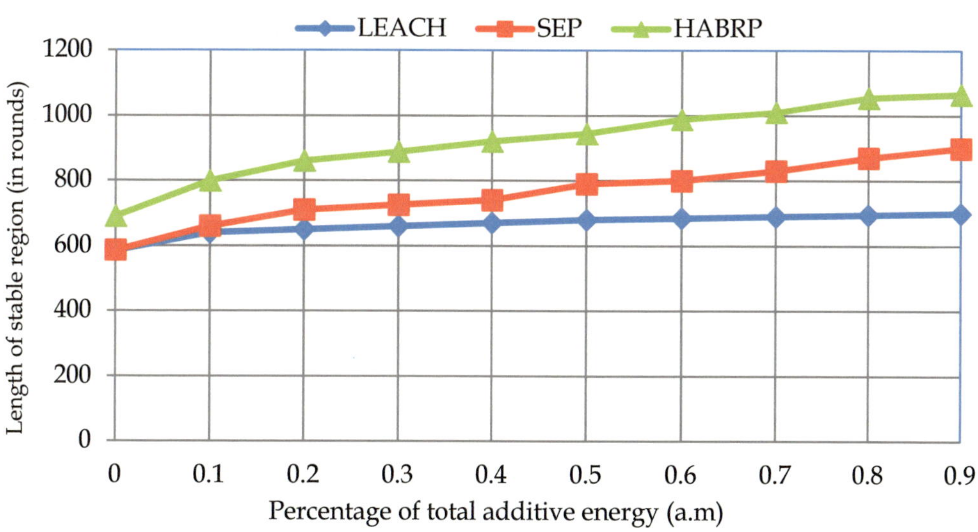

Figure 18. Sensitivity of HABRP, LEACH and SEP to degree of heterogeneity in large scale networks with 900 nodes in an 300mx300m field.

In Fig.16. the simulation result shows, that the network lifetime decrease in large area network and the period that the first dead node appears is earlier than those of previous cases. The phenomenon is caused by the fact that the cluster heads waste the considerable amount of energy for transmitting their data to the far away base station. In HABRP cluster head would transmit data to base station through gateway to eliminate that the cluster head far away from the base station dissipate their energy much faster than those close to the BS. HABRP outperforms LEACH and SEP for different values of total additive energy $a * m$. Because in LEACH and SEP, all cluster heads transmits aggregated data to the BS directly.

6.3.7. Improvement of stability period

The comparison results are shown in Table2. show that HABRP is more energy-efficient and the stability period is extended than LEACH in both homogeneous and heterogeneous WSNs.

	a * m	LEACH	HABRP	Improvement
FND (First Node Dies)	0	892	1335	49,66%
	0.1	908	1704	87,66%
	0.3	1020	1820	78,34%
	0.5	1076	1888	75,46%
	0.7	1100	1966	78,72%
	0.9	1145	2111	84,36%

Table 2. Improvement of HABRP compared to LEACH with $a = 1$

6.4. Result analysis

From our simulations, we observed the followings:

- HABRP can achieve twice the energy savings than LEACH and SEP protocol.
- HABRP outperforms LEACH and SEP due to balanced energy dissipation of individual node through out the network and extends network lifetime.
- For all base station locations we simulated, as the base station moves further away from the network, the energy efficient performance of HABRP improves compared to LEACH and SEP.
- In heterogeneous WSNs, HABRP provides an extended lifetime of approximately twice LEACH protocol and the stability period of the HABRP was prolonged than LEACH and SEP in heterogeneous settings.
- Energy dissipation is balanced between normal nodes and advanced nodes in the HABRP compared to LEACH and SEP.
- Balancing the energy consumption, reducing the phenomenon of rapid death of the cluster head caused by excessive energy consumption, also preventing the situation of instability period caused by one cluster head failure to work, ensure that the network work normally.
- Using gateway and cluster head, it saves excessive energy consumption for long-distance transmission, increased energy utilization of the entire network.

The simulation results show that HABRP protocol could suitably form clusters and effectively prolonging the survival time of the entire networks .

7. Conclusion

Energy efficient routing is paramount to extend the stability and lifetime of the wireless sensor networks. Routing in sensor networks is very challenging due to several characteristics that distinguish them from traditional communications and wireless ad-hoc networks since several restrictions, e.g., limited energy supply, computing power, and bandwidth of the wireless links connecting sensor nodes. The major difference between the WSN and the traditional wireless network is that sensors are very sensitive to energy consumption. Introducing clustering into the networks topology has the goal of reducing the number of message that need to be delivered to the sink in large-scale WSNs.

In this chapter, we have proposed an Hierarchical Adaptive Balanced energy efficient Routing Protocol (HABRP) for wireless sensor networks. The energy efficiency and ease of deployment make HABRP a desirable and robust protocol for wireless sensor networks. In order to improve the lifetime and performance of the network, routing in HABRP works in rounds and each round is divided into two phases, the Set-up phase and the Steady State phase. During the set-up phase some high-energy nodes called NCG nodes are elected gateways, other choised cluter heads and the clusters are organized. During the steady-state phase, data are transmitted from the cluster members nodes to the cluster head to agregate data and transmit it to the base station through a chosen gateways that requires the minimum communication energy to reduce the energy consumption of cluster head and decrease probability of failure nodes.

Simulation results shows that the HABRP improves the stable region of the clustering hierarchy and decrease probability of failure nodes and increase the lifetime of the network due to balanced energy dissipation of individual node through out the network and extends network lifetime. Balancing the energy consumption, reducing the phenomenon of rapid death of the cluster head caused by excessive energy consumption, also preventing the situation of instability period caused by one cluster head failure to work, ensure that the network work normally.

Finally, HABRP is scalable and achieves better performance compared to SEP and LEACH in both heterogeneous and homogenous environments.

Author details

Said Ben Alla, Abdellah Ezzati and Ahmed Mohsen
Science and Technical Faculty Hassan 1 University, Settat, Morocco

8. References

W. Heinzelman, A. Chandrakasan and H. Balakrishnan. (2000). Energy-Efficient Communication Protocol for Wireless Microsensor Networks. Proceedings of HICSS '00.

K. Akkaya, M. Younis, (2005). A survey on routing protocols for wireless sensor networks, Ad Hoc Networks 3 (3) 325–349.

S. Lindsey and C. S. Raghavendra, (2002). PEGASIS: Power Efficient GAthering in Sensor Informatio Systems. Proceedings of the IEEE Aerospace Conference, Big Sky, Montana, March.

G. Smaragdakis, I. Matta, A. Bestavros, 2004. SEP: A Stable Election Protocol for clustered heterogeneous wireless sensor networks. Second International Workshop on Sensor and Actor Network Protocols and Applications,

L. Qing, Q. Zhu, M. Wang, (2006). Design of a distributed energy efficient clustering algorithm for heterogeneous wireless sensor networks. ELSEVIER, Computer communications 29, pp 2230- 2237.

X. Du, F. Lin. (2005). Improving Routing in Sensor Networks with Heterogeneous Sensor Nodes. Proceedings of IEEE 61st Vehicular Technology Conference, VTC, Vol. 4, pp. 2528-2532.

LI Han, (2010). An Energy Efficient Routing Algorithm for Heterogeneous Wireless Sensor Networks. Proceedings of th International Conference on Computer Application and System Modeling, pp. V3-612- V3-616, IEEE.

Q. Xuegong, M. Fuchang, Ch.Yan, Y. Weizhao (2009). The Protocol of Cluster Multi-Hop Transmission Based on Heterogeneous Wireless Sensor Networks. Proceedings of International Conference on Computational Intelligence and Software Engineering, CiSE. pp. 1-4, IEEE.

V. Katiyar, N. Chand, S. Soni, (2011). A Survey on Clustering algorithms for Heterogeneous Wireless Sensor Networks. Internationnal journal Advanced Networking and Applications Vol. 02, Issue: 04, pp. 745-754,

M. Yarvis, N. Kushalnagar and H. Singh (2005).Exploiting heterogeneity in sensor networks, IEEE INFOCOM, 2005.

W.B.Heinzelman et al. (2002),An application-specific protocol architecture for wireless microsensor networks, IEEE Transactions on Wireless Communications 1 (4) 660–670.

R. Sheikhpour, S. Jabbehdari, A. Khadem-Zadeh (2011). Comparaison of Energy Efficient Clustering Protocols in Heterogeneous Wireless Sensor Networks. International Journal of Advanced Science and Technology. Vol.36.

Chapter 16

Street Lighting System Based on Wireless Sensor Networks

Rodrigo Pantoni, Cleber Fonseca and Dennis Brandão

Additional information is available at the end of the chapter

http://dx.doi.org/10.5772/48718

1. Introduction

An urban network, according to the document RFC 5548 (2009) - Routing Requirements for Urban Low-Power and Lossy Networks, is defined as:

"Sensing and actuating nodes placed outdoors in urban environments so as to improve people's living conditions as well as to monitor compliance with increasingly strict environmental laws. These field nodes are expected to measure and report a wide gamut of data (for example, the data required by applications that perform smart-metering or that monitor meteorological, pollution, and allergy conditions). The majority of these nodes are expected to communicate wirelessly over a variety of links such as IEEE 802.15.4, low-power IEEE 802.11, or IEEE 802.15.1 (Bluetooth), which given the limited radio range and the large number of nodes requires the use of suitable routing protocols".

According to Gungor et al. (2010), low-range WSNs are being widely recognized as a promising technology due to results obtained for smart metering, in particular the IEEE 802.15.4 standard, standardized by IEEE (Institute of Electrical and Electronics Engineers) in 2006 (IEEE 802.15.4), for its robustness, financial costs, low power consumption and simplicity.

Furthermore, there are two task groups that foresee the extension of IEEE 802.15.4 protocol in order to suit the requirements from smart grids. One of them (IEEE 802.15 WPAN TASK GROUP 4G, 2011) is preparing a protocol to support large networks, geographically diverse, with minimal infrastructure and millions of nodes; the other (IEEE 802.15 WPAN TASK GROUP 4E, 2011) is including characteristics extracted from CWPAN (Chinese WPAN standard – Chinese Wireless Personal Area Network), which specify power-saving methods for networks with high latency tolerance.

This paper presents an application for urban networks using the IEEE 802.15.4 standard, which is used for monitoring and control electric variables in a public lighting scenario.

INTECH
open science | open minds

© 2012 The Author(s). Licensee InTech. This chapter is distributed under the terms of the Creative Commons Attribution License http://creativecommons.org/licenses/by/3.0), which permits unrestricted use, distribution, and reproduction in any medium, provided the original work is properly cited.

This application consists of an urban network, where sensors (nodes) are coupled to the lamp posts or lighting points in order to control the control the lighting for these points, and capture important information from diagnostics, operation and failures.

This type of application is expected to improve quality in public lighting service and rationalize power consumption through smart sensors and supervision screens. The next section describes specific aspects involving the quality of public lighting service and the current stage of its structure in Brazil. Section 3 describes the system requirements, which its understanding is fundamental to develop the control and monitoring application and routing protocols. In section 4 the solutions regarding the control and monitoring application are detailed. Section 5 is presented the routing protocols strategies. In the last section, the conclusions are indicates the conclusions.

2. Specific aspects for the application in Brazil

An application for public lighting involves some specific aspects that should be highlighted. Economic aspects and aspects of sustainability are relevant to a public lighting system. The current public lighting system in Brazil needs to be modernized and must rely on sustainable energy sources and energy rationing.

Moreover, by automating the lighting system, it is possible to provide the population with a better service in resolving problems. According to Elektro (2011), the concessionaire responsible for the public lighting has no means to obtain the information related to problems, such as a lighting point that is out of operation (either because of a problem with the lamp or the device). Currently, a public lighting failure is usually reported to the concessionaire by a customer complaining on the phone or at the concessionaire's web page, as indicated at the Cemig web site (ATENDIMENTO CEMIG, 2011). Another way to detect a problem is patrolling the structure, which is made without a defined logistic. For this reason, it is desirable that the system to be developed is able to report the status of the devices installed on public streets, and also provide emergency alarms.

This work aligns with the work that began in 2009 in the USA: a comprehensive program in which electricity distribution companies, device manufactures, telecommunication companies and Information Technology companies have joined the Department of Energy (DOE) and the National Institute of Standards and Technology (NIST, U.S. agency for standardization) to define standards for smart grids and assure interoperability for protocols and devices in various areas, such as telecommunications, Information Technology and energy. That work has produced a document (NIST, 2009), which contains a survey of more than 80 standards, including some directly related to this project: transmitting messages over wireless networks, using networks based on IP (Internet Protocol), ZigBee (ZIGBEE, 2007), using the standard specification of GPS signals (Global Positioning System) (GPS NAVSTAR, 1995) and W3C Simple Object Protocol (SOAP, Web Services standard).

Looking at the work developed in the U.S., in June 2010, the Brazilian government, through the Brazilian Electricity Regulatory Agency (ANEEL), opened a public hearing for strategic projects on the "Brazilian Smart Grid Program" (ANEEL, 2010a). This program focuses on presenting project proposals in order to join efforts to coordinate and generate new technological knowledge of great relevance for the Brazilian electricity sector. More specifically, in accordance with the document generated by the mentioned public hearing (ANEEL, 2010a), this study contributes to the research of "telecommunication infrastructure".

In addition, the regulatory resolution nº 414, of September 9, 2010 (ANEEL, 2010b) from ANEEL establishes changes to the maintenance policy for the Brazilian public lighting structure. This document states that, from September 2012, municipal governments shall be responsible for maintaining the public lighting structure, instead of the concessionaires, which shall be responsible only for power supplying.

Finally, control and monitoring public lighting, through lamps or dimming ballasts, decreases light pollution, caused by poor adjustment of luminosity and number of lamps lit in specific directions, depending on the angle, intensity, etc (Figure 1). Besides, light pollution should be studied aiming to improve roads, streets and avenues conditions, in order to provide better maneuverability for drivers, and brightness for unsafe places, such as alleys and public squares (MIZON, 2001) (SCHWARZ, 2003).

Figure 1. Light Pollution

The current public lighting structure in Brazil consists of a set of lighting points arranged in a certain manner. Each point has a housing with a device called photocell relay, which is responsible for turning the lights on or off according to a brightness sensor. Then, at nightfall, the relay turns the light on at the lighting point. This limited structure may compromise consumption efficiency, quality of service to the community, and other factors, for example, problems in the lamp, power supply, the photocell relay, etc.

To solve the problems mentioned above in an automatic way, allowing fast and efficient preventive and corrective maintenance, an automation system may be applied. The basis for such a system is implementing the communication between parts involved, like devices, computer software, among others. Therefore, the photocell relay mechanism can be attached to an Information Technology structure, through device and sensor networks, supervision workstations and applications.

The stage of development of public lighting projects in Brazil is restricted to rationing energy and improving lighting efficiency, by replacing sodium or mercury lamps by LED lamps, installing photovoltaic panels for sustainable lighting, powered by solar energy, and, finally, luminaires to reduce lighting focus dispersion (CPFL CAMPINAS, 2011).

LED lamps offer better color reproduction; they do not emit ultraviolet and infrared rays, they propagate less heat, and they attract fewer insects. They also show good efficiency and longer life, estimated at approximately 50,000 hours of operation. Sodium vapor lamps, most commonly used today, last up to 32,000 hours; mercury vapor lamps may last 12,000 hours, and metal halide lamps, commonly used in facades of buildings, 10,000 hours.

The Companhia Energética de Minas Gerais (Cemig), an electric energy utility company in the State of Minas Gerais, Brazil, together with the municipal government of Belo Horizonte, began testing LED lamps. Some luminaires have been already installed on poles near San Francisco Church (CEMIG, 2009). Cemig provides annual reports and sustainability reports that can be accesses at the company's website (CEMIG, 2011).

In addition, driven by the investments for the World Cup 2014, Cemig's project "Minas Solar 2014" seeks to install photovoltaic panels to power public lighting, aiming to improve public infrastructure and soccer stadiums (CHIARETTI, 2011).

Cemig signed a cooperation agreement with the U.S. government, through the U.S. Trade and Development Agency (USTDA), to fund their smart grid project. The U.S. agency will provide US$ 710,000 for the Brazilian company, to be applied on the study that analyzes the feasibility of implementing smart grids, which has been developed by the company together with Light (SMART ENERGY ONLINE, 2011).

CPFL Energia, an electric energy utility company in Brazil, together with the Development Foundation from the University of Campinas (FUNCAMP), the Foundation for Research and Advisory Service to the Industry from the Federal University of Itajubá (UNIFEI), HYTRON Hydrogen technology, FUSION, and the AQUA GENESIS Institute for Studies and Projects on Energy, Hydrogen and Environment, is currently investing in a project of a solar photovoltaic and/or wind hybrid system to generate electricity, which can operate connected to the distribution network or isolated, and the goal is to evaluate the integration of distributed generation at low voltage, with and without energy storage. Considering the increasing use of local alternative sources of energy generation, low power systems, renewable energy sources, and technological availability that allow immediate applications, utilization of solar photovoltaic and wind energy, with and without energy storage, stands out above other options (PORTAL INOVAÇÃO CPFL, 2011a).

A project that involves communication, however not between sensors (PORTAL DE INOVAÇÃO CPFL, 2011b), called "failure diagnostic system for public lighting points", is being developed by CPFL in partnership with companies MATRIX and PLAYMUSIC. The development stage is restricted to alterations in the master part device, aiming to determine more appropriate, and also less vulnerable to transportation, means to perform maintenance on lighting points. Maintenance concerns handling from the field team, to make it practical to use. Furthermore, the project involves adapting the device to industrial production.

Closer to the project proposed in this work, which involves sensors communication in a system, the Smart Substation project from CPFL, in partnership with the Foundation for the Technological Development in Engineering (MUSP – FDTE) and CONTREL, proposes a smart system by adapting or inserting monitored devices with built-in intelligence in maintenance, security, qualimetry and control, connecting to the respective remote management centers. Nowadays, relays, sensors, qualimeters, and other digital devices operate underutilized, less integrated and do not repay their high costs. Back office processes (operation, planning, security, qualimetry, maintenance, etc.) are still detached and without intelligence (PORTAL INOVAÇÃO CPFL, 2011c).

Information for this project was provided by Elektro and the Eldorado Institute, in order to modernize the infrastructure of public lighting through WSN communication and control and monitoring workstations, which allow rationing energy, sustainability, automated detection of physical or logical fails, alarms related to emergencies, supervision of lighting points, and controlling actions (turning the lamp on or off and adjusting brightness, for example) for lighting points.

3. System requirements

The recent document proposed by IETF (RFC 5548, 2009), which describes requirements for routing on urban networks, also affirms that these networks must attend convergent traffic, where one node (usually a gateway or sink) receives messages from several low frequency meters (a maximum of one measurement per hour, and a minimum of one measurement per day). The number of nodes must be in the order of 10^2 to 10^7, distributed in areas varying from hundreds of meters to one square kilometer, considering that nodes are commissioned in groups, and the battery shelf life is usually in the order of 10 to 15 years. The frequency band must be ISM (Industrial, Scientific and Medical), and the nodes will probably have from 5 to 10 immediate neighbors to communicate. The routing protocol must enable the network to be autonomous and organize itself, requiring a minimum energetic cost for maintenance functions. The protocol must also ensure that any diagnostic or failure information from the nodes is communicated without interfering on the network operational mode, respecting the time limits.

The project of a Wireless Sensor Network depends on the application where the network will be used, and it should be based on functional and non-functional requirements. Functional requirements for this type of project are:

- Points supervision: node status, whether it is connected to the network or not; battery power; lifetime estimation for battery and lamp; and LEDs luminosity level;
- Control: switch on/off the luminosity level lamp; switch on/off a lamp post, a selected segment, a street, neighborhood, city, etc;

- Automatic programmed actions: for example, the city lights should be turned on at 7 PM;
- Actuate the nodes through a remote tool: besides the automatic actions, the operator should be able to select a region through a remote tool to actuate;
- Diagnostic and alarms: trigger an event when a network, hardware or lamp failure occurs;
- Automation of information storage: mechanisms automating the storage process of related street lighting information (geodesic coordinates supplied for electrical companies) should be applied to simplify and streamline the process.

To meet the functional requirements, a control and monitoring application is required to report system information to the user (Figure 2). This application is able to monitor operating conditions, for example, notifying the user about the end of life of a lamp if it exceeds a particular limit of hours, and graphically displaying city maps on a supervision screen, with the lighting points in operation. Figure 2 shows an example of status of lighting points, which are represented in different colors: lighting points in yellow indicate lit points; lighting points in blue indicate unlit points; lighting points in red indicate the boundaries of a specific region in which the application must act. Regarding control, the application sends commands to activate or deactivate lighting points as pre-programmed or when there is a request from the system operator.

As non-functional requirements, portability and robustness are considered essential factors in the application project. For more information, section 4 covers the details about the application project.

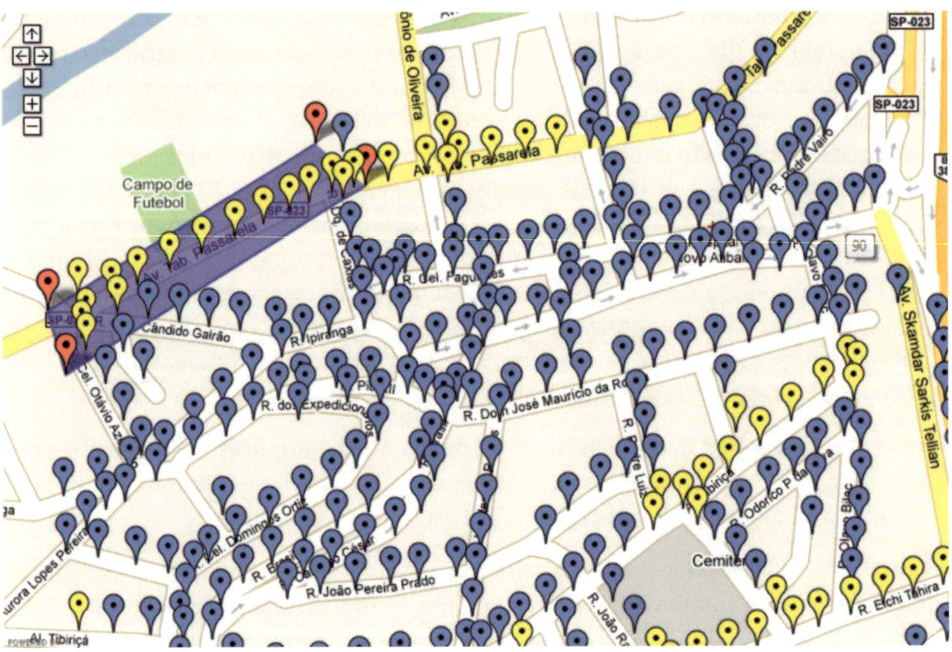

Figure 2. Control and Monitoring Tool Display

To understand the non-functional requirements, the list below briefly describes parameters relevant to the application, in order of importance (KHALIFA et al., 2011):

- Scalability: when proposing an approach to sensor networks, it is important to consider that the project should apply to small (up to 25 nodes), medium (25 to 10^3 nodes), or large networks (10^2 to 10^7 nodes);
- Fault tolerance: depending on the physical conditions, it may be necessary to adopt sensor nodes with special characteristics. In addition, due to difficult accessibility in some environments, it may be unmanageable to replace damaged nodes or batteries. Therefore, it is necessary to provide sensor nodes with adaptive capacities to deal with unstable environments. In the type of application described in this work, nodes must support high temperatures and heavy rain. Considering the failures related to sensor nodes, the routing algorithm proposed in this work must be fault-tolerant because the messages must find a reliable path to reach the destination;
- Delivery Guarantee: regarding the unstable wireless communication nature, because of external interference and/or electronic noise, it is necessary to create a point-to-point acknowledgment mechanism, which is a consensus among researches in the wireless network area. End-to-end confirmation messages are implemented in higher layers, preferably in the transport layer (TCP);
- Lifetime: sensor nodes depend on batteries, which require the use of efficient protocols in energetic terms. However, for this project, most sensor nodes are powered by energy cables that are already used in current street lighting structures. Avoiding the use of batteries is an ecological issue, because devices' maintenance involves a considerable amount of trash. Batteries would be used only in areas where there is no energy cable structure, such as rural areas. Therefore, the lifetime estimation should be medium to high, but it should not be prioritized in detriment of other requirements, especially not to compromise the delivery rate;
- Latency: this parameter reflects the time interval the network has to inform the observer about a specific phenomenon. Some applications can be sensible to latency, therefore it is necessary to respond in a short time interval (equal or less than one second), for example, in control process supervision. On the other hand, other applications do not depend on a quick response time (for example, monitoring bridge stabilization). In relation to street lighting application, the acceptable latency should be on a time interval in the order of tens of seconds.

Routing protocols are used in two types of package traffic in Wireless Sensor Networks: the first type directs packages from the sink node to the network nodes (i); the second type directs packages from the network nodes to the sink node (ii). Different terminologies may be used for these two traffic types: Tian and Georganas [1] and She et. al [2] use the terms *downstream traffic* for (i) and *upstream traffic* for (ii); Watteyne et al. [3] use the terms *divergent traffic* for (i) and *convergent traffic* for (ii); in other works, such as [4], authors name (i) as *gradient-descendent traffic* and (ii) as *gradient-descent traffic*, if routing is based on gradient. This works uses the terms *divergent traffic* for (i) and *convergent traffic* for (ii).

Figure 3 represents the two types of package traffic, differing by the communication flow in relation to the sink node.

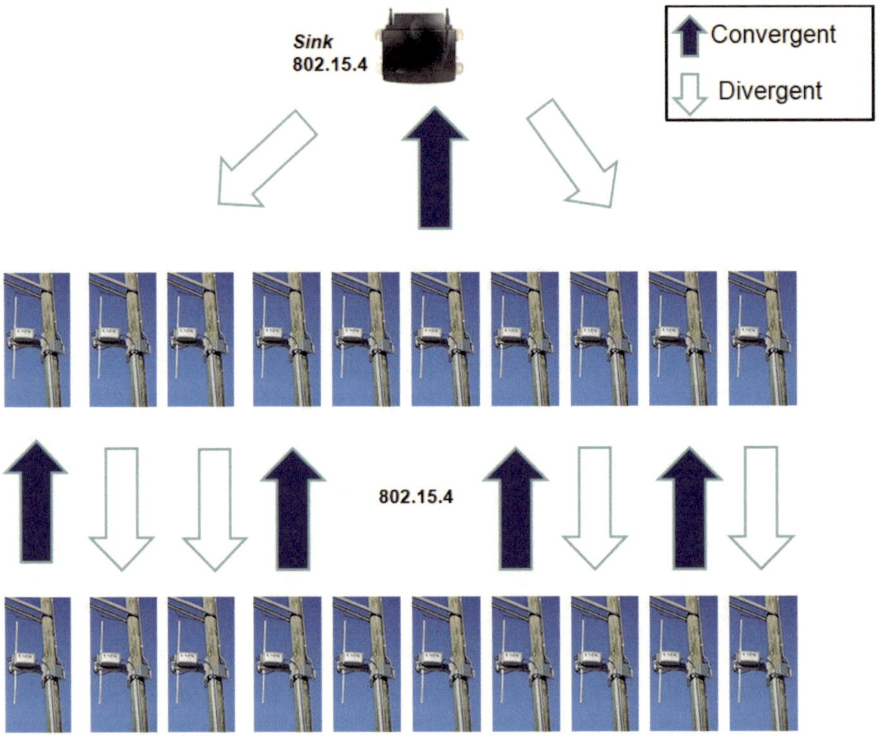

Figure 3. Typical WSN simplified architecture

In order to control and monitor public lighting systems, which is the goal of this work, the following message types based on RFC 5548 are needed, according to the mentioned requirements, subdivided in convergent and divergent traffic.

Convergent traffic:

- Message to notify the system about failures (which can occur on a lamp, hardware, etc);
- Message to update the system about status (on/off) and other information. Periodicity is a maximum of one message per hour, and each node has its own cycle or clock;
- Message to notify the system that some neighboring node is out of radio range;
- Specific diagnostic response (battery power, LEDs luminosity level, battery lifetime estimation, and lamp lifetime estimation).

Divergent traffic:

- Actuate the nodes luminosity (activation, dimerization, etc);
- Specific diagnostic request (battery power, LEDs luminosity level, battery lifetime estimation, and lamp lifetime estimation, etc).

4. Control and monitoring application

The proposed architecture is based on the cell model. The cell should provide access to supervision and control of a number of sensor nodes for the management applications. By definition, the sensor node is the device coupled to the lighting point, which is able to communicate via a wireless network.

A cell is composed by a compartment, an industrial computer, a GPS antenna, a communication controller, and sensor nodes. Numerous cells can be formed to completely cover an urban center.

The industrial computer is installed inside the compartment and it is connected to a communication controller, communicating via a serial protocol, while the communication controller communicates to the sensor nodes via wireless network.

The GPS (Global Positioning System) antenna is considered the only time basis among several cells, and then alarms that occurred in different cells can be ordered. Management applications (Configuration/Topology, remote operation server and monitoring) are installed on the Operation computer and they communicate to the cells via Ethernet network.

Remote access is done through an Internet browser from the remote operation application computer to the remote operation server. The Google Maps library (GOOGLE MAPS FAMILY, 2011) is used to plot the lighting points on the maps (Figure 4).

Figure 4. Screen obtained with Google Maps library

Figure 5 shows the system architecture and illustrates cell 01, which is responsible for grouping some sensors, and also shows the capability of adding new cells until all lighting points from the city are mapped.

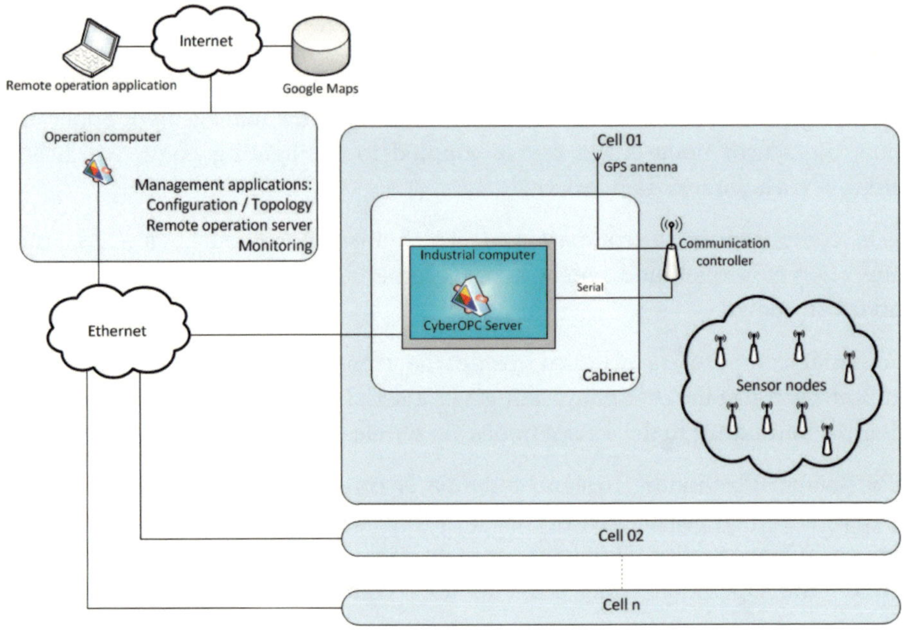

Figure 5. Application Architecture

The communication standard adopted between the operation and the industrial computer is CyberOPC. CyberOPC is an academic research project that proposes an open communication system, based on HTTP (Hyper Text Transfer Protocol), specially developed for remote control and supervision of industrial systems over public IP (Internet Protocol) networks (TORRISI, 2011).

Communication works as follows: management applications installed on the Operation computer request control and supervision data via the Ethernet network to the industrial computer installed on the compartment. The message is received by the computer and transferred to the communication controller connected to the serial port.

The communication controller transmits the message via the wireless network, identifying the request. The sensor nodes receive the notifications, process the request and transmit the response to the controller via the wireless network. The response is received by the controller via the wireless network, and retransmitted to the computer via the serial port. The industrial computer sends the response to the operation computer via the Ethernet network.

For detailed information about this control and monitoring application, refer to Fonseca (2011).

5. Routing protocols

Due to the robustness, low cost, low power consumption and simplicity, this work is based on the IEEE 802.15.4 standard. For the devices to form large scales networks, as described in

RFC 5548 (2009), the protocol from this standard must operate on a mesh-type topology, so that the devices communicate in multiple hops, based on the low range communication (around 100 meters) with distributed routing to transmit information to their final destinations.

However, the large number of nodes and the peculiar nature of operation of wireless sensors urban networks require the project of routing and medium access control (MAC) protocols to be specifically developed.

Protocols that use only global disseminations to communicate, through broadcast messages (known as flooding protocols), may cause high package traffic, which can lead to a high number of collisions and, thus, reduce the delivery rate. In addition, they demand high energy consumption to exchange information.

On the other hand, *unicast* protocols should be stateless, which means that packages cannot store information about the route taken to avoid loops or package delivering problems. This is due to a limitation of information that the package of lower layers from the IEEE 802.15.4 standard has. Moreover, protocols must considerer the minimum energy consumption and a satisfactory delivery rate.

The public lighting concessionaire has the geodesic coordinates of the lighting points in their digital maps, which help in the structure management and maintenance. The routing protocols can then implement mechanisms that use such information.

Furthermore, devices must have mechanisms that minimize energy consumption, such as periodically switching the radio off, since devices would be idle most of the time.

This section intends to only cite the proposed protocols. They were divided according to two different types of traffic (divergent and convergent), and evaluated to demonstrate which are more suitable for the mentioned urban network.

5.1. Divergent traffic

The proposed protocol for divergent traffic consists of four parts: it is based on the GPSR protocol (KARP and KUNG, 2000), which has two modes, greedy and perimeter (i); the second part is that when it reaches a particular target region, indicated by the geodesic coordinates, it spread the message, thus operating in *geocast* mode (ii); the third part is that it is able to reverse the direction of the perimeter mode, in order to go around blank spaces competently (iii); finally, the last part consists of retransmitting to different neighbors in case a fail occurs, using the original GPSR criterion to chose neighbors (iv). Originally, GPSR uses the right-hand rule to define the node to retransmit the package, choosing the neighbor by the smaller angle. Then, to reverse the direction, the logic was reversed, using the left-hand rule. Figure 6 shows the mode of operation of the protocol for divergent traffic.

For detailed information about the protocol used for divergent traffic, including flowcharts and analyses, refer to Pantoni and Brandão (2011).

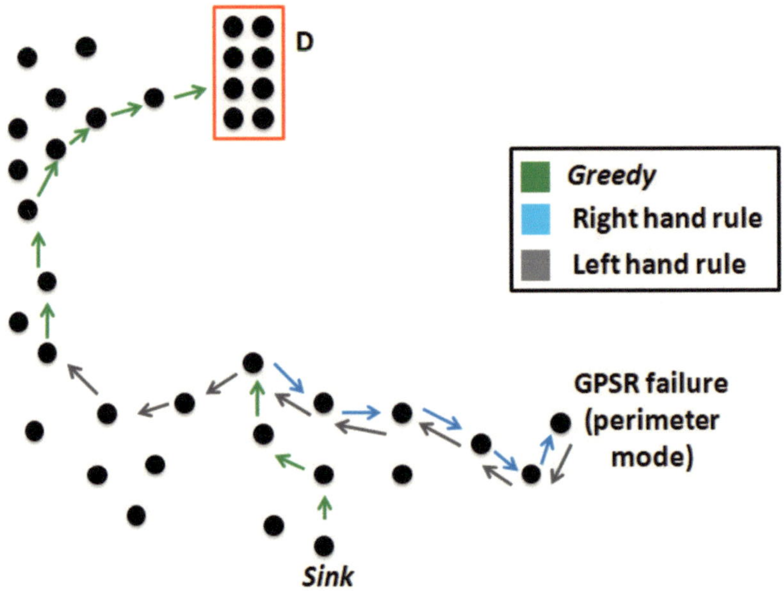

Figure 6. Operation of the routing protocol for divergent traffic

5.2. Convergent traffic

The main idea of the routing algorithm based on gradient is to forward the package towards the sink node choosing the neighbors with the lowest height, where height represents the number of hops over nodes in relation to the sink node.

In order to enable the routing process, the sink node broadcasts a message with its height equals to zero (H=0) in the initialization phase. The nodes that receive the "gradient initialization" package assign the incremented value to their height variable and, later on, broadcast the message to other nodes on the network, in order to assign heights to all network nodes, creating height levels (H) on the network, as indicated in Figure 7.

The logic of the protocol consists of transmitting a package to only one neighbor, according to the lowest height. The proposed protocol differs from the previous works because it considers other aspects to forward the package in case the neighbors have identical heights. In addition, in case acknowledgment messages identify a failure in transmission, the package is retransmitted to other neighbors, and the retransmissions are limited to only one transmitter node per data flow.

The protocol is based on choosing the neighbors for retransmitting the message according to the number of attempts made. The next hop is selected based on a list that is sorted in descending order by the longest distance and populated with all neighbors, except the neighbor from the last hop. On the first transmission, among the neighbors with the lowest heights, the most distant node from the current node will be chosen; in other words, the first element from the list. If it is the first retransmission, the second most distant element will be chosen, and so on. If no nodes with the same lowest height are chosen, the next group to be

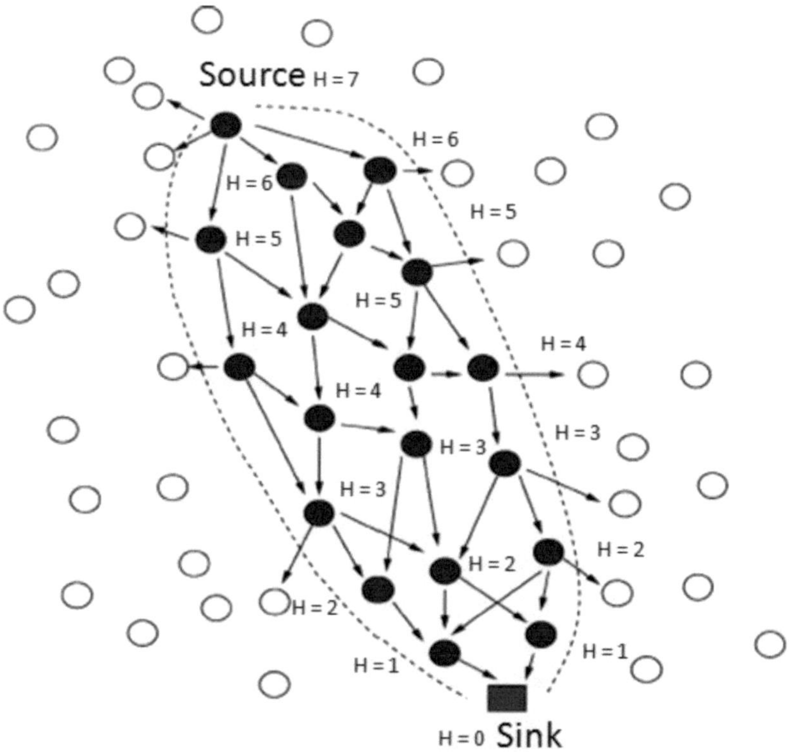

Figure 7. Height mapping for gradient routing protocol. Source: Ye et al. (2005)

sorted according to the distance will be selected based on the succeeding lowest height among the neighbors.

For detailed information about the protocol used for convergent traffic, including flowcharts and analyses, refer to Pantoni and Brandão (2012).

6. Conclusions

The public lighting system provides automation for control process, diagnostics and alarms from possible failures in the structure. Therefore, an improvement in public lighting service is expected, plus a rationalization of energy consumption.

The two task forces related to the IEEE 802.15.4 standard (IEEE 802.15 WPAN TASK GROUP 4G, 2011; IEEE 802.15 WPAN TASK GROUP 4E, 2011) assure a protocol extension to the lower layers in order to increasingly meet urban networks requirements, as well as smart grid applications.

Although this work comprises a specific application for public lighting, it can also be applied to other applications, such as monitoring energy and water consumption meters, measurement of climatic factors in order to implement weather forecast systems, with humidity sensors, temperature sensors, etc.

Author details

Rodrigo Pantoni
Federal Institute of São Paulo, Brazil

Cleber Fonseca and Dennis Brandão
University of São Paulo, Brazil

7. References

ANEEL (2010a). Agência Nacional de Energia Elétrica. Chamada N 011/2010 Projeto Estratégico: "Programa Brasileiro de Rede Elétrica Inteligente". Brasília, Jul. 2010.

ANEEL (2010b). RESOLUÇÃO NORMATIVA Nº 414, set. 2010. In: *ANEEL Web Site*, 19 Sept, 2011, Available at: <http://www.aneel.gov.br/cedoc/ren2010414.pdf>.

ATENDIMENTO CEMIG (2011). In: *Cemig Web Site*, 12 Sept, 2011, Available at: <http://www.cemig.com.br/Atendimento/Paginas/default.aspx>.

CEMIG (2009). In: *Cemig Web Site*, 12 Sept, 2011, Available at: <http://www.jusbrasil.com.br/politica/3212811/cemig-testa-em-belo-horizonte-iluminacao-publica-com-leds>.

CEMIG (2011). In: *Cemig Web Site*, 12 Sept, 2011, Available at: <http://www.cemig.com.br/ACemig/paginas/Relatorios. aspx>.

Chiaretti, D. (2011). In: *MP Web Site*, 12 Sept, 2011, Available at: <https://conteudoclippingmp.planejamento.gov.br /cadastros/noticias/2011/5/17/alemaes-vao-investir-em-energia-solar-no-pais/>.

GOOGLE MAPS FAMILY (2011). In: *Google API Web Site*, 11 Sept, 2011, Available at: <http://code.google.com/intl/pt-BR/apis/maps/>.

CPFL CAMPINAS (2011). In: *Jornal Local Web Site*, 22 Sept, 2011, Available at: <http://jornalocal.com.br/site/cidades/cpfl-paulista-entrega-na-sexta-feira-nova-iluminacao-publica-para-avenidas-de-campinas/>.

ELEKTRO (2011). In: *Elektro Web Site*, 25 Sept, 2011, Available at: <http://www.elektro.com.br>.

GPS NAVSTAR (1995). *Global positioning system standard positioning service signal specification.* Second Edition, 1995.

Gungor V. C.; Lu, B.; Hancke, G. P. (2010). Opportunities and challenges of wireless sensor networks in smart grid. *IEEE Transactions on Industrial Electronics*, vol. 57, n. 10, p. 3557 – 3564, Oct, 2010.

IEEE 802.15 WPAN TASK GROUP 4E (2011). In: *IEEE Task Group Web Site*, 15 Sept, 2011, Available at: <http://www.ieee802.org/15/pub/TG4e.html>

IEEE 802.15 WPAN TASK GROUP 4G (2011). In: *IEEE Task Group Web Site*, 15 Sept, 2011, Available at: <http://www.ieee802.org/15/pub/TG4g.html>

IEEE 802.15.4 (2006). *Wireless Medium Access Control (MAC) and Physical Layer (PHY) Specifications for Low-Rate Wireless Personal Area Networks (WPANs),* IEEE Computer Society, Revision of IEEE Std 802.15.4-2003, 2006.

INSTITUTO ELDORADO (2012). In: *Eldorado Institute Web Site,* 15 Sept, 2011, Available at: <http://www.eldorado.org.br>

Karp, B; Kung, H. T. (2000) GPSR: greedy perimeter stateless routing for wireless networks. In: International Conference On Mobile Computing And Networking, Boston, Massachusetts, 2000. p. 243-254.

Lim, J.; Shin, K.G. (2005). Gradient-Ascending Routing via Footprints in Wireless Sensor Networks. Proceedings of the 26th IEEE International Real-Time Systems Symposium (RTSS'05), 2005.

Mizon, B. (2001). *Light Pollution: Responses and Remedies*. Springer, 2001. 255 p.

NIST (2009). NIST Framework and Roadmap for Smart Grid Interoperability Standards. National Institute of Standards and Technology, Sept. 2009. In: NIST Web Site, 22 Dec, 2011, Available at: <http://www.nist.gov/public_affairs /releases/upload/ smartgrid_interoperability_final.pdf>.

Pantoni, R. P.; Brandao, D. (2011). A confirmation-based geocast routing algorithm for street lighting systems. *Computers & Electrical Engineering,* v. 37, n. 6, p 1147-1159, 2011.

Pantoni, R. P.; Brandao, D. (2012). A gradient based routing scheme for wireless street lighting sensor networks. *Journal of Network and Computer Applications,* 2012.

PORTAL INOVAÇÃO CPFL (2011a). Hybrid System. In: *CPFL Web Site,* 15 Sept, 2011, Available at: <http://www.cpfl.com.br/parceiros_inovacao_tecnologica/ Detalhes_Projetos_23>.

PORTAL INOVAÇÃO CPFL (2011b). Diagnostic System to light point failure. In: *CPFL Web Site,* 15 Sept, 2011, Available at: <http://www.cpfl.com.br/ parceiros_inovacao_tecnologica/Detalhes_Projetos_20.aspx>.

PORTAL INOVAÇÃO CPFL (2011c). Smart Substation. In: *CPFL Web Site,* 15 Sept, 2011, Available at: <http://www.cpfl.com.br/parceiros_inovacao_tecnologica/ Detalhes_Projetos_21.aspx>.

RFC 5548 (2009). Routing Requirements for Urban Low-Power and Lossy Networks. In: *RFC Web Site,* 30 Dec, 2011, Available at: <http://datatracker.ietf.org/doc/ rfc5548.txt>.

Schwarz, H.E (2003). *Light Pollution: The Global View*. Springer, 2003. 324 p.

She, H.; Lu, Z.; Jantsch, A.; Zheng, L.R.; Zhou, D. Traffic Splitting with Network Calculus for Mesh Sensor Networks. Proceedings of Future Generation Communication and Networking (FGCN 2007), 2007.

Tian, D.; Georganas, N. D. (2003) Energy efficient routing with guaranteed delivery in wireless sensor networks. Proceedings of IEEE Wireless Communications and Networking, vol. 3, 2003, pp. 1923–1929.

Watteyne, T; Pister, K.; Barthel, D.; Dohler, M.; Augeblum, I. (2009) Implementation of Gradient Routing in Wireless Sensor Networks. Proceedings of GLOBECOM'09 Proceedings of the 28th IEEE conference on Global telecommunications, 2009.

Ye, F.; Zhong, G.; Lu, S.; Zhang, L. (2005). GRAdient Broadcast: A Robust Data Delivery Protocol for Large Scale Sensor Networks, *ACM Wireless Networks*, vol. 11, pp. 285–298, 2005.

ZIGBEE (2007). *ZigBee PRO Specification*, ZigBee Alliance.

Chapter 17

Energy Efficiency in the ICT - Profiling Power Consumption in Desktop Computer Systems

Giuseppe Procaccianti, Luca Ardito, Antonio Vetro' and Maurizio Morisio

Additional information is available at the end of the chapter

http://dx.doi.org/10.5772/48237

1. Introduction

Energy efficiency is finally becoming a mainstream goal in a limited world where consumption of resources cannot grow forever. ICT is both a key player in energy efficiency, and a power drainer. The Climate Group reported that the total footprint of the ICT sector was 830 MtCO2e and that the ICT was responsible for 2% of global carbon emissions [13]. Even if energy efficient IT technologies were developed and implemented, this figure would still grow up at a rate of 6% per year until 2020. Recently, much of the attention in green IT discussions focuses on data centers. However, it is foreseen that data centers will only add up to less than 20 percent of the total emissions of ICT in 2020. The majority (57 percent) will come from PCs, peripherals, and printers, as shown in Figure 1 [13].

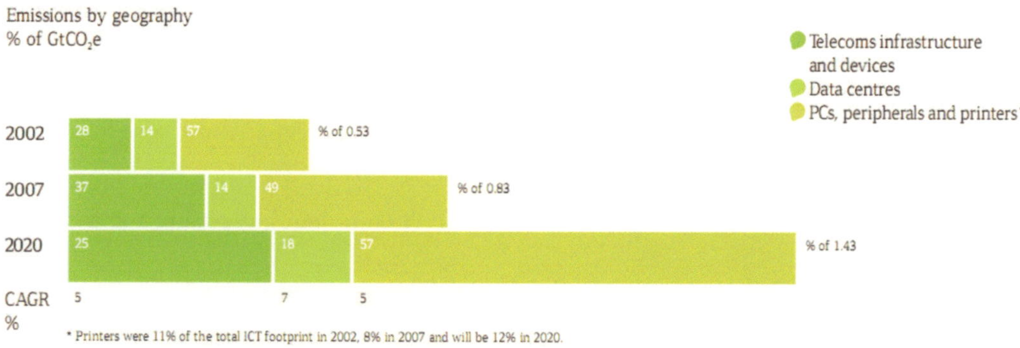

Figure 1. The 2020 global footprint by subsector

This is because of the enormous number of machines used by individuals and businesses: it is estimated there will be 4 billion PCs in the world by 2020. So the vast number of PCs is going to dominate ICT energy consumption. Finding precise figures of the current energy consumption of computer systems and ICT equipment is essential, in order to understand how to reduce their power consumption and improve their energy efficiency. Today these figures are incomplete and not precise.

INTECH
open science | open minds

© 2012 The Author(s). Licensee InTech. This chapter is distributed under the terms of the Creative Commons Attribution License http://creativecommons.org/licenses/by/3.0), which permits unrestricted use, distribution, and reproduction in any medium, provided the original work is properly cited.

The management of system components can be done either in hardware or software. When we buy a device and it is not programmable, we can not do anything to limit its energy consumption. The designers have already made choices in terms of selection of components and in terms of resource management. On the other hand, if a system can be programmed, choices made by developers will affect the management of energy the device consumes. Looking at embedded systems, all the responsibilities in terms of management of energy resources are dependent on the hardware management and on the firmware. Firmware optimizations have immediate effects that can be verified directly by measuring the current the device consumes. If we consider a general purpose device, the hardware and the operating system have an important role in global energy management, but it is not the only one. On this type of device is it possible to install a multitude of programs that will impact on the management of energy resources. For example, if a third party software uses a particular peripheral incorrectly, it could increase its energy demand even when not needed.

The underlying idea is that power consumption depends not only on hardware features, but also (and probably mostly) on software usage and software internal characteristics. For instance, a more complex software will require more CPU cycles, or a single long write operation on disk can be less power consuming rather than several small write operations. The energetic impact of software will also be analyzed in order to understand the effect of human behaviour on power consumption: for example we can check whether customized power management profiles are more efficient than default ones. Usually, if a person is aware of how much he is consuming, he is able to find his own solution that is the most appropriate for his device usage, and to save more power.

This chapter deals with the matter of finding relationships between software usage and power consumption. For this purpose, two experiments have been designed, consisting in running benchmarks[1] on two common desktop machines, simulating some typical scenarios and then measuring the energy consumption in order to make some statistical analysis on results.

This chapter is organized as follows:

- in Section 2, we present some related works in literature dealing with Green Software and Energy-Aware issues.
- Section 3 describes the experiment design process in all of its steps.
- in Section 4 we present the results of the experiments.
- in Section 5 we discuss the results of the experiments in detail, exposing the facts that became evident.
- in Section 6, we try to give some conclusions and present our future works.

2. Related work

In 2011, a post appeared on the MSDN Blog[2] : it concerned the energy consumption measurement of internet browsers. Authors measured power consumption and battery life of

[1] *A computer benchmark is typically a computer program that performs a strictly defined set of operations (a workload) and returns some form of result (a metric) describing how the tested computer performed.* [11] In our benchmark the *workload* is a set of usage scenarios and the *metric* is the power consumption.

[2] Browser Power Consumption - Leading the Industry with IE 9, `http://blogs.msdn.com/b/ie/archive/2011/03/28/browser-power-consumption-leading-the-industry-with-internet-explorer-9.aspx`

a common laptop across six scenarios and different browsers. They allowed each scenario to run for 7 minutes and calculated the average power consumption over that duration. Authors ran *IE9*, *Firefox*, *Opera* and *Safari* for each scenario, maintaning a fixed running duration, and then they made a comparison of the obtained results. Notable differences arose regarding power consumption and laptop battery life. Namely, using IE9 the battery lasted 3:45 hours, while using Opera the battery ran out of power after 2:45 hours. Hence, software can impact on energy consumption, as we also found in our previous work [15], where we monitored three servers for a whole year, observing that one of them consumed up to 75% more when used for graphical operations.

Kansal et al. [8] presented a solution for VM power metering. Since measuring the power consumption of a Virtual Machine is very hard and not always possible, authors built power models to get power consumption at runtime. This approach was designed to operate with low runtime overhead. It also adapts to changes in workload characteristics and hardware configuration. Results showed 8% to 12% of additional savings in virtualized data centers. Another related work is PowerScope [6]: this tool uses statistical sampling to profile the energy usage of a computer system. Profiles are created both during the data collection stage and during the analysis stage. During the first stage, the tool samples both the power consumption and the system activity of the profiling computer and then generates an energy profile from this data without profiling overhead. During data collection, authors use a digital multimeter to sample the current drawn by the profiling computer through its external power input. After that, they modified Odyssey platform for mobile computing. When there is a mismatch between predicted demand and available energy, Odyssey notifies applications to adapt. This is one of the first examples of *Energy-Aware* software.
In 1995, the first attempts were made to profile the energy performance of a computer. Lorch [9] in his M.S. thesis explained that there are two aspects to consider while measuring the breakdown of power consumption on a portable computer: I) Measuring how much power is consumed by each component, II) Profiling how often each component is in each state.

Other works about profiling and measuring energy consumption are related to embedded systems. For instance, JouleTrack [12] runs each instruction or short sequences of instruction in a loop and measure the current/power consumption. The user can upload his C source code to a Web Server which compiles, links and executes it on an ARM simulator. Program outputs, assembly listing and the run-time statistics (like execution time, cycle counts etc.) are then available and passed as parameters to an engine which estimates the energy consumed and produces graphs of different energy variables. Results showed that the error of predictions was between 2% and 6%. The concept of energy-awareness is based upon a complete knowledge on how and where energy is consumed on a device. Carroll and Heiser [3] present a detailed analysis of power consumption in a mobile device, focusing on the hardware subsystems, through common and realistic usage scenarios. Results show that the GSM module and the display are the most power-consuming components: for example, a GSM phone call on OpenMoko Neo Freerunner, HTC Dream G1 and Google Nexus One consumes 1135 mW, 822 mW and 846 mW respectively.

Usually, an accurate power consumption analysis of mobile or embedded devices is component-based. However, instantaneous information about discharge current and remaining battery capacity is not always available, because most devices do not have built-in sensors to collect these data. PowerBooter [17] has been proposed as a technique to build a battery-based model automatically. Authors motivate this decision by considering that different mobile devices of the same category show different power consumption, and a specific power consumption model for each device is difficult to obtain. Thus, instead of using external metering instrumentation to detect power consumption, only the internal battery voltage sensor is used, which is found across many modern smartphones. Authors indicate that for a 10-second interval, the PowerBooter technique has an accuracy of about 4.1% within measured values.

From a software engineering point of view, most contributions are devoted in developing frameworks and tools for energy metering and profiling. The authors of PowerBooter also propose an on-line power estimation tool called PowerTutor [17]. It implements the PowerBooter model in order to profile power consumption of applications, basing upon their component usage. Another example, which makes use of external metering devices, is ANEPROF [4], which authors define as a real-measurement-based energy profiler able to reach function-level granularity. It is developed for Android OS-based devices, thus it is aimed at profiling Java applications. It is based on JVM event profiling, using software probes to record runtime events and system calls. Authors had to address several design issues, such as overhead control and proper time synchronization. Power consumption profiling is made through correlation of real-time power measurements done by an external DAQ, connected to a ARM Computer-on-Module running Android 2.0. Authors also provide profiling data of four popular applications (Android Browser, GMail, Facebook, Youtube). The accuracy of ANEPROF depends on the hardware meter used. Its CPU overhead is stated to be less than 5%. Finally, SEMO [5] is a smart energy monitoring system, developed for Android, which provides also application-level consumption monitoring. This system is composed of three components: an *inspector*, which monitors the information on the battery, warning users when the battery reaches a critical condition; a *recorder*, which basically logs the actual charge of the battery and the running applications, and an *analyzer*, which calculates the energy consumption rate for each application and ranks them according to it.

Another alternative for energy measurement is low-level power-analysis using instruction-level models [14]. These models provide accurate power estimates for small kernels of code. An example of this kind of model is presented in Equation 1 where [10] Energy is the total energy dissipation of the program.

$$Energy = \sum_{i}(BC_i) + \sum_{i,j}(SC_{i,j}\, N_{i,j}) + \sum_{k}(OC_k) \tag{1}$$

The first part is the summation of the base energy cost of each instruction (BC_i is the base energy cost and N_i is the number of times instruction i is executed). The second part accounts for the circuit state ($SC_{i,j}$ is the energy cost when instruction i is followed by during the program execution). The third part accounts for energy contribution OC_k of other instruction effects such as stalls and cache misses during the program execution.

The study presented here is instead focused on the analysis of power consumption data, and it is designed to find out usage patterns of IT devices' energy consumption and to identify situations in which there is a waste of energy. Webber et al. [16] also collected data on devices, focusing on the after-hours power state of networked devices in office buildings: they showed that most of devices are left powered on during night, concluding that this is the first cause of energy waste.

Usage analysis is a crucial step to optimize the energy consumption: this task is even more necessary within data centers where the number of computers is large. In this field, Bein et al. [2] tried to improve the energy efficiency of data centers: they studied the cost of storing vast amounts of data on the servers in a data center and they proposed a cost measure together with an algorithm that minimizes such cost.

3. Study design

3.1. Goal description and research questions

The aim of this research is to assess the impact of software and its usage on power consumption in computer systems. The goal is defined through the Goal-Question-Metric (GQM) approach. [1]. This approach, applied to the experiment, led to the definition of the model presented in Table 1. The first research question investigates whether and how much software impacts power consumption. The second research question investigates whether a categorization of usage scenarios with respect to functionality is also valid for power consumption figures. The third research question tries to find a quantifiable relationship between power consumption and actual usage of the computer system, by selecting four metrics relative to the main system resources (CPU, Disk, Memory and Network).

Goal	*Evaluate* *for the purpose of* *with respect to* *from the viewpoint of* *in the context of*	software usage assessing its energetic impact power consumption the System User Desktop applications
Question 1	Does software impact power consumption?	
Metric	Consumed Power (Watts)	
Question 2	Is it possible to classify software usage scenarios basing upon power consumption?	
Metric	Consumed Power (Watts)	
Question 3	What is the relationship between usage and power consumption?	
Metrics	CPU Usage (percentage) Memory Usage (reads/writes) Disk Usage (reads/writes) Network Usage (Packets/sec) Consumed Power (Watts)	

Table 1. The GQM Model

3.2. Usage scenarios

The following usage scenarios, described in detail, will provide the basis for the analysis. The scenarios have been designed trying to simulate common operations for a desktop user, and they provide benchmarks (see Section 1) for the different resources of the computer system. This way, we will obtain useful information on the relationship between resource usage and power consumption.

0 - Idle. This scenario aims at evaluating power consumption during idle states of the system. In order to avoid variations during the runs, most of OS'automatic services were disabled (i.e. Automatic Updates, Screen Saver, Anti-virus and such).

1 - Web Navigation. This scenario depicts one of the most common activities for a basic user - Web Navigation. During the simulation, the system user starts a web browser, inputs the URL of a web page and follows a determined navigation path. Google Chrome has been chosen as the browser for this scenario because of its better performance on the test system, which allowed us to increase navigation time. The website chosen for this scenario is the homepage of the SoftEng research group `http://softeng.polito.it`, so that the same contents and navigation path could be maintained during all the scenario runs.

2 - E-Mail. This scenario simulates sending and receiving E-Mails. For this scenario's purpose, a dedicated E-Mail account has been created in order to send and receive always the same message. In this scenario, the system user opens an E-Mail Client, writes a short message, sends it to himself, then starts checking for new messages by pushing on the send/receive button. Once the message has been received, the user reads it (the reading activity has been simulated with an idle period), then deletes the messages and starts over.

3 - Productivity Suite. This scenario evaluates power consumption during the usage of highly-interactive applications, such as office suites. For this scenario, Microsoft Word 2007 has been chosen, the most used Word Processor application. During the scenario execution, the system user launches the application and creates a new document, filling it with content and applying several text editing/formatting functions, such as enlarge/shrink Font dimension, Bold, Italics, Underlined, Character and background colors, Text alignment and interline, lists. Then the document is saved on the machine's hard drive. For each execution a new file is produced, thus the old file gets deleted at the end of the scenario.

4 - Data Transfer (Disk). This scenario evaluates power consumption during operations that involve the File System, and in particular the displacement of a file over different positions of the hard drive, which is a very common operation. For this scenario's purpose, a data file of a relevant size (almost 2 GB) has been prepared in order to match the file transfer time with the prefixed scenario duration (5 minutes). The scenario structure is as follows: the system user opens an Explorer window, selects the file and moves it to another location. It waits for file transfer to end, then closes Explorer and exits.

5 - Data Transfer (USB). As using portable data storage devices has become a very common practice, this scenario has been developed to evaluate power consumption during a file transfer from the system hard drive to an USB Memory Device. This scenario is very similar to the previous one, exception given for the file size (which is slightly lower, near 1.8 GB) and the file destination, which is the logical drive of the USB Device.

6 - Image Browsing/Presentation. This scenario evaluates power consumption during another common usage pattern, which is a full-screen slide-show of medium-size images, which can simulate a presentation as well as browsing through a series of images. In this scenario, the system user opens a PDF File composed of several images, using the Acrobat Reader application. It sets the Full-Screen visualization, then manually switches through the images every 5 seconds (thus simulating a presentation for an audience).

7 - Skype Call (Video Disabled). For an average user, the Internet is without any doubt the most common resource accessed via a Computer System. Moreover, as broadband technologies become always more available,it has been thought to be reductive not to consider usage scenarios that make a more intensive use of the Internet than Web Navigation and E-Mails. Thus, the Skype scenario has been developed. Skype is the most used application for Video Calls and Video Conferences among private users. For this scenario's purposes, a Test Skype Account was created, and the Skype Application was deployed on the test machine. Then, for each run, a test call is made to another machine (which is a laptop situated in the same laboratory) for 5 minutes, which is the prefixed duration of all scenarios.

8 - *Skype Call (Video Enabled).* This scenario is similar to scenario 7, but the Video Camera is enabled during the call. This allows to evaluate the impact of the Video Data Stream both on power consumption and on system resources.

9 - *Audio Playback.* This scenario aims to evaluate power consumption during the reproduction of an Audio content. For this scenario's purpose, an MP3 file has been selected, with a length of 5 minutes, to reproduce through a common multimedia player. Windows Media Player has been chosen, as it is the default player in Microsoft systems, and thus one of the most diffused.

10 - *Video Playback.* Same as above, but in this case the subject for reproduction is a Video File in AVI format, same duration.

11 - *Peer-to-Peer Data Exchange.* As for the Skype scenarios, it has been made the decision to take into account also a Peer-to-Peer scenario, which has proven to be a very common practice among private users. For this scenario, BitTorrent was selected as a Peer-to-Peer application, because of its large diffusion and less-variant usage pattern if compared to other Peer-to-Peer networks with more complex architectures. During this scenario, the system user starts the BitTorrent client, opens a previously provided .torrent archive, related to an Ubuntu distribution, and starts the download, which proceeds for 5 minutes.

In Table 2 all the scenarios are summarized with a brief description of each of them. The last column reports the category which scenarios belong to, from a functional point of view, according to the following:

- *Idle* (Scenario 0) is the basis of the analysis, evaluates power consumption during the periods of inactivity of the system.
- *Network* (Scenarios 1,2,7,8,11) represents activities that involve networking and Internet.
- *Productivity* (Scenario 3) is related to activities of personal productivity.
- *File System* (Scenarios 4,5) concerns activities that involve storage devices and File System operations.
- *Multimedia* (Scenarios 6,9,10) represents activities that involve audio/video peripherals and multimedia contents.

3.3. Variable selection

In order to answer the Research Questions, it is necessary to specify the independent variables that will characterize the experiment. As anticipated in the previous section, four metrics have been selected to evaluate the system usage. These metrics were measured by means of software logging (as will be explained in the *Instrumentation* section) considering the following values:

Nr.	Title	Description	Category
0	Idle	No user input, no applications running, most of OS'automated services disabled.	Idle
1	Web Navigation	Open browser, visit a web-page, operate, close browser.	Network
2	E-Mail	Open e-mail client, check e-mails, read new messages, write a short message, send, close client.	Network
3	Productivity Suite	Open word processor, write a small block of text, save, close.	Productivity
4	Data Transfer (disk)	Copy a large file from a disk position to another.	File System
5	Data Transfer (USB)	Copy a large file from an USB Device to disk.	File System
6	Presentation	Execute a full-screen slide-show of a series of medium-size images.	Multimedia
7	Skype Call (no video)	Open Skype client, execute a Skype conversation (video disabled), close Skype.	Network
8	Skype Call (video)	Open Skype client, execute a Skype conversation (video enabled), close Skype.	Network
9	Audio Playback	Open a common media player, play an Audio file, close player.	Multimedia
10	Video Playback	Open a common media player, play a Video file, close player.	Multimedia
11	Peer-to-Peer	Open a common peer-to-peer client, put a file into download queue, download for 5 minutes, close.	Network

Table 2. Software Usage Scenarios Overview

- CPU
  - CPU Time Percentage, intended as time spent by the CPU doing active work in a second
  - CPU User Time Percentage, intended as time spent by the CPU executing user instructions (i.e. applications) in a second
  - CPU Privileged Time Percentage, intended as time spent by the CPU executing system instructions (services, daemons) in a second
  - CPU Deferred Procedure Calls Percentage, intended as time spent by the CPU executing DPC in a second
  - CPU Interrupt Time Percentage, intended as time spent by the CPU serving interrupts in a second

- CPU C1 Time Percentage, intended as time spent by the CPU in low-power (C1) State
- CPU C2 Time Percentage, intended as time spent by the CPU in low-power (C2) State
- CPU C3 Time Percentage, intended as time spent by the CPU in low-power (C3) State

- Memory
  - Memory Page Writings per second
  - Memory Page Readings per second
  - Memory Available (KiloBytes) per second
- Hard Disk
  - Physical Disk Transfers (Read/Write) per second
  - Logical Disk Transfers (Read/Write) per second
- Network
  - Network Packets per second as seen by the Network Interface Card

The dependent variable selected for the experiment is P i.e. the instant power consumption (W). Therefore, P_n is the average power consumption during Scenario $n = 1..11$ and $P_{idle|net|prod|file|MM}$ is the average power consumption of (respectively) Idle, Network, Productivity, File System and Multimedia scenarios.

3.4. Hypotheses formulation

Basing upon the GQM Model, the Research Questions can be formalized into Hypotheses. In order to formally express Research Question 3, $\rho(x, y)$ expresses the correlation coefficient between variables x and y. β represents a significant correlation value, which will be defined later in this Section.

- *RQ 1: Does Software impact Power Consumption?*

  $H1_0$: $P_{idle} \geq P_n, n \in [1, 11]$
  $H1_a$: $P_{idle} < P_n, n \in [1, 11]$

- *RQ 2: Is it possible to classify software usage scenarios basing upon power consumption?*

  $H2_0$: $P_{idle} = P_{net} = P_{prod} = P_{file} = P_{MM}$
  $H2_a$: $P_{idle} \neq P_{net} \neq P_{prod} \neq P_{file} \neq P_{MM}$

- *RQ 3: What is the relationship between usage and power consumption?*

  $H3_0$: $\rho(I_{CPU}, P) = \rho(I_{Memory}, P) = \rho(I_{Disk}, P) = \rho(I_{Network}, P) = 0$
  $H3_a$: $max[\rho(I_{CPU}, P), \rho(I_{Memory}, P), \rho(I_{Disk}, P), \rho(I_{Network}, P)] > \beta$

3.5. Instrumentation

Every scenario has been executed automatically by means of a GUI Automation Software for 5 minutes, obtaining 30 runs per scenario, each composed of 300 observations (one per second) of the instant power consumption value (W).

The test machines selected are two desktop PCs of different generations. In Table 3, the Hardware/Software configuration of the machines is presented. As can be seen, the difference

	Desktop 1 (old generation)	Desktop 2 (new generation)
CPU	AMD Athlon XP 1500+	Intel Core i7-2600
Memory	768 MB DDR SDRAM	4 GB DDR3 SDRAM
Display Adapter	ATI Radeon 9200 PRO 128 MB	ATI Radeon HD 5400
HDD	Maxtor DiamondMax Plus 9 80GB Hard Drive	Western Digital 1 TB
Network Adapter	NIC TX PCI 10/100 3Com EtherLink XL	Intel 82579V Gigabit Ethernet
OS	Microsoft Windows XP Professional SP3	Windows 7 Professional SP1

Table 3. HW/SW Configuration of the test machine

in terms of hardware is relevant; this will allow us to make some evaluations about how power consumption varied over the years, with the evolution of hardware architectures.

Different software and hardware tools have been used to do monitoring, measurement and test automation. The Software tool adopted is Qaliber[3], (see Figure 2) which is mainly a GUI Testing Framework, composed of a Test Developer Component, that allows a developer to write a specific test case for an application, by means of "recording" GUI commands, and a Test Builder Component, which allows to create complex usage scenarios by combining the use cases. One of the most important features of Qaliber is its possibility to log system information during scenario execution, using Microsoft's Performance Monitor Utility. By defining a specific Counter Log, adding all the variables of interest, it is possible to tell Qaliber to start Performance Monitor simultaneously with the Scenario, thus allowing a complete monitoring of all the statistics needed for this analysis.

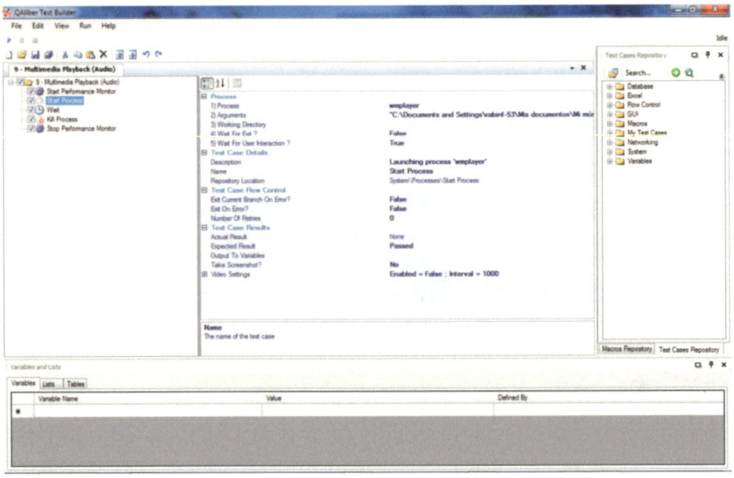

Figure 2. Qaliber Test Builder screenshot

[3] Qaliber - GUI Testing Framework, `http://www.qaliber.net/`

The measurement of power consumption was done through two different devices. For the old-generation PC, PloggMeter[4] (see Figure 3) device was used. This device is capable of computing Active and Reactive Power, Voltage, Current Intensity, $Cos\varphi$. The data is stored within the PloggMeter's 64kB memory and can be downloaded in a text file format via Zigbee wireless connection to a Windows enabled PC or Laptop or viewed as instantaneous readings on the installed Plogg Manager software. The device drivers were slightly modified to adapt the PloggMeter recording capability to this analysis' purposes, specifically to decrease the logging interval from 1 minute (which is too wide if compared to software time) to 1 second.

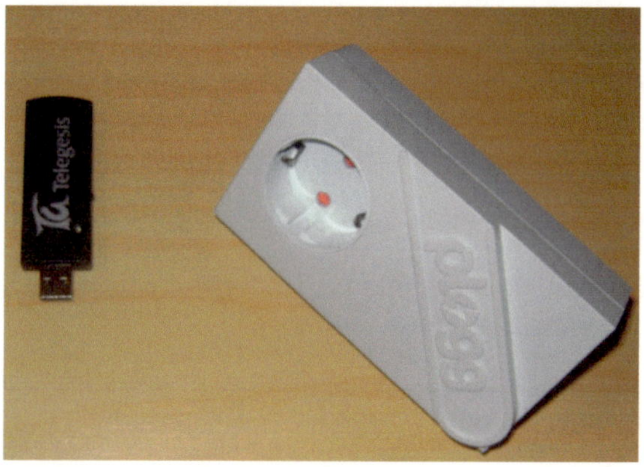

Figure 3. The PloggMeter device

For the new-generation PC, WattsUp PRO ES[5] (see Figure 4) device was used. This device is capable of measuring current power consumption (Watts), power factor, line voltage and other metrics. The data is stored within the device internal memory, and then downloadable via USB interface. The sampling rate resolution is 1 second.

Figure 4. The WattsUp Pro ES device

[4] Youmeter - Plogg Technologies, `http://www.plogginternational.com/products.shtml`

[5] WattsUp Pro ES, `https://www.wattsupmeters.com/secure/products.php?pn=0&wai=0&spec=2`

3.6. Analysis methodology

The goal of data analysis is to apply appropriate statistical tests to reject the null hypothesis. The analysis will be conducted separately for each scenario in order to evaluate which one has an actual impact on power consumption.

In order to extract a Power Consumption profile for each Usage Scenario, a set of descriptive statistics was derived from the experimental data. For a single scenario, a total of 30 runs were executed, each composed of 300 observations (one per second) of the power consumption value. Thus, the calculations for the descriptive statistics were made using two approaches: firstly, the average of each run is extracted, obtaining a short vector of 30 elements, which was used as the subject of our analysis. This method allowed to speed up the calculations, and because of the decreased sampling rate, the data was less variant and showed an almost regular distribution.

Afterwards, the same analysis on the full datasets was applied, which means a total of 9000 observations. Comparing the results from these two approaches, focusing on the Index of Dispersion and the variance, the variability of a single scenario can be appreciated, which was also a useful tool for validating the experiment.

First of all, the null hypothesis $H1_0$ will be tested for each scenario. Then the scenarios will be grouped into categories and $H2_0$ will be tested for each category.

First of all, data distribution must be analysed, in order to determine the appropriate testing method for each hypothesis. The data distribution analysis was conducted using the Shapiro-Wilk normality test. Since its results pointed out that the data was not normally distributed, non parametric tests were adopted, in particular the Mann-Whitney test [7] for testing $H2_0$, and the Spearman's rank correlation coefficient (also known as Spearman's ρ) for testing $H3_0$.

The first hypothesis $H1_0$ is clearly directional, thus the one-tailed variant of the test will be applied. The second and third hypotheses $H2_0$, $H3_0$ are not directional, therefore the two-sided variant of the tests will be applied.

We will draw conclusions from our tests based on a significance level $\alpha = 0.05$, that is we accept a 5% risk of type I error – i.e. rejecting the null hypothesis when it is actually true. Moreover, since we perform multiple tests on the same data – precisely twice: first overall and then by category – we apply the Bonferroni correction to the significance level and we actually compare the test results versus a $\alpha_B = 0.05/2 = 0.025$. As regards Spearman's ρ significance, using 298 degrees of freedom (since 300 observations per scenario are available) the significance level of the ρ coefficient is 0.113. Thus, correlations coefficients resulting higher than this value can be considered as significant positive or negative correlations.

3.7. Validity evaluation

The threats of experiment validity can be classified in two categories: **internal** threats, derived from treatments and instrumentation, and **external** threats, that regard the generalization of the work.

There are three main internal threats that can affect this analysis. The first concerns the *measurement sampling*: measurements were taken with a sampling rate of 1 second. This interval is a compromise between the power metering device capability and the software logging service. However, it could be a wide interval if compared to software time.

Subsequently, *network confounding factors* could arise: as several usage scenarios involving network activity and the Internet are included in our treatments, the unpredictability of the network behaviour could affect some results. Another confounding factor is represented by *OS scheduling operations*: the scheduling of user activities and system calls is out of the experiment control. This may cause some additional variability in the scenarios, especially for those that involve the File System.

In addition, the two machines on which our tests are performed are different in terms of hardware and software configuration. This is done on purpose, because we wanted to test devices which could represent common machines used in home and office scenarios, for both generations. Thus, installing an old version of an operating system on a new machine or viceversa would have altered this assumption. However, this will introduce another confounding factor, but still, will provide useful information regarding the evolution of these systems, even if no specific research hypotheses can be verified about the comparison.

Finally, the main external threat concerns a possible *limited generalization* of the results: this is due to the fact that the experiment was conducted on only two different test machines, which is a limited population to be representative of a whole category.

4. Results

4.1. Preliminary data analysis

We present in Table 4 and Table 5 the following descriptive statistics about measurements for each scenario. Tables reports in this order mean (Watts), median (Watts), standard error on the mean, 95% confidence interval of the mean, variance, standard deviation (σ), variation coefficient (the standard deviation divided by the mean), index of dispersion (variance-to-mean ratio, VMR).

Power consumptions show an excursion of about 11 W for both PCs, even if the baseline is quite different (an average of 87 W in Idle scenario for the Old PC, 51 W for the New PC). Moreover, the very low variability indexes ensure that the different samples for each scenario are homogeneous.

4.2. Hypothesis testing

The results of hypotheses testing of the research questions are exposed in this section.

The testing of hypothesis H_1 and H_2 are exposed in Table 6 and 7. These table report the scenarios tested, the p-value of Mann-Whitney test and the estimated difference of the medians between Idle scenario and the other ones.

Figure 5 represents the bar plot of the power consumption increase (in watts), with respect to idle, of each scenario. Figure 6 shows the box plot of scenario categories for each PC. As regards hypothesis H_3, which evaluates correlations between resource usage and power consumption, more steps are needed. First of all, Table 8 reports the results of the Data Distribution Analysis. Then, in Table 9 and Table 10, are presented the results of the correlation

	Old-Generation PC							
	Mean	Median	S.E.	C.I.	Variance	σ	Var.Co.	VMR
0 - Idle	86.81	86.69	0.007	0.013	0.424	0.650	0.007	0.005
1 - Web	89.09	88.57	0.011	0.022	3.372	1.836	0.021	0.038
2 - E-Mail	88.03	87.11	0.024	0.047	5.195	2.279	0.026	0.059
3 - Prod	90.12	89.40	0.025	0.500	5.862	2.421	0.027	0.065
4 - Disk	94.12	97.21	0.048	0.095	21.12	4.595	0.049	0.224
5 - USB	96.41	97.10	0.024	0.046	5.047	2.246	0.023	0.052
6 - Image	91.97	91.48	0.041	0.081	15.474	3.934	0.043	0.168
7 - Skype	91.87	91.69	0.015	0.029	1.981	1.407	0.015	0.022
8 - SkypeV	95.40	95.75	0.020	0.040	3.844	1.960	0.020	0.040
9 - Audio	88.14	87.94	0.013	0.025	1.429	1.195	0.013	0.016
10 - Video	88.61	88.57	0.009	0.017	0.677	0.823	0.009	0.008
11 - P2P	88.46	88.25	0.010	0.019	0.842	0.917	0.010	0.009

Table 4. Scenarios Statistics Overview: Old-Generation PC

	New-Generation PC							
	Mean	Median	S.E.	C.I.	Variance	σ	Var.Co.	VMR
0 - Idle	51.39	51.20	0.007	0.015	0.507	0.712	0.013	0.009
1 - Web	54.05	53.90	0.014	0.028	1.883	1.372	0.025	0.035
2 - E-Mail	53.40	53.40	0.011	0.021	1.123	1.059	0.019	0.021
3 - Prod	53.09	52.70	0.016	0.032	2.369	1.539	0.029	0.044
4 - Disk	60.24	62.10	0.037	0.072	12.38	3.518	0.058	0.205
5 - USB	61.29	61.90	0.023	0.046	4.901	2.214	0.036	0.080
6 - Image	52.75	52.50	0.011	0.023	1.214	1.102	0.021	0.023
7 - Skype	56.23	56.30	0.016	0.032	2.420	1.555	0.027	0.043
8 - SkypeV	62.13	62.90	0.036	0.070	11.428	3.380	0.054	0.184
9 - Audio	52.87	52.70	0.006	0.012	0.315	0.561	0.010	0.006
10 - Video	54.14	54.00	0.007	0.013	0.420	0.648	0.012	0.008
11 - P2P	54.32	54.50	0.008	0.016	0.609	0.780	0.014	0.011

Table 5. Scenarios Statistics Overview: New-Generation PC

test using Spearman's method, with a 95% confidence interval, applied to every couple $(watt, variable)$ for each scenario. Only the significant coefficients are listed.

4.2.1. Question 1: Does software impact power consumption?

$H1 : P_{idle} \neq P_n \forall n \in [1, 11]$.

4.2.2. Question 2: Is it possible to classify software usage scenarios basing upon power consumption?

$H2 : P_{idle} \neq P_{net} \neq P_{prod} \neq P_{file} \neq P_{MM}$

	Old-Gen PC		*New-Gen PC*	
Scenario Comparison	**p-value**	**Est. Diff**	**p-value**	**Est. Diff**
0 - Idle vs. 1 - Web Navigation	< 0.0001	-1.87	< 0.0001	-2.60
0 - Idle vs. 2 - E-Mail	< 0.0001	-0.52	< 0.0001	-2.10
0 - Idle vs. 3 - Productivity Suite	< 0.0001	-2.71	< 0.0001	-1.50
0 - Idle vs. 4 - IO Operation (Disk)	< 0.0001	-10.41	< 0.0001	-10.80
0 - Idle vs. 5 - IO Operation (USB)	< 0.0001	-10.41	< 0.0001	-10.60
0 - Idle vs. 6 - Image Browsing	< 0.0001	-4.69	< 0.0001	-1.20
0 - Idle vs. 7 - Skype Call (No Video)	< 0.0001	-5.10	< 0.0001	-5.00
0 - Idle vs. 8 - Skype Call (Video)	< 0.0001	-9.05	< 0.0001	-11.50
0 - Idle vs. 9 - Audio Playback	< 0.0001	-1.25	< 0.0001	-1.50
0 - Idle vs. 10 - Video Playback	< 0.0001	-1.87	< 0.0001	-2.80
0 - Idle vs. 11 - Peer-to-Peer Data Exchange	< 0.0001	-1.66	< 0.0001	-3.30

Table 6. Hypotheses *H*1 Test Results

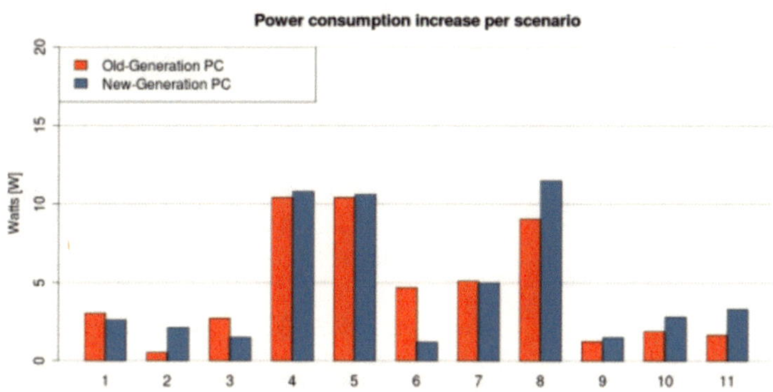

Figure 5. Bar Plot of per-scenario Power Consumption increase with respect to Idle

	Old-Gen PC		*New-Gen PC*	
Scenario Comparison	**p-value**	**Est. Diff**	**p-value**	**Est. Diff**
Idle vs. Network	< 0.0001	-2.08	< 0.0001	-3.20
Idle vs. Productivity	< 0.0001	-2.71	< 0.0001	-1.50
Idle vs. File System	< 0.0001	-10.41	< 0.0001	-10.60
Idle vs. Multimedia	< 0.0001	-1.67	< 0.0001	-1.60
Network vs. Productivity	< 0.0001	-0.31	< 0.0001	1.70
Network vs. File System	< 0.0001	-6.97	< 0.0001	-6.80
Network vs. Multimedia	< 0.0001	0.31	< 0.0001	1.60
Productivity vs. File System	< 0.0001	-6.87	< 0.0001	-9.10
Productivity vs. Multimedia	< 0.0001	0.73	< 0.0001	-0.20
File System vs. Multimedia	< 0.0001	8.53	< 0.0001	8.60

Table 7. Hypothesis *H*2 Test Results

4.2.3. Question 3: What is the relationship between usage and power consumption?

$$H3_a :\neq max[\rho(I_{CPU}, P), \rho(I_{Memory}, P), \rho(I_{Disk}, P), \rho(I_{Network}, P)] > \beta$$

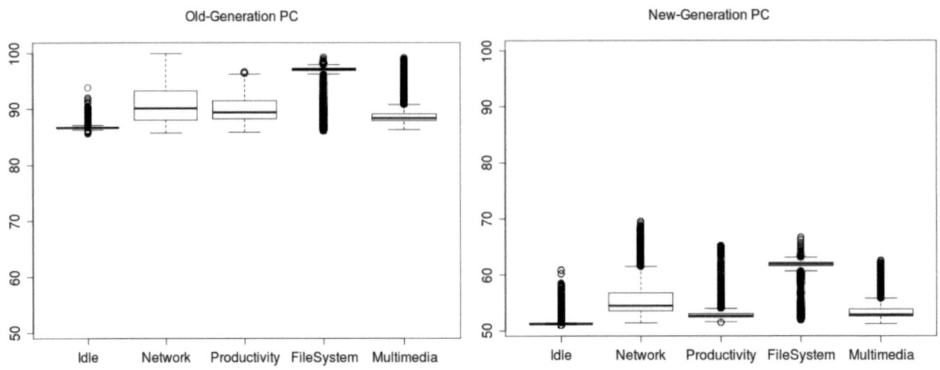

Figure 6. Box Plots of Scenario Categories

	Old-Gen PC		*New-Gen PC*	
Scenario	**Data Distr.**	**Max p-val.**	**Data Distr.**	**Max p-val.**
0 - Idle	Not Normal	1.5e-63	Not Normal	2.2e-39
1 - Web Navigation	Not Normal	4.4e-36	Not Normal	1.1e-20
2 - E-Mail	Not Normal	9e-73	Not Normal	1.2e-19
3 - Productivity Suite	Not Normal	1e-45	Not Normal	9.4e-29
4 - IO Operation (Disk)	Not Normal	1.2e-46	Not Normal	8.7e-51
5 - IO Operation (USB)	Not Normal	6.4e-52	Not Normal	2.5e-29
6 - Image Browsing	Not Normal	1.1e-35	Not Normal	6.7e-22
7 - Skype Call (No Video)	Not Normal	8.2e-30	Not Normal	3e-67
8 - Skype Call (Video)	Not Normal	1.3e-35	Not Normal	5.2e-36
9 - Audio Playback	Not Normal	7.9e-54	Not Normal	5.2e-44
10 - Video Playback	Not Normal	1.6e-44	Not Normal	6.6e-81
11 - Peer-to-Peer Data Exchange	Not Normal	8.9e-36	Not Normal	2.2e-35

Table 8. Data Distribution Analysis

Old-Generation PC				
Scenario Title	**Variable**	**p-value**	ρ	**R2**
2 - E-Mail	CPUC1Time.	< 0.0001	-0.36	13 %
4 - IO Operation (Disk)	CPUTime.	< 0.0001	0.35	12 %
4 - IO Operation (Disk)	CPUC1Time.	< 0.0001	-0.35	12 %
5 - IO Operation (USB)	CPUTime.	< 0.0001	0.47	22 %
5 - IO Operation (USB)	CPUC1Time.	< 0.0001	-0.47	22 %
7 - Skype Call (No Video)	CPUC1Time.	< 0.0001	-0.39	15 %
8 - Skype Call (Video)	CPUTime.	< 0.0001	0.63	40 %
8 - Skype Call (Video)	CPUUserTime.	< 0.0001	0.53	28 %
8 - Skype Call (Video)	CPUC1Time.	< 0.0001	-0.7	49 %
11 - Peer-to-Peer	MemoryKByteAvailable	< 0.0001	-0.34	12 %

Table 9. Spearman's ρ Coefficient between Power and Resource variables

5. Discussion

The collected data shows several facts, and gives the answers for the Research Questions.

	New-Generation PC			
Scenario Title	**Variable**	**p-value**	ρ	**R2**
2 - E-Mail	CPUUserTime.	< 0.0001	0.42	17 %
2 - E-Mail	CPUPrivTime.	< 0.0001	0.43	18 %
3 - Productivity Suite	CPUUserTime.	< 0.0001	0.33	11 %
4 - IO Operation (Disk)	PhysicalDiskTransfers	< 0.0001	0.45	20 %
4 - IO Operation (Disk)	LogicalDiskTransfers	< 0.0001	0.45	20 %
4 - IO Operation (Disk)	MemoryPages	< 0.0001	0.44	19 %
4 - IO Operation (Disk)	MemoryKByteAvailable	< 0.0001	-0.54	29 %
4 - IO Operation (Disk)	CPUC3Time.	< 0.0001	-0.59	35 %
4 - IO Operation (Disk)	CPUTime.	< 0.0001	0.55	31 %
4 - IO Operation (Disk)	CPUUserTime.	< 0.0001	0.58	34 %
4 - IO Operation (Disk)	CPUPrivTime.	< 0.0001	0.39	15 %
6 - Image Browsing	CPUUserTime.	< 0.0001	0.34	12 %
7 - Skype Call (no video)	NetworkPkts	< 0.0001	0.62	39 %
7 - Skype Call (no video)	MemoryKByteAvailable	< 0.0001	-0.45	20 %
7 - Skype Call (no video)	CPUC3Time.	< 0.0001	-0.66	43 %
7 - Skype Call (no video)	CPUTime.	< 0.0001	0.52	27 %
7 - Skype Call (no video)	CPUUserTime.	< 0.0001	0.63	39 %
8 - Skype Call (Video)	NetworkPkts	< 0.0001	0.67	46 %
8 - Skype Call (Video)	MemoryKByteAvailable	< 0.0001	-0.62	39 %
8 - Skype Call (Video)	CPUC3Time.	< 0.0001	-0.88	77 %
8 - Skype Call (Video)	CPUTime.	< 0.0001	0.87	76 %
8 - Skype Call (Video)	CPUUserTime.	< 0.0001	0.9	81 %
9 - Audio Playback	MemoryKByteAvailable	< 0.0001	-0.34	12 %
11 - Peer-to-peer	NetworkPkts	< 0.0001	0.45	20 %
11 - Peer-to-peer	MemoryKByteAvailable	< 0.0001	-0.42	18 %
11 - Peer-to-peer	CPUPrivTime.	< 0.0001	0.35	12 %

Table 10. Spearman's ρ Coefficient between Power and Resource variables

5.1. Question 1: Does software impact power consumption?

As observed in Table 6, in both our test machines, every usage scenario consumes more power than the Idle scenario. This difference is even more evident in the New-Generation PC, where we obtain our highest increase percentage (up to 20%), as can be observed in Figure 7.

5.2. Question 2: Is it possible to classify software usage scenarios basing upon power consumption?

As regards our second RQ, scenarios classification, results are not homogeneous: for instance, in Figure 6 it can be observed that Network category has a very wide range if compared to the others. Moreover, the comparison not always gives a clear distinction between the profiles. This suggests that a classification based on functionality can be inadequate for power consumption. Another classification may arise from the analysis of every single scenario. As can be seen from Tables 4, 5 and 6, the most power-consuming scenarios are those that involve File System, followed by Skype (both with and without Video Enabled) and Image Browsing. From the hardware point of view, these scenarios are also the most expensive in terms of system resources. Thus, classifying our scenarios basing upon resource utilization can be a more accurate way to estimate their power consumption. For instance, the power

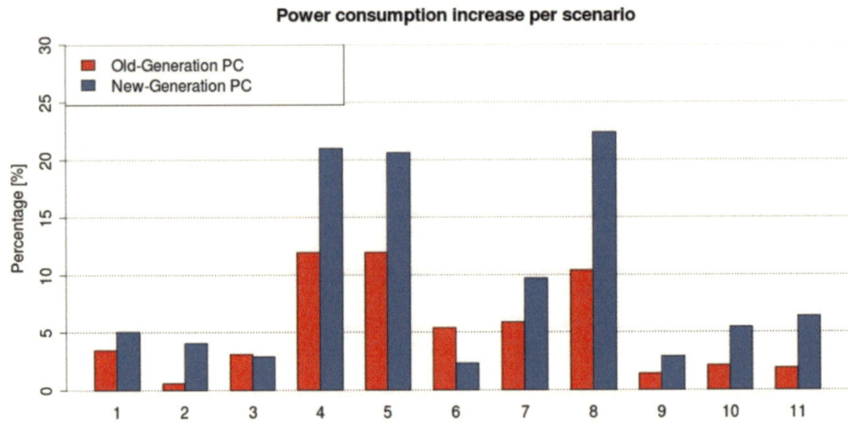

Figure 7. Per-scenario Power Consumption increase with respect to Idle (in percentage)

consumption profile of Skype is very different (about 4-5 Watts in average) with and without enabling the Video Camera.

Another interesting question that arises from the analysis is, in case of applying these Scenarios in groups, if their power consumption would follow a linear composition rule (thus summing up the values). That is, for example, supposing a composed Usage Scenario S that involves a Skype Call, a Web Navigation and a Disk Operation performed simultaneously, their linear composition would give, on our Old-Gen PC, an estimated Power Consumption per second of

$$P_{idle} + \Delta P_S = 86.81W + 21.33W = 108.14W$$

introducing a 25% overhead on power consumption. On the New-Gen PC, the estimated Power Consumption would be

$$P_{idle} + \Delta P_S = 51.39W + 24.90W = 76.29W$$

which gives a 48% overhead on power consumption.

5.3. Question 3: What is the relationship between usage and power consumption?

Taking a look at the results of the correlation analysis, represented by the third Research Questions, more conclusions can be made. First of all, we can observe that the coefficients related to the New-Gen PC are higher with respect to the Old-Gen PC. This may suggest that as hardware evolves, the software usage is even more significant for determining the power consumption of the system. This assumption is confirmed by Figure 7, where we can observe that the percentage increase of the New-Gen PC is higher, in most cases, with respect to the Old-Gen.

However, it is remarkable that, for both machines, the variables that show higher correlation coefficients are undoubtedly those related to CPU Usage and Memory Usage. High coefficients are also present in the Hard Disk Index, but only in those scenarios that, clearly, involve File System operations. This means that those resources have a greater influence upon power consumption related to the others selected for the analysis. Further researches should probably focus upon these two variables.

As expected, power consumption has always a *negative* correlation with the time spent by the CPU in the C1 and C3 states, which are power saving, low-activity states, and with the available memory, which means that using more memory has a positive correlation with power, which is a reasonable and correct behaviour. This is also a confirmation that the analysis was conducted with the right premises.

Moreover, as expected, the scenarios who exhibit higher correlations are those who use more resources, such as Skype and IO scenarios. In particular, the Skype scenario with video enabled has a strong correlation with the CPU usage, probably because the real-time video elaboration makes the CPU the dominant resource for power consumption.

6. Conclusions and future works

This experiment assessed quantitatively the energetic impact of software usage. It consisted in building up common application usage scenarios (e.g.: Skype call, Web Navigation, Word writing) and executing them independently to collect power consumption data. Each single scenario introduced an overhead on power consumption, which may raise up to 20% for recent systems: if their power consumption would follow a linear composition rule, the impact could be even higher.

The relationship between usage and power consumption was also analysed in terms of correlation between resource usage. Although a clear linear relationship did not arise, the analysis showed that some resources drive power consumption more than others, such as memory and CPU usage. Using a precise control over how an application consumes these resources, it can be possible to predict its power consumption, thus including dedicated countermeasures in the Software Design Phase – which, by itself, is the essence of Energy-Aware Programming.

Our experiment also gives us the indication that modern desktop systems, although being more energy efficient in standby and idle states, due to their higher scalability, are even more sensible to the impact of software usage on power consumption. This indicates that research should focus on reducing this impact, as it will be always more significant as time goes by.

Moreover, results set the basis for future works and research projects. A more accurate correlation analysis will be conducted, focusing on the more relevant resources and taking into account also different kinds of relationships (not just linear). Moreover, we will focus our attention on battery-powered mobile devices, where software power consumption is a key issue. Our idea is that re-factoring applications by considering a more efficient resource utilization, the impact of software on power consumption could be easily reduced.

Author details

Giuseppe Procaccianti, Luca Ardito, Antonio Vetro' and Maurizio Morisio
Politecnico di Torino, Italy

7. References

[1] Basili, V. R., Caldiera, G. & Rombach, H. D. [1994]. The goal question metric approach, *Encyclopedia of Software Engineering*, Wiley.

[2] Bein, D., Bein, W. & Phoha, S. [2010]. Efficient data centers, cloud computing in the future of distributed computing, *Proceedings of the 2010 Seventh International Conference on Information Technology: New Generations*, ITNG '10, IEEE Computer Society, Washington, DC, USA, pp. 70–75.

[3] Carroll, A. & Heiser, G. [2010]. An analysis of power consumption in a smartphone, *Usenix technical conference*, Boston, MA, USA, pp. 1–14.

[4] Chung, Y., Lin, C. & King, C. [2011]. ANEPROF: Energy Profiling for Android Java Virtual Machine and Applications, *Parallel and Distributed Systems, International Conference on* 0: 372–379.

[5] Ding, F., Xia, F., Zhang, W., Zhao, X. & Ma, C. [2012]. Monitoring energy consumption of smartphones, *CoRR* abs/1201.0218.

[6] Flinn, J. & Satyanarayanan, M. [1999]. Powerscope: A tool for profiling the energy usage of mobile applications, *Mobile Computing Systems and Applications, IEEE Workshop on* 0: 2.

[7] Hollander, M. & Wolfe, D. A. [1973]. *Nonparametric Statistical Methods*, John Wiley & Sons, New York.

[8] Kansal, A., Zhao, F., Liu, J., Kothari, N. & Bhattacharya, A. A. [2010]. Virtual machine power metering and provisioning, *Proceedings of the 1st ACM symposium on Cloud computing*, SoCC '10, ACM, New York, NY, USA, pp. 39–50.

[9] Lorch, J. [1995]. A complete picture of the energy consumption of a portable computer, *Technical report*.

[10] Macii, E., Pedram, M. & Somenzi, F. [1998]. High-level power modeling, estimation, and optimization, *Computer-Aided Design of Integrated Circuits and Systems, IEEE Transactions on* 17(11): 1061 –1079.

[11] Marcu, M., Vladutiu, M., Moldovan, H. & Popa, M. [2007]. Thermal benchmark and power benchmark software, *ArXiv e-prints* .

[12] Sinha, A. & Chandrakasan, A. [2001]. Jouletrack-a web based tool for software energy profiling, *Design Automation Conference, 2001. Proceedings*, pp. 220 – 225.

[13] The Climate Group [2008]. Smart 2020: Enabling the low carbon economy in the information age, *Technical report*, GeSi.

[14] Tiwari, V., Malik, S. & Wolfe, A. [1994]. Power analysis of embedded software: a first step towards software power minimization, *Very Large Scale Integration (VLSI) Systems, IEEE Transactions on* 2(4): 437 –445.

[15] Vetro', A., Ardito, L., Morisio, M. & Procaccianti, G. [2011]. Monitoring IT power consumption in a research center: 7 facts, *in* IEEE CPS (ed.), *Proceedings of Energy 2011*.

[16] Webber, C. A., Roberson, J. A., McWhinney, M. C., Brown, R. E., Pinckard, M. J. & Busch, J. F. [2006]. After-hours power status of office equipment in the usa, *Energy* 31(14): 2823 – 2838.

[17] Zhang, L., Tiwana, B., Qian, Z., Wang, Z., Dick, R. P., Mao, Z. M. & Yang, L. [2010]. Accurate online power estimation and automatic battery behavior based power model generation for smartphones, *Proceedings of the eighth IEEE/ACM/IFIP international conference on Hardware/software codesign and system synthesis*, CODES/ISSS '10, ACM, New York, NY, USA, pp. 105–114.

Chapter 18

Energy Efficiency in Cooperative Wireless Sensor Networks

Glauber Brante, Marcos Tomio Kakitani and Richard Demo Souza

Additional information is available at the end of the chapter

http://dx.doi.org/10.5772/47780

1. Introduction

Wireless Sensor Networks (WSNs) are composed by a large number of sensor nodes, which are usually small in size and are deployed inside or close to some phenomenon of interest. Moreover, since usually there is no need for regular or predefined deployment, the sensors can be placed over irregular or inaccessible areas. Therefore, according to [2], it is expected that the sensors possess self-organizing capabilities. Such attributes provide to the WSNs a large number of applications such as medical, military, and commercial. For instance, in medical applications wireless sensor networks can be used in patient monitoring systems. In the military, the fast set-up and self-organizing characteristics of the sensors make WSNs interesting for communication applications, security, monitoring, and terrain recognition. Commercial applications can include inventory management, product quality control and monitoring disaster areas.

The nodes in a WSN are typically equipped with limited power sources such as batteries, whose recharge or replacement may not always be possible or of economical interest. Moreover, batteries capacity presented a modest increase in the last decades when compared to the gains obtained in computational capacity and wireless throughput, which motivates the study of the energy efficiency of these devices. The wireless throughput has grown by roughly one million times and the computational capacity has had an increase of 40 million times since 1957, while the average nominal battery capacity has increased only 3.5 percent per year over the last two decades, as shown in [11, 24]. Thus, according to [9], due to these power source limitations, the overall energy consumption and energy efficiency have great importance and are major concerns in the design and analysis of wireless sensor networks.

Another challenge faced by WSNs is the wireless environment itself. The wireless channel is a difficult and unpredictable communication medium. A signal transmitted through wireless is subjected to many factors, such as noise, random fluctuations in time (usually referred to as fading), attenuation due to moving objects, etc. Therefore, a reliable system design comes at the expense of a significant amount of power, required to transmit a block of data from

INTECH
open science | open minds

© 2012 The Author(s). Licensee InTech. This chapter is distributed under the terms of the Creative Commons Attribution License http://creativecommons.org/licenses/by/3.0), which permits unrestricted use, distribution, and reproduction in any medium, provided the original work is properly cited.

the sensors to the sink. According to [14], one of the most promising techniques to overcome such limitations of the wireless medium is to exploit diversity techniques. Time diversity, frequency diversity and spatial diversity are among the most common strategies used in wireless transmissions. For instance, the use of error correction codes is an example of time diversity, introducing a level of correlation among the symbols to be transmitted. Orthogonal Frequency-Division Multiplexing (OFDM) and spread spectrum techniques are examples of frequency diversity. Recently, spatial diversity, through the use of multiple antennas, has been on the focus of many works, as for instance in [3, 12, 30].

However, for the spatial diversity gains to be obtained in practice it is necessary that the antennas are sufficiently spaced at the transmitter and receiver. Small-sized devices such as sensor nodes do not dispose of sufficient area to place multiple antennas appropriately spaced. Another practical way to obtain spatial diversity is through the use of cooperative communications. Cooperative communications are based on the channel model introduced by [31], which was originally composed by three nodes: one source of information, the destination of the communication, and a relay node. The relay node is responsible for helping the communication between the source and the destination, so that it may be possible to establish a more reliable communication, or to reduce the transmission power. Thus, exploiting the broadcast nature of the wireless channel, the relay may be able to overhear the transmission from the source in a first time instant, and then retransmit this information to the destination in a second time instant.

At the time that it was proposed by Van der Meulen in 1971, the relay channel was of more theoretical interest. However, due to technological advances of wireless communications in the last decades, a renewed interest in cooperative communications appeared motivated by the recent works of [20, 26], showing that cooperation is a strong practical candidate to improve robustness and help in reducing the energy consumption of wireless networks.

Motivated by these recent advances in the cooperative communications field, and by the importance of reducing the energy consumption of wireless devices, the energy efficiency of some transmission schemes for wireless sensor networks is analyzed in this chapter. The goal is to outline the best strategy in terms of energy efficiency given the characteristics of a network. Such characteristics can include, for instance, the amount of error allowed at the receiver, or the maximum delay in the communication between two nodes. Moreover, in order to approximate the theoretical results to a practical sensors network scenario, the following analysis seeks to model the network in a realistic way. As the nodes in a WSN are often at close distances to each other, the severity of the wireless fading must be taken into account, since when nodes are closer, a better wireless channel is expected. Moreover, another factor that cannot be ignored in the energy efficiency analysis of WSNs is the consumption of the internal circuitry of the devices. As shown in [10, 25, 28], in networks where the nodes are distant, the transmit power dominates over the consumption of the RF circuits. However, when the nodes are closer, the circuitry consumption becomes relevant.

In the following, Section 2 reviews some important concepts of cooperative communications. This section starts by showing the gains that can be obtained with spatial diversity, followed by the introduction of the relay channel as a practical way to obtain spatial diversity. Moreover, some cooperative protocols are presented at the end of the section. In the sequence, Section 3 shows some applications of such concepts to wireless sensor networks. Various WSN scenarios are analyzed in terms of the energy consumption of the devices. The section starts with simple examples considering only three nodes, as in the case of the classical relay channel in [31], and is further generalized to multiple nodes randomly distributed over a field. Then, Section 4 presents the final comments of this chapter.

2. Cooperative communications

The modern cooperative communications concept is based on the classical relay channel model in [31], which allows different nodes to share resources in order to achieve a more reliable transmission. Relay channel is usually referred to systems where the relay is a dedicated device, without information of its own to transmit. On the other hand, the term cooperative communication is used when the relay is another user or node in the same network, which also has information to transmit. The main objective of this approach is to achieve spatial diversity gains, which usually are obtained by adding more antennas to the nodes. However, in cooperative communications spatial diversity is obtained through the shared use of the source and the relay antennas. Thus, even if each device has only one antenna, spatial diversity can be obtained, what in this case is usually referred to as cooperative diversity.

2.1. Spatial diversity

The spatial diversity exploits the use of multiple antennas at the transmitter and/or receiver in order to create independent paths for transmitting the same information, allowing the system to:

- increase the transmission rate without increasing the bandwidth, as in [12, 33];
- improve the link quality, and therefore decrease the transmission error probability as in [3, 30];
- combine the two previous alternatives in a hybrid option, as in [13, 39].

When only the receiver is equipped with multiple antennas, as illustrated in Figure 1(a), diversity combining techniques such as Maximal Ratio Combining (MRC) can be applied. The case that only the transmitter has multiple antennas is illustrated in Figure 1(b). In this scenario, one of the most effective techniques is the Alamouti scheme, which establishes a space-time coding for the symbols to be transmitted. And when there is a combination of multiple antennas at both the transmitter and the receiver the system is known as MIMO (Multiple-Input Multiple-Output), as shown in Figure 1(c).

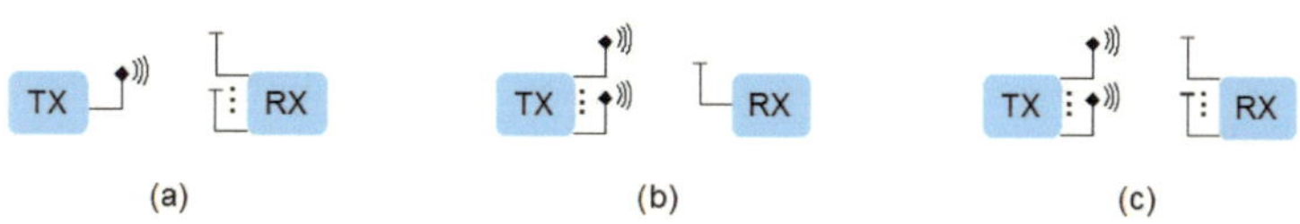

Figure 1. Spatial diversity through the use of multiple antennas: (a) at the receiver; (b) at the transmitter; (c) at both (MIMO system).

A comparison of different spatial diversity techniques is shown in Figure 2. The figure compares the bit error rate (BER) as a function of the signal-to-noise ratio (SNR), which is defined as E_b/N_0, where E_b is the energy per information bit, and N_0 is the power spectral density of the noise. In this particular example, the wireless channels are independent and modeled by a Rayleigh distribution, and the system operates with a Binary Phase Shift Keying (BPSK) modulation. It can be noted from the figure that all strategies using multiple antennas outperform the case when the transmitter and the receiver have only one antenna each (1TX and 1RX). The main conclusion that can be obtained from this figure is that the spatial diversity significantly increases the system performance. Moreover, the gain increases with the number of available antennas. It is important to note that the diversity gain can be observed by the change in the slope of the curves in relation to the case where there is only one transmitting antenna and one receiving antenna.

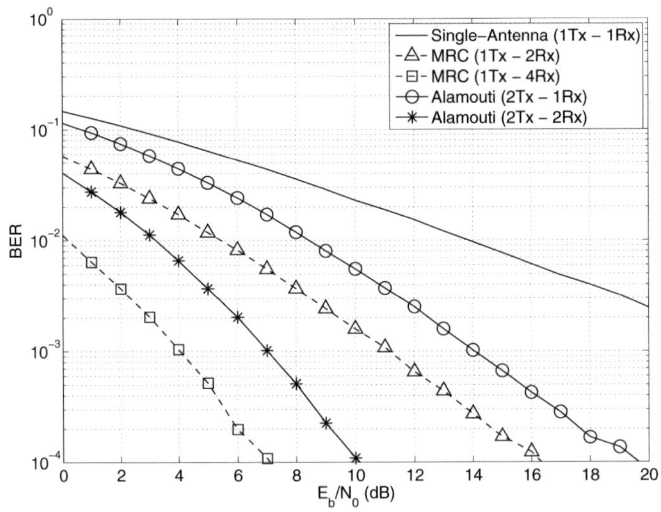

Figure 2. Bit error rate performance of different spatial diversity techniques.

However, it is necessary that the antennas are sufficiently spaced at the transmitter and at the receiver to obtain spatial diversity gain in practical applications. But small size devices, such as mobile phones or nodes in a WSN, may not dispose of enough area for the placement of multiple antennas properly spaced. Moreover, it is not expected from the user to accept a considerable increase in his device size to obtain a better performance. It is from this scenario that the cooperative communication emerged, aiming to obtain spatial diversity gains through sharing the resources of different devices that use the same wireless channel, as shown in [20, 26].

2.2. Relay channel

The relay channel, as proposed by [31], consists of three nodes: the source (S), the relay (R), and the destination (D), as shown in Figure 3. The function of the relay is to assist the source, forwarding the information to the destination using a different path, thus proving spatial diversity. The relay can either employ the same codebook as the source, acting as a repeater, or use a different codebook, also providing code diversity in this case. Moreover, this assisting node can either be a dedicated relay that has no specific information to transmit, or a system user. The term cooperative communication is usually employed when the relay is a user of the system, which also has information to transmit to the destination. Thus, the source and the relay act as partners to transmit the information from both, so the nodes act both as source, or as relay.

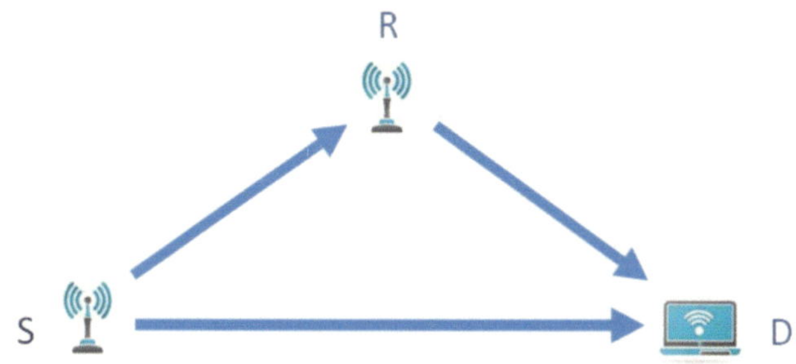

Figure 3. Source (S), relay (R) and destination (D) in a cooperative scenario.

The relay channel can be classified according to some of its characteristics, as the way that the bidirectional communication is performed, the multiple access method employed, the availability or not of a feedback channel, and the deployment scheme, as follows:

1. Based on the way that the bidirectional communication is made, it can be classified as:
   - *Full-Duplex*:
     The nodes are able to transmit and receive simultaneously. From a theoretical point of view, it is the model that provides the greatest channel capacity. However, there may be a large power difference between the transmitted signal and the received signal, up to a hundred dBs, making the isolation of these signals on the transceiver a very difficult task. Thus, its practical implementation is still considered a great challenge.
   - *Half-Duplex*:
     The transmission and the reception by each node are multiplexed in time, i.e., the nodes are not able to send and receive simultaneously. Although this model offers a smaller capacity if compared to full-duplex systems, it has a good trade-off between performance and complexity, being widely used in wireless scenarios.
2. Based on the multiple access method employed:
   - *Superposition*:
     The source and the relay broadcast their information at the same time and at the same frequency, i.e., in a superposed way. Thus, a larger complexity is required by the destination, since it must be able to decode the information from each user. The main advantage of this model is that there is no spectral efficiency loss if compared to the direct (non-cooperative) transmission.
   - *Orthogonal*:
     The source and the relay transmissions are multiplexed in time, in frequency, or in code (TDMA, FDMA, or CDMA). Taking the TDMA system as an example, the communication in the relay channel is performed in two different time slots. In the first time slot, the source broadcasts its information, and in the second time slot the relay forwards the information from the source to the destination. Thus, if compared to the direct transmission, there is loss of spectral efficiency due to this two steps communication process.
3. Based on the availability of a feedback channel:

- *No Feedback Channel*:
  The communication is only performed in the source-destination and relay-destination directions. The system provides no feedback from the destination.
- *With a Feedback Channel*:
  If there is a communication channel in both directions, the destination can exchange information with the source and the relay nodes to optimize the communication. The relay channel model that makes use of a return channel is able to apply classical retransmission techniques, and is denoted in [38] as a generalization of Automatic Repeat reQuest (ARQ) protocols.

4. Based on the deployment scheme and the mobility:
   - *Ad-Hoc Relays*:
     In general, the ad-hoc relays consist of other users that have information to send, share the same wireless environment, and are willing to cooperate. These relays usually face similar channel conditions to those faced by the source in relation to the destination.
   - *Infra-Structured Relays*:
     They are relays that operate in a dedicated manner, i.e., devices that are fixed and do not have information of their own to transmit. Furthermore, it can be assumed that the relay will be under better channel conditions in relation to the destination, since these devices are installed by the service provider with the appropriate positioning and antennas height as to optimize the communication.

2.3. Cooperative communication protocols

Some of the most known cooperative protocols for the wireless channel were presented in [20], as the Amplify-and-Forward (AF) and Decode-and-Forward (DF) protocols, while some other options can be found in [19, 21, 23]. The main idea of the AF protocol is that the relay only amplifies the received signal from the source, in order to compensate for the effects of the source-relay channel, and then the information is forwarded to the destination, as illustrated by Figure 4(a). In the DF protocol, illustrated by Figure 4(b), the relay tries to retrieve the information sent by the source, converts it to information bits, re-encodes them, modulates, and then forwards the message to the destination.

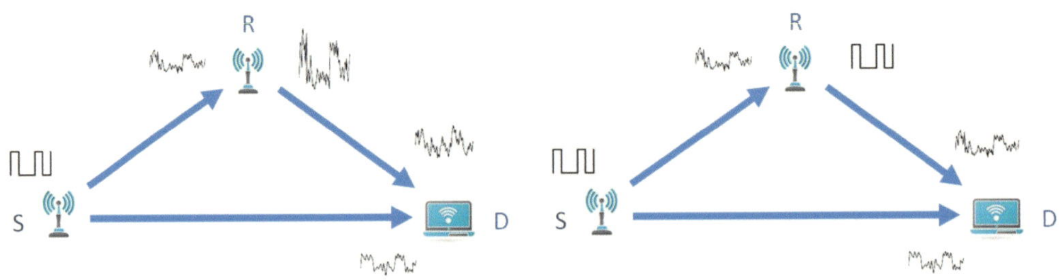

(a) Amplify-and-Forward. (b) Decode-and-Forward.

Figure 4. Cooperative protocols.

The DF protocol has at least three major variants: Fixed DF (FDF), Selective DF (SDF), and Incremental DF (IDF). In the FDF protocol the relay always acts in the communication, i.e., the message sent by the source is always forwarded to the destination, regardless the fact that the decoding at the relay was successfully performed or not. The disadvantage of this protocol is that the error propagation can be very large, reducing the system performance. The SDF protocol, on the other hand, establishes a condition for the relay to act. The objective is to allow the relay to detect whether the estimated message corresponds to the original message from the source or not. Thus, the information is only forwarded to the destination by the relay if the estimation is error free. Finally, the IDF protocol exploits the feedback channel from the the destination, so that the relay acts only if requested.

Another cooperative protocol that is less employed, but not less relevant, is the Compress-and-Forward (CF). In the CF protocol, the relay quantizes and compresses the message from the source, and then forwards it to the destination. Although [19] shows that the CF protocol can have better performance than the AF and DF protocols in some situations, CF is less employed due to the difficulty of implementation in practical systems.

2.4. Coded cooperation

With the use of error-correcting codes by the source, the information symbols vector, which represents a non-deterministic and uncorrelated sequence, is converted into a codeword vector, which adds redundancy to the symbols of the original word, introducing a correlation degree in the symbols to be sent. Thus, the decoding process in the receiver may be able to recover symbols that were incorrectly received due to the channel attenuation effect and noise at the receiver. Different error correction methods have been proposed over the years, among which the block codes, the convolutional codes, the turbo codes, and the Low-Density Parity-Check (LDPC) codes can be cited. Among these codes, the turbo and the LDPC codes are the ones that have the performance closer to the theoretical limit predicted by [27], and greatly outperform the convolutional and the block codes, although the former require in general more complex decoding algorithms.

The coded cooperation can be classified into two main types: Repetition Coding (RC) and Parallel Coding (PC). In repetition coding the same encoder is used in both the source and the relay, sending the same information and parity symbols, as shown in Figure 5(a). This technique has the advantage of simplicity in decoding, since the receiver uses the same circuit to decode the received words from the source and from the relay. In the parallel coding, the source and the relay encoders are specially designed to send different parities, increasing the coding robustness and, therefore, the complexity of the encoding and the decoding designs. Figure 5(b) illustrates this strategy.

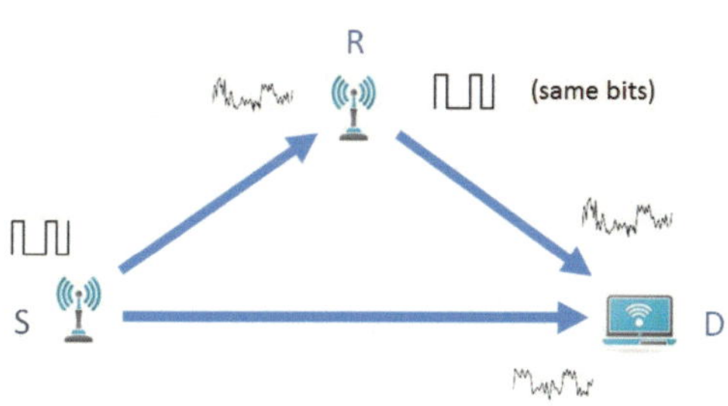

(a) Repetition Coding.

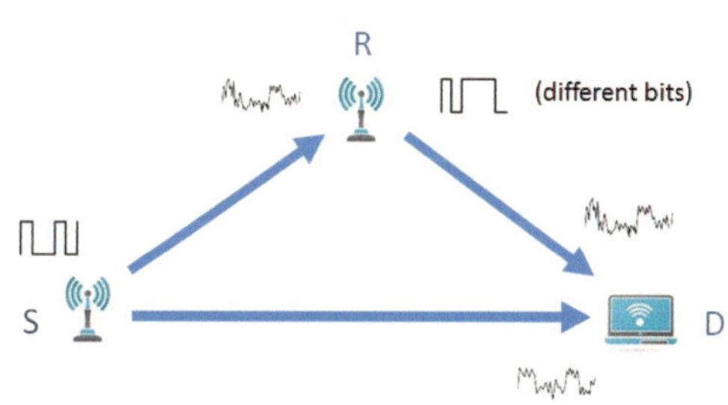

(b) Parallel Coding.

Figure 5. Coded cooperation.

Several schemes using repetition coding can be found in the literature, as in [29, 36]. Parallel coding had as pioneers the authors of [37], followed by [15, 36]. The results from these works show that the parallel coding outperforms the repetition coding in terms of error probability, specially when irregular LDPC or turbo codes are used. However, these codes need to be specially designed for the relay channel, substantially increasing the complexity. Another important factor is that the decoding process at the destination also becomes more complex.

3. Application of cooperative protocols to wireless sensor networks

This section focuses on the application of some cooperative communication concepts to WSNs and its impact on the energy efficiency of the system. Therefore, some non-cooperative and cooperative transmission schemes are analyzed in terms of their total energy consumption. Moreover, aiming at practical telecommunication scenarios, several characteristics of a real wireless network are taken into account for a more accurate performance measure.

In the sequence, some important concepts are presented in Section 3.1, and the transmission techniques are discussed in Sections 3.2 and 3.3. Three relevant nodes in a WSN are considered: one source node S, one destination node D, and one relay node R, where the source tries to communicate with the destination, and the relay is at an intermediate position. Some numerical examples are given in Section 3.4. Moreover, since practical scenarios may be composed of many sensors, the extension of this simple analysis to multiple nodes is discussed in Section 3.5 and further generalized in Section 3.6. Finally, a comparison among different cooperative protocols is given in Section 3.7.

3.1. Concepts

Typically, the data collected by each sensor in a WSN is transmitted to a fusion center (FC), where estimates are formed based on the aggregated data from the ensemble of the sensors, and where the end user can access such data. Depending on the application of the WSN, the data transmission from the sensor to the FC can be made by radio, infrared, optical, etc. In the case of a wireless communication using radio frequency (RF) circuits, according to [14], a transmission from a node i to a node j can be written as:

$$\mathbf{y}_j = \sqrt{P_i\,\gamma_{ij}}\,h_{ij}\,\mathbf{x} + \mathbf{n}_{ij}, \tag{1}$$

where $\mathbf{y}_j$ represents the signal received at the node j, while $\mathbf{x}$ is the original message transmitted by node i. From this equation it is possible to notice that the signal received at j depends on the power used by node i to transmit, denoted by P_i, the path-loss γ_{ij} between i and j, which will be detailed in the sequence, and on the characteristics of the wireless medium h_{ij}. In addition, the received signal is corrupted by the communication noise $\mathbf{n}_{ij}$, which is typically modeled as additive white Gaussian noise (AWGN), with variance $N_0/2$ per dimension, where N_0 is the thermal noise power spectral density per hertz.

The path-loss between i and j is a factor that expresses the attenuation of the signal propagating through the wireless channel. The path-loss depends on the distance between the transmitter and the receiver and is given by:

$$\gamma_{ij} = \frac{G\lambda^2}{(4\pi)^2 d_{ij}^{\alpha} M_l N_f}, \tag{2}$$

where G includes the gain of the antennas of the transmitter and receiver, λ corresponds to the wavelength of the signal being transmitted, M_l represents the link margin, and N_f is the noise figure at the receiver. Note that all these terms are constant, and the unique term that varies is d_{ij}, which is the distance between the nodes i and j. Finally, α represents the path-loss exponent, which usually assumes values between 2 and 4 depending on the type of the environment. For instance, $\alpha = 4$ is usually assumed for very dense urban areas. For a more detailed explanation on this subject, the work of [14] is suggested as reference.

To describe the behavior of the wireless medium, several probabilistic models can be used depending on the characteristics of this environment. One of the most adopted models is the Rayleigh distribution, which is mostly suitable for non line-of-sight (NLOS) communications, meaning that the transmitter and the receiver have no direct link to each other, and communication is achieved through signals reflected in different directions. Nevertheless, WSNs often experience at least a portion of LOS between the nodes, specially in dense networks. With the nodes closer to each other, there exists a higher probability of an available direct communication path between two nodes. Another statistical model that can be used in such conditions is the Nakagami-m distribution. In such distribution the severity of the fading can be adjusted by the parameter m. Lower values of m represent a channel with little or no LOS, while higher values of m are representative of some relevant LOS. Experimental results in [34] show that $m = 1$ suits NLOS scenarios (where Nakagami-m is equal to the Rayleigh distribution in this case) and $m = 2$ models a scenario with some LOS.

Then, an important concept in the transmission between i and j is the Signal-to-Noise Ratio (SNR) in the i-j link, defined as:

$$SNR_{ij} = |h_{ij}|^2 \cdot \frac{\gamma_{ij} P_i}{N}, \tag{3}$$

where $N = N_0 \cdot B$ is the noise power spectral density, with B being the system bandwidth.

Finally, another important concept in the wireless transmission is the outage probability. An outage event between i and j occurs when the SNR at the node j falls below a threshold β which allows error-free decoding. The term β can be calculated based on the capacity of the channel given in [14] resulting in $\beta = 2^{\Delta} - 1$, where Δ is the system spectral efficiency. Then, the outage probability depends on the probabilistic model used for the wireless channel, so that in the case of Nakagami-m fading is given by:

$$\mathcal{O}_{ij} = \frac{\Psi\left(m, \frac{mN(2^{\Delta}-1)}{\gamma_{ij} P_i}\right)}{\Gamma(m)}, \tag{4}$$

where $\Psi(.,.)$ is the incomplete gamma function and $\Gamma(.)$ is the complete gamma function. At high SNR, according to [32], the outage probability in (4) can be approximated as:

$$\mathcal{O}_{ij} \simeq \frac{1}{\Gamma(m+1)} \left[\frac{mN(2^{\Delta} - 1)}{\gamma_{ij} P_i}\right]^m. \tag{5}$$

The energy efficiency is analyzed in terms of the total energy consumption per bit of the wireless transmission. In the case of WSNs, the following aspects must be taken into account to compute the energy consumption:

- the power P_i required by node i to transmit the data, which depends on the distance between i and j;
- the additional power wasted by the power amplifier, which is proportional to P_i;
- the power consumed by the RF circuitry of the transmitter and of the receiver;
- the bit rate of the communication.

It is noteworthy that, since the focus is on WSNs, where the nodes are typically equipped with narrow-band single-carrier transceivers, the power consumed by internal signal processing is very small when compared to the circuitry power consumption, and therefore can be neglected in this energy consumption analysis. If broadband multi-carrier transceivers were considered, as for instance in [5], then the power consumption of the baseband processing should also be taken into account. Moreover, in a more general wireless network concept, some control messages may be exchanged by the nodes in order to acknowledge if the packets have been correctly received or not. However, as shown in [8], the impact of these control messages in the overall energy consumption is also negligible since these messages are usually much smaller than the message of interest **x**.

Then, the total energy consumption per bit in a transmission from i to j can be expressed as:

$$E_{ij} = \frac{P_{PA,ij} + P_{TX} + P_{RX}}{R_b}, \tag{6}$$

where $P_{PA,ij} = \frac{\xi}{\eta} P_i$ is the power consumed by the power amplifier, which depends on the peak-to-average ratio ξ of the employed modulation scheme and on the drain efficiency η of the power amplifier, P_{TX} and P_{RX} are the RF circuitry power consumption for transmitting and receiving, respectively, and $R_b = \Delta \cdot B$ corresponds to the bit rate in bits/s. A representative model for the RF circuitry is given in [10], illustrated by Figure 6, which represents the state of the art for current hardware for sensor technologies, as also depicted in [9]. From the figure, the following components can be identified for the transmit circuit: digital-to-analog converter, mixer, transmission filter and frequency synthesizer, with the respective power consumptions given by P_{DAC}, P_{mix}, $P_{fil_{tx}}$ and P_{syn}, totalizing:

$$P_{TX} = P_{DAC} + P_{mix} + P_{fil_{tx}} + P_{syn}. \tag{7}$$

At the receiver side, the following components can be identified: frequency synthesizer, low noise amplifier, mixer, intermediate frequency amplifier, and analog-to-digital converter, with the respective power consumptions of P_{syn}, P_{LNA}, P_{mix}, P_{IFA}, $P_{fil_{rx}}$, and P_{ADC}, totalizing:

$$P_{RX} = P_{syn} + P_{LNA} + P_{mix} + P_{IFA} + P_{fil_{rx}} + P_{ADC}. \tag{8}$$

In the sequence, some wireless transmission schemes are presented. Specifically, a simple single-hop scheme is analyzed in Section 3.2, while cooperative amplify-and-forward is analyzed in Section 3.3.

3.2. Traditional non-cooperative transmission

Single-hop (SH) is the simplest communication scheme involving only two nodes, with a direct transmission from S to D, as illustrated by Figure 7. The total energy consumed per bit of SH can be simply obtained by replacing i and j by S and D in (6):

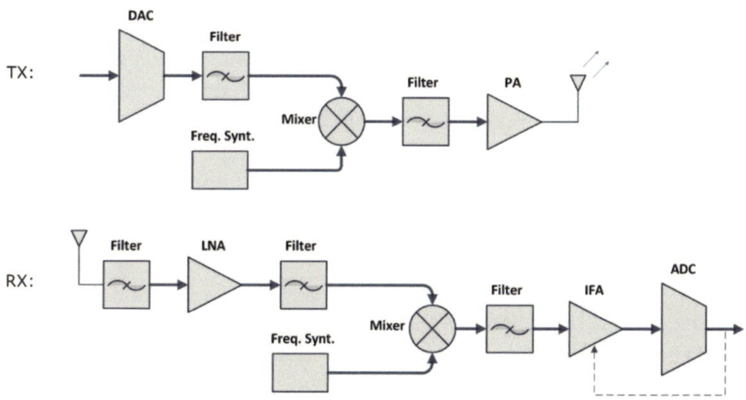

Figure 6. Block diagram for the TX and RX circuits.

$$E_{SH} = \frac{P_{PA,SD} + P_{TX} + P_{RX}}{R_b}. \quad (9)$$

Note that to minimize the energy consumption $P_{PA,SD}$ must be minimized, since P_{TX} and P_{RX}

Figure 7. Single-hop transmission scheme.

are fixed and depend on the current technology. In order to do so, the following methodology can be applied:

1. a target outage probability $\mathcal{O}^\star$ is established at the destination. In other words, $\mathcal{O}^\star$ represents the maximum amount of frame error rate that the system may accept.
2. based on the outage probability of the scheme, which for the case of SH is given by (5) while replacing i and j by S and D, the optimal transmit power can be found as the minimal power that still reaches the outage threshold $\mathcal{O}^\star$.

Such strategy has been widely exploited in the literature, and for some more detailed examples the works of [7, 8, 17, 25] are given as references.

3.3. Cooperative amplify-and-forward transmission

Many cooperative protocols can be applied to WSNs in order to improve the throughput performance, or to reduce the energy consumption of the network. For instance, Selective and Incremental Decode-and-Forward have been analyzed in [7, 8, 17, 25, 28]. Nevertheless, motivated by the simplicity of analog schemes and since no decoding is required at the relay node in this case, Amplify-and-Forward is considered in this section.

In the cooperative transmission, two time slots are reserved for the communications process. In the first time slot the source broadcasts its message, which is received by the destination and also overheard by the relay. Then, in the second time slot, the relay amplifies the received message and forwards it to the destination. At the receiver, a combination between the two received signals is made, which increases the performance. However, note that the cooperative transmission presents an inherent spectral efficiency loss when compared to SH, since the end-to-end throughput is reduced to half due to the communication in two time slots. Such spectral efficiency loss can compromise the performance of some systems. In order to avoid this, the nodes in the AF scheme must transmit with a higher spectral efficiency. Thus, the nodes are assumed to operate with a spectral efficiency two times higher than that in SH. The main concern here is to obtain the same end-to-end throughput in both transmission schemes.

Therefore, since the spectral efficiency is multiplied by two, an outage event occurs when the received SNR falls below a threshold of $\beta' = 2^{2\Delta} - 1$. Then, the outage probability of each i-j link becomes:

$$\mathcal{O}_{ij} \simeq \frac{1}{\Gamma(m+1)} \left[\frac{mN(2^{2\Delta} - 1)}{\gamma_{ij} P_i} \right]^m. \tag{10}$$

In addition, another important aspect in analyzing the energy consumption of AF is the exploitation of a feedback channel. The energy consumption of AF differs if a feedback channel is present or not. For instance, when a feedback is not available, the relay will always retransmit the message from the source in the second time slot, independently on the result of the first transmission. In such case, the total energy consumption of AF can be expressed by:

$$E_{AF} = \frac{P_{PA,S} + P_{TX} + 2P_{RX}}{2R_b} + \frac{P_{PA,RD} + P_{TX} + P_{RX}}{2R_b}, \tag{11}$$

where the first term corresponds to the transmission from S to R and D, and the second term corresponds to the transmission from R to D. Moreover, note that all terms are divided by two, since with a spectral efficiency multiplied by two, each individual transmission is two times faster. It is also noteworthy that, since both the relay and the destination listen to the source transmission in the first time slot, additional energy is consumed by the receive hardware (represented by $2P_{RX}$). In (11), $P_{PA,S}$ and $P_{PA,RD}$ represent the power consumed by the source and by the relay, respectively, and can be obtained based on the outage probability of the cooperative scheme.

On the other hand, Incremental AF (IAF) exploits a feedback channel from the destination so that the relay retransmits only if the destination could not decode the message from the source in the first time slot. This clearly leads to an energy improvement when compared to AF without feedback, since the transmission from the relay may not be always necessary. The energy consumption of IAF can be expressed as:

$$E_{IAF} = \frac{P_{PA,S} + P_{TX} + 2P_{RX}}{2R_b} + p_{SD} \cdot \frac{P_{PA,RD} + P_{TX} + P_{RX}}{2R_b}, \tag{12}$$

where the term p_{SD} represents the probability of incorrect decoding at the destination of the message from the source after the first time slot.

3.4. Numerical examples

In this section we discuss the energy efficiency of a WSN with three nodes and using the AF protocol. First, consider the required transmission power for single-hop and amplify-and-forward schemes, $P^\star_{SH}$ and $P^\star_{AF}$. The rest of the system parameters are given by

Table 1 and the relay is assumed to be at the intermediate position between the source and the destination. Figure 8 shows the required transmit power for each of the transmission schemes for both NLOS and LOS scenarios, where many important conclusions can be obtained. For instance, a significant difference in the required transmit power is observed for NLOS and LOS scenarios. The power consumed by SH is around 22 times smaller in LOS than in NLOS, and 5 times smaller for AF. In addition, the gains of the cooperative transmission becomes evident in Figure 8, where AF consumes up to 11 times less transmit power than the non-cooperative scheme.

Link Margin	$M_l = 40$ dB
Noise Figure	$N_f = 10$ dB
Antenna Gain	$G = 5$ dBi
Carrier Frequency	$f_c = 2.5$ GHz
Noise Power Spectral Density	$N_0 = -174$ dBm
Bandwidth	$B = 10$ KHz
Path-Loss Exponent	$\alpha = 2.5$
Spectral Efficiency	$\Delta = 2$ b/s/Hz
Target Outage Probability	$\mathcal{O}^{\star} = 10^{-3}$

Table 1. System Parameters.

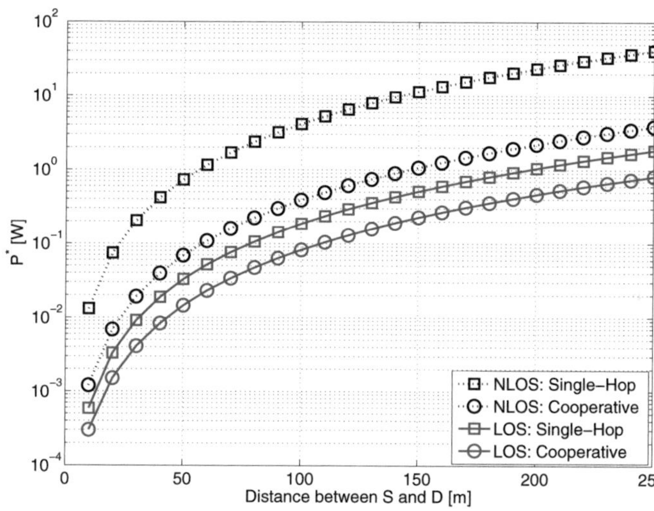

Figure 8. Transmit power for SH and AF schemes in Nakagami-m fading.

A more insightful comparison is given by the total consumed energy per bit for each scheme, E_{SH}, E_{AF} and E_{IAF}. In order to model the circuitry energy consumption, the same parameters as in [10] are used and are listed in Table 2. Figure 9 shows the obtained results where it is interesting to notice that SH is more energy efficient than AF at short transmission ranges. When the distance between the nodes increases, SH is outperformed. This fact is explained by the energy consumption of the circuitry of the additional node involved in AF. When the distance between the nodes is small, the circuitry consumption dominates in the total energy consumption, and therefore SH presents the best performance. On the other hand, while the distance increases, transmit power becomes more relevant and the cooperation outperforms the other schemes. Considering the NLOS scenario of Figure 9(a), AF is more energy efficient than SH when the *S-D* distance is longer than 12 m. On the other hand, observing the LOS scenario of Figure 9(b), it is possible to notice that these values increase considerably, with AF being more energy efficient for distances grater than 52 m, which is four times greater than the distances for the NLOS scenario. Finally, the most interesting conclusion is that IAF outperforms all the other schemes at any transmission range, which shows that a significant performance gain can be obtained with cooperation when a feedback channel is available.

Mixer	$P_{mix} = 30$ mW
TX/RX Filters	$P_{fil_{tx}} = P_{fil_{rx}} = 2.5$ mW
Frequency Synthesizer	$P_{syn} = 50$ mW
Low Noise Amplifier	$P_{LNA} = 20$ mW
Intermediate Frequency Amplifier	$P_{IFA} = 3$ mW
Analog-to-Digital Converter	$P_{ADC} = 6.7$ mW
Digital-to-Analog Converter	$P_{DAC} = 15.4$ mW
Drain Efficiency of the Amplifier	$\eta = 0.35$

Table 2. RF Circuitry Power Consumption.

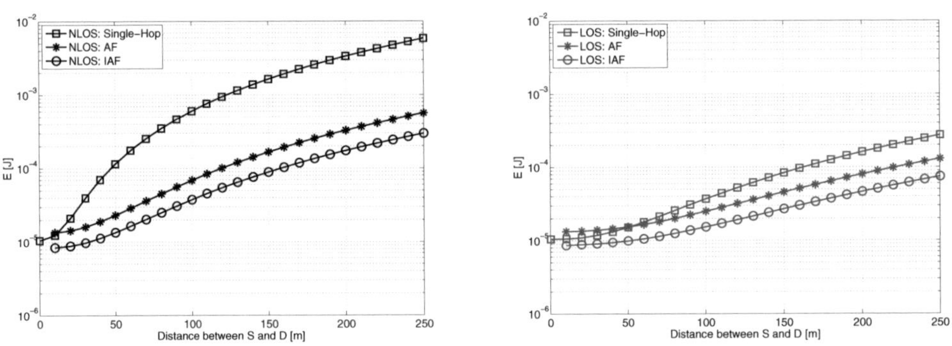

(a) NLOS ($m = 1$). (b) LOS ($m = 2$).

Figure 9. Total energy consumed per bit for SH, AF without feedback and IAF in Nakagami-m fading.

3.5. Multiple relays

As it is usual in WSNs, multiple nodes can be available in the network and the cooperative concept can be extended so that there is not only one, but multiple relays. The performance of the cooperative schemes increases with multiple relays since a larger number of independent paths will be available and, consequently, the probability that one of these relays is in good conditions increases. On the other hand, the complexity of the cooperative protocols increases since some criterion for choosing which relay will cooperate must be defined.

Two different approaches for relay selection are discussed in [4]. These two algorithms are named *reactive* and *proactive* relay selection, illustrated by Figure 10. In the reactive algorithm of Figure 10(a) the relay is chosen after the source transmission, and all relays have to listen to the source, what may increase the network energy consumption. On the other hand, the proactive algorithm of Figure 10(b) selects the relay before the source transmission, such that only the *a priori* selected relay has to listen to the source. In practice, the reactive algorithm is easier to implement, since it is distributed and no global information about the channel quality of the other nodes is required. Specific details about the implementation of each algorithm will not be discussed in this chapter, and the works of [1, 6, 22] are suggested to the interested reader.

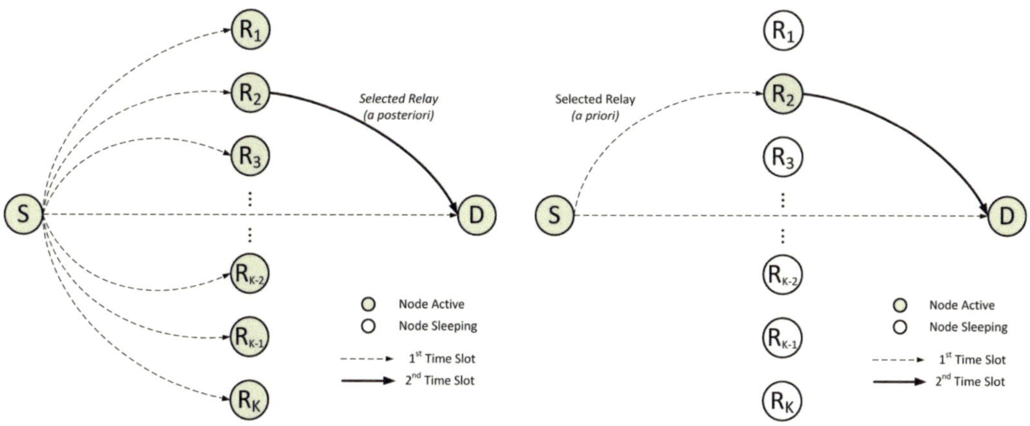

(a) Reactive relay selection. (b) Proactive relay selection.

Figure 10. Relay selection algorithms.

As an example, consider a network composed by one source S, one destination D, and K relay nodes denoted by R_k, $1 \leq k \leq K$, where all relays lie at the intermediate position between S and D. This simplification is only to allow the mathematical tractability of the problem, which may bring important insights into the energy consumption of relay selection algorithms. The relays operate under the IAF protocol, exploiting the presence of the feedback channel from the destination.

In terms of energy consumption, the total consumption depends on the employed relay selection algorithm. In the case of the proactive algorithm, only the *a priori* selected relay and the destination overhear the transmission from the source in the first time slot. Thus:

$$E_{IAF}^{(pro)} = \frac{P_{PA,S} + P_{TX} + 2P_{RX}}{2R_b} + p_{SD} \cdot \frac{P_{PA,R_kD} + P_{TX} + P_{RX}}{2R_b}. \tag{13}$$

Note that this equation is very similar to (12) in terms of energy consumption, however, the power required to transmit the data decreases with a higher number of relays, since the outage probability decreases with K.

On the other hand, in the reactive algorithm, besides the source, the destination and the selected relay, all other $K-1$ relays overhear the transmission from the source in the first time slot. Therefore:

$$E_{IAF}^{(re)} = \frac{P_{PA,S} + P_{TX} + (K+1)P_{RX}}{2R_b} + p_{SD} \cdot \frac{P_{PA,R_kD} + P_{TX} + P_{RX}}{2R_b}, \tag{14}$$

which clearly indicates a higher energy consumption with respect to the proactive algorithm.

The optimal transmit power $P_{PA}^{\star}$ for $K \in \{0,1,2,4,8\}$ is shown in Figure 11 for a NLOS scenario. From the figure it can be observed that IAF requires less transmit power than SH and that the transmit power decreases with K. As the transmit power depends only on the outage probability, reactive and proactive algorithms lead to the same results. In the LOS scenario, similar conclusions are obtained.

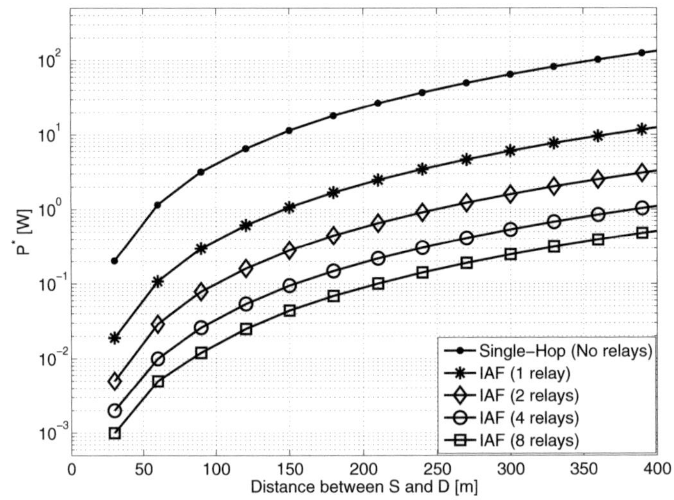

Figure 11. Optimal transmit power required in NLOS for multi-relays WSNs.

The total energy consumption is presented in Figure 12, also for NLOS. Regarding the reactive algorithm, it can be observed from Figure 12(a) that IAF with $K = 2$ relays is more energy efficient than with $K = 1$ when $d_{SD} \geq 49$ m, IAF with $K = 4$ outperforms $K = 1$ when $d_{SD} \geq 72$ m, and IAF with $K = 8$ outperforms $K = 1$ when $d_{SD} \geq 100$ m. It can also be observed that the energy savings in long transmission ranges do not increase linearly with K. For instance, reactive IAF with $K = 8$ is more energy efficient than reactive IAF with $K = 4$ only when $d_{SD} \geq 264$ m, and by a very small margin. In the LOS scenario, the energy savings are even less significant.

It can be also seen from Figure 12(b) that the proactive algorithm takes a better advantage from a larger number of relays. In this case, IAF with $K = 2$ relays is always more energy efficient than with $K = 1$, $K = 4$ outperforms $K = 1$ when $d_{SD} \geq 38$ m, and $K = 8$ outperforms $K = 1$ when $d_{SD} \geq 53$ m. Moreover, reactive IAF with $K = 8$ is more energy efficient than reactive IAF with $K = 4$ already with $d_{SD} \geq 150$ m, which is a considerable decrease in the energy consumption when compared to the reactive algorithm. This is due to the *a priori* relay selection, since all other relays remain in sleep mode during the source transmission. However, this algorithm depends on a fixed (or reduced mobility) topology, where the channel is constant for a long period, allowing for a pre-selection strategy. In addition, while the transmission range increases, the energy consumption of reactive IAF approaches that of proactive IAF, since the transmit power dominates over the RF circuitry energy consumption.

A more detailed comparison, which also considers the impact of a nonlinear discharge model that the batteries of the sensors may have, is also given in [6].

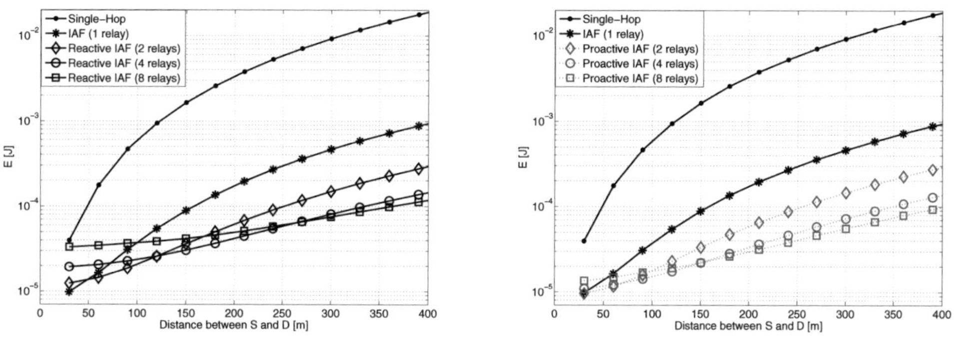

(a) Reactive algorithm. (b) Proactive algorithm.

Figure 12. Total consumed energy per bit for multi-relays WSNs in NLOS.

3.6. Generalized wireless sensor networks

Simpler scenarios composed by three or more nodes distributed over a line segment are very useful due to the mathematical tractability of the problem. Nevertheless, the sensors in a WSN are usually distributed over a certain region in order to monitor some phenomenon of interest, which characterizes a two-dimensional topology with a random deployment of the sensors. Therefore, one should question if the results obtained over a line segment are still valid for a more general and realistic network scenario.

To investigate a larger scenario, consider that a number of sensors are randomly distributed over a certain area of interest. All the sensor nodes can act as source by gathering information from the environment and sending it to the destination node, which is positioned at the center[1]. Moreover, any sensor node can be selected to operate as relay. Since multiple nodes are available, relay selection is employed. Here, two strategies are compared: proactive relay selection, and random relay selection. Random relay selection is the simplest selection algorithm, as analyzed in [35], and the choice for the proactive algorithm is due to its good performance in terms of energy efficiency, as shown in Section 3.5.

A total of 121 sensor nodes are randomly deployed over a square area and the energy efficiency of SH and AF are analyzed for different distances between the nodes. Figure 13 plots the most energy efficient scheme as a function of the distance between the nodes in the square area. For instance, a result of 0.8 means that such a scheme is more energy efficient for 80% of the nodes in that scenario. Random relay selection is presented in Figure 13(a), and proactive relay selection in Figure 13(b). Moreover, a LOS scenario is considered. Note that at shorter distances, due to the circuitry consumption provided by the additional transmission of AF, SH is the most energy efficient transmission scheme. However, as transmit power increases with distance, AF presents better efficiency and outperforms SH when the distance between the nodes increases.

When random relay selection of Figure 13(a) is compared to proactive relay selection of Figure 13(b), it is possible to notice that the advantage of AF increases when the best relay is able to be selected, as the percentage of nodes operating with AF is higher when proactive relays

[1] Note that assuming D at the center is a general case. For instance, considering D at a corner can be seen as a particular case, by dividing the area it into quadrants.

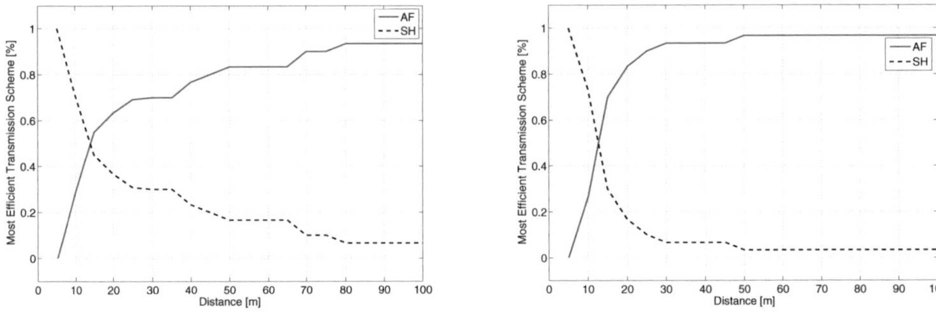

(a) Random relay selection.

(b) Proactive relay selection.

Figure 13. The most efficient transmission scheme, considering SH and AF, for different distances between nodes in LOS.

selection is employed. In addition, if a return channel is available, IAF is the most energy efficient method for all distances. Under NLOS, AF is the most energy efficient scheme for all cases, regardless of the availability of a return channel or not. The energy consumption in LOS is 3.5 times lower than in NLOS, and the availability of a feedback channel also presents a significant impact on the energy consumption, as IAF consumes up to six times less than AF without feedback. This results corroborate with the findings of Section 3.4, showing that the mathematical predictions obtained for simpler scenarios of a few nodes are representative of more general cases of wireless sensor networks.

3.7. Other cooperative protocols

The energy efficiency analysis carried out so far assumes that the cooperation may occur using the Amplify-and-Forward protocol, whose use is motivated by its low complexity. Nevertheless, other cooperative protocols exist and could be also applied to WSNs, as described in Section 2.3. For instance, the Decode-and-Forward protocol is also of practical interest. In DF, the relay no longer operates in the analog mode, but the message received from the source is decoded by the relay, re-encoded, and then forwarded to the destination.

From a practical point of view, AF is very interesting due to its simplicity and DF may be more robust to transmission errors. Moreover, one important characteristic of DF is that different channel codes can be used at the source and relay, which is known as parallel coding (PC), in opposition to repetition coding (RC) when source and relay use the same channel code. The difference among these protocols in terms of energy consumption comes from the difference in the outage probability of each scheme. For the the derivation of the outage probability of these three schemes, the work of [16, 18] are recommended references.

The goal of the following analysis is to compare the energy efficiency of each one of these cooperative techniques: AF and DF. Since in WSNs the nodes are usually assumed to have the same hardware configurations, only DF with repetition coding is considered[2].

---

[2] DF with parallel coding requires different encoders at source and relay, such that the relay forwards the message from the source with a different codebook, increasing the error correction capability of the network. However, the hardware complexity increases with this protocol. Nevertheless, for a more detailed comparison including DF with PC, the work of [16] is suggested for the interested reader.

As the outage probability of these schemes behaves differently according to the relative position of the relay with respect to the source, the energy consumption of each scheme is analyzed with respect to the relative position of the relay, which is defined as $d_r = d_{SR}/d_{SD}$. Figure 14 illustrates the energy efficiency of these schemes when d_r is between 0.1 and 0.9, with the distance between the source and the destination being of $d_{SD} = 50$ m. Note that when the relay is close to the source (when d_r is small) both AF and DF present similar performance in terms of energy consumption. On the other hand, when R is not so close to the source (when $d_r > 0.4$ according to Figure 14), the AF method outperforms DF. The SH consumed energy is also shown as comparison, but note that it is constant since the SH performance is not a function of d_r. These results show that AF can be a very good option for WSNs.

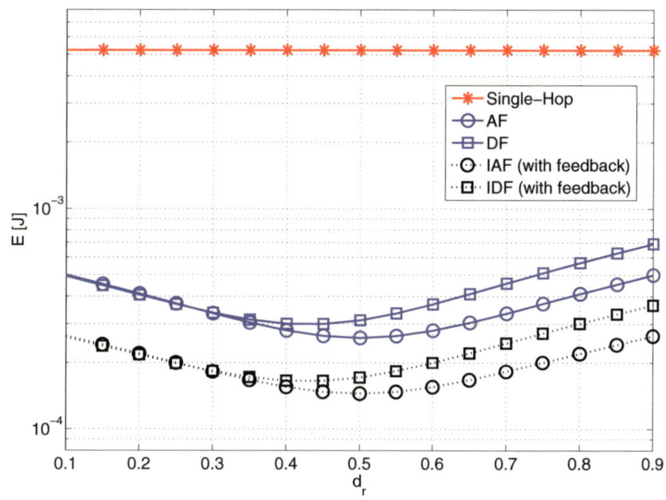

Figure 14. Total consumed energy per bit of AF and DF with RC for $d_{SD} = 50$ m.

4. Final comments

Although sensor nodes have existed for decades, the modern development of tiny sensor nodes is due to recent advances in hardware miniaturization, making possible to produce silicon footprints with more complex and lower powered microcontrollers. As a consequence, a large number of modern applications makes use of such devices. However, many challenges are still to be faced in the development of these systems. Nowadays, one major concern in the sensors industry is to develop low cost sensors with low energy consumption.

Coupled with the hardware development for WSNs, many advances have been reached in the telecommunications industry in the last years. Since WSNs are composed by many nodes, usually close to each other, the broadcast nature of the wireless medium can be exploited by the use of cooperative techniques. As shown in Section 2.1, spatial diversity is a promising technique to improve system performance, and cooperative communications, discussed in Section 2.2, is a practical way to achieve spatial diversity with small-sized devices, where the use of multiple antennas may not be possible. The use of cooperative protocols, as those presented in Section 2.3, proved to be effective in terms of reducing the power required by each node.

When the energy efficiency of WSNs is analyzed, some important characteristics of the network must be taken into account in order to obtain a fair comparison. For instance, as shown in Section 3, the maximum amount of error tolerated by the receiver and the characteristics of the wireless environment can have significant impact on the conclusions, and therefore must be carefully taken into account. Moreover, since WSNs usually deal with short range communications, the energy consumption of the transmit and receive circuits must also be taken into account. As shown in the examples of Section 3.4, in networks where there is no feedback from the destination, simpler transmission schemes such as single-hop are more energy efficient at short transmission ranges, since less nodes are involved in the communication (as well as less circuitry energy is being consumed). On the other hand, cooperation is able to save an important amount of energy when the transmission range increases. Nevertheless, if a feedback channel is available, the advantage of using cooperative schemes becomes evident at any transmission range.

The basic idea of energy efficiency in cooperative WSNs is presented in Section 3.3, while Section 3.5 extends such concept to a more interesting scenario. When multiple nodes are available, multiple sensors are potential candidates to act as a relay, and therefore some relay selection criterion can be established. However, from an energy efficiency point of view, to select one relay may be a challenging task. An important amount of energy is spent when multiple nodes are involved in a relay selection process, since these nodes consume energy if they must overhear the source transmission. Two relay selection algorithms are discussed in Section 3.5, however, energy efficient relay selection schemes are still a quite open research area.

Finally, in order to validate the results of Sections 3.4 and 3.5, the study is further generalized in Section 3.6. In this section, a WSN composed of multiple nodes is considered, with the nodes randomly distributed over a finite area. The results match with the predictions obtained over simplified networks, confirming the relevance of the analysis. In addition, Section 3.7 compares the performance of cooperation with analog relaying, by employing the amplify-and-forward protocol, to that of digital relaying, by employing the decode-and-forward protocol with repetition coding. The comparison shows a similar performance of both protocols when the relay is very close to the source, with a performance advantage of AF when the relay moves towards the destination.

Author details

Glauber Brante, Marcos Tomio Kakitani and Richard Demo Souza
Federal University of Technology - Paraná (UTFPR), Curitiba, Brazil

5. References

[1] Abdulhadi, S., Jaseemuddin, M. & Anpalagan, A. [2010]. A survey of distributed relay selection schemes in cooperative wireless ad hoc networks, *Springer Wireless Personal Communications* pp. 1–19.

[2] Akyildiz, I. F., Su, W., Sankarasubramaniam, Y. & Cayirci, E. [2002]. Wireless sensor networks: A survey, *Computer Networks* 38(4): 393 – 422.

[3] Alamouti, S. [1998]. A simple transmit diversity technique for wireless communications, *IEEE Journal on Selected Areas in Communications* 16(8): 1451 – 1458.

[4] Bletsas, A., Shin, H. & Win, M. [2007]. Cooperative communications with outage-optimal opportunistic relaying, *IEEE Transactions on Wireless Communications* 6(9): 3450 –3460.

[5] Bougard, B., Lenoir, G., Dejonghe, A., Van der Perre, L., Catthoor, F. & Dehaene, W. [2007]. Smart MIMO: an energy-aware adaptive MIMO-OFDM radio link control for next-generation wireless local area networks, *EURASIP Journal on Wireless Communications and Networking* 2007: 1–12.

[6] Brante, G., Souza, R. D. & Vandendorpe, L. [2012]. Battery-aware energy efficiency of incremental decode-and-forward with relay selection, *IEEE Wireless Communications and Networking Conference (WCNC)*, pp. 1–5.

[7] Brante, G., Kakitani, M. & Souza, R. D. [2011a]. On the energy efficiency of some cooperative and non-cooperative transmission schemes in wsns, *45th Annual Conference on Information Sciences and Systems (CISS)*, pp. 1 –6.

[8] Brante, G., Kakitani, M. T. & Souza, R. D. [2011b]. Energy efficiency analysis of some cooperative and non-cooperative transmission schemes in wireless sensor networks, *IEEE Transactions on Communications* 59(10): 2671 –2677.

[9] Chen, G., Hanson, S., Blaauw, D. & Sylvester, D. [2010]. Circuit design advances for wireless sensing applications, *Proceedings of the IEEE* 98(11): 1808 –1827.

[10] Cui, S., Goldsmith, A. & Bahai, A. [2005]. Energy-constrained modulation optimization, *IEEE Trans. Wireless Communications* 4(5): 2349–2360.

[11] Dohler, M., Heath, R., Lozano, A., Papadias, C. & Valenzuela, R. [2011]. Is the PHY layer dead?, *IEEE Communications Magazine* 49(4): 159 –165.

[12] Foschini, G. J. [1996]. Layered space-time architecture for wireless communication in a fading environment when using multi-element antennas, *Bell Labs Tech. Journal* 2: 41–59.

[13] Freitas-Jr., W. C., Cavalcanti, F. R. P. & Lopes, R. R. [2005]. Hybrid transceiver schemes for spatial multiplexing and diversity in MIMO systems, *SBrT/IEEE Journal of Communication and Information Systems* 20(3): 63–76.

[14] Goldsmith, A. [2005]. *Wireless Communications*, 1st edn, Cambridge University Press.

[15] Hu, J. & Duman, T. M. [2007]. Low density parity check codes over wireless relay channels, *IEEE Transactions on Wireless Communications* 6(9): 3384–3394.

[16] Kakitani, M. T., Brante, G., Souza, R. D. & Imren, M. A. [2012]. Energy efficiency of amplify-and-forward, repetition coding and parallel coding in short range communications, *35th International Conference on Telecommunications and Signal Processing (TSP)*, Prague, Czech Republic.

[17] Kakitani, M. T., Brante, G., Souza, R. D. & Munaretto, A. [2011]. Comparing the energy efficiency of Single-Hop, Multi-Hop and incremental Decode-and-Forward in Multi-Relay wireless sensor networks, *22nd IEEE Personal, Indoor and Mobile Radio Communications (PIMRC)*, Toronto, Canada.

[18] Khormuji, M. & Larsson, E. [2009]. Cooperative transmission based on decode-and-forward relaying with partial repetition coding, *IEEE Transactions on Wireless Communications* 8(4): 1716–1725.

[19] Kramer, G., Gastpar, M. & Gupta, P. [2005]. Cooperative strategies and capacity theorems for relay networks, *IEEE Transactions on Information Theory* 51(9): 3037–3063.

[20] Laneman, J. N., Tse, D. N. C. & Wornell, G. W. [2004]. Cooperative diversity in wireless networks: Efficient protocols and outage behavior, *IEEE Transactions on Information Theory* 50(12): 3062–3080.

[21] Li, Y. [2009]. Distributed coding for cooperative wireless networks: An overview and recent advances, *IEEE Communications Magazine* 47(8): 71–77.

[22] Nosratinia, A. & Hunter, T. [2007]. Grouping and partner selection in cooperative wireless networks, *IEEE Journal on Selected Areas in Communications* 25(2): 369 –378.

[23] Nosratinia, A., Hunter, T. E. & Hedayat, A. [2004]. Cooperative communication in wireless networks, *IEEE Communications Magazine* 42(10): 74–80.

[24] Pentikousis, K. [2010]. In search of energy-efficient mobile networking, *IEEE Communications Magazine* 48(1): 95–103.

[25] Sadek, A. K., Yu, W. & Liu, K. J. R. [2009]. On the energy efficiency of cooperative communications in wireless sensor networks, *ACM Transactions on Sensor Networks* 6(1): 1–21.

[26] Sendonaris, A., Erkip, E. & Aazhang, B. [2003]. User cooperation diversity - part I: System description, *IEEE Transactions on Communications* 51(11): 1927–1938.

[27] Shannon, C. E. [1948]. A mathematical theory of communications, *Bell Systems Tech. Journal* 27: 379–423; 623–656.

[28] Stanojev, I., Simeone, O., Bar-Ness, Y. & Kim, D. H. [2009]. Energy efficiency of non-collaborative and collaborative hybrid-ARQ protocols, *IEEE Transactions on Wireless Communications* 8(1): 326 –335.

[29] Stefanov, A. & Erkip, E. [2004]. Coperative coding for wireless networks, *IEEE Transactions on Communications* 52(9): 1470–1476.

[30] Tarokh, V., Jafarkhani, H. & Calderbank, A. R. [1999]. Space-time block codes from orthogonal designs, *IEEE Transactions on Information Theory* 45: 1456–1467.

[31] Van der Meulen, E. C. [1971]. Three-terminal comunication channels, *Advanced Applied Probability* 3: 120–154.

[32] Wang, Z. & Giannakis, G. [2003]. A simple and general parameterization quantifying performance in fading channels, *IEEE Transactions on Communications* 51(8): 1389 – 1398.

[33] Wolniansky, P., Foschini, G., Golden, G. & Valenzuela, R. [1998]. V-BLAST: an architecture for realizing very high data rates over the rich-scattering wireless channel, *Proc. URSI International Symposium on Signals, Systems, and Electronics*, pp. 295–300.

[34] Yacoub, M. [2007]. The κ-μ distribution and the η-μ distribution, *IEEE Antennas and Propagation Magazine* 49(1): 68 –81.

[35] Zarifi, K., Abuthinien, M., Ghrayeb, A. & Affes, S. [2009]. Relay selection schemes for uniformly distributed wireless sensor networks, *IEEE Wireless Communications and Networking Conference (WCNC)*, pp. 1 –6.

[36] Zhang, Z. & Duman, T. M. [2005]. Capacity-approaching turbo coding and iterative decoding for relay channels, *IEEE Transactions on Communications* 53(11): 1895–1905.

[37] Zhao, B. & Valenti, M. C. [2003]. Distributed turbo coded diversity for relay channel, *IEE Electronics Letters* 39(10): 786–787.

[38] Zhao, B. & Valenti, M. C. [2005]. Practical relay networks: A generalization of hybrid-ARQ, *IEEE Journal on Selected Areas in Communications* 23(1): 7–18.

[39] Zheng, L. & Tse, D. [2003]. Diversity and multiplexing: a fundamental tradeoff in multiple antenna channels, *IEEE Transactions on Information Theory* 49: 1073–1096.